MATERIALS
SCIENCE
of
DNA

MATERIALS
SCIENCE
of
DNA

Edited by
Jung-Il Jin
James Grote

CRC Press
Taylor & Francis Group
Boca Raton London New York

CRC Press is an imprint of the
Taylor & Francis Group, an **informa** business

CRC Press
Taylor & Francis Group
6000 Broken Sound Parkway NW, Suite 300
Boca Raton, FL 33487-2742

First issued in paperback 2017

© 2012 by Taylor & Francis Group, LLC
CRC Press is an imprint of Taylor & Francis Group, an Informa business

No claim to original U.S. Government works

ISBN-13: 978-1-4398-2741-3 (hbk)
ISBN-13: 978-1-138-19918-7 (pbk)

Library of Congress Cataloging-in-Publication Data

Materials science of DNA / editors, Jung-Il Jin, James Grote.
    p. ; cm.
  "A CRC title."
  Includes bibliographical references and index.
  ISBN 978-1-4398-2741-3 (hardcover : alk. paper)
  I. Jin, Jung-Il. II. Grote, James Gerard, 1955-
  [DNLM: 1. DNA--chemistry. 2. DNA--genetics. 3. Nanostructures. 4. Nanotechnology. QU 58.5]

579.24--dc23   2011039583

Visit the Taylor & Francis Web site at
http://www.taylorandfrancis.com

and the CRC Press Web site at
http://www.crcpress.com

Contents

Preface...vii

The Editors..ix

Contributors xi

Chapter 1   Materials Science of DNA: An Introduction.. 1

                      Jung-Il Jin

Chapter 2   Nanostructures and Nanomaterials via DNA-Based
                      Self-Assembly... 13

                      Yuanqin Zheng and Zhaoxiang Deng

Chapter 3   Intercalation of Organic Ligands as a Tool to Modify the
                      Properties of DNA.. 49

                      Heiko Ihmels and Laura Thomas

Chapter 4   DNA and Carbon-Based Nanomaterials: Preparation and
                      Properties of Their Composites.. 77

                      Thathan Premkumar and Kurt E. Geckeler

Chapter 5   Electrical and Magnetic Properties of DNA 121

                      Chang Hoon Lee, Young-Wan Kwon, and Jung-Il Jin

Chapter 6   DNA Ionic Liquid.. 163

                      Naomi Nishimura and Hiroyuki Ohno

Chapter 7   DNA-Surfactant Thin-Film Processing and Characterization......... 179

                      Emily M. Heckman, Carrie M. Bartsch, Perry P. Yaney,
                      Guru Subramanyam, Fahima Ouchen, and James G. Grote

Chapter 8   Applications of DNA to Photonics and Biomedicals 231

                      Naoya Ogata

Chapter 9 DNA-Based Thin-Film Devices...255

Carrie M. Bartsch, Joshua A. Hagen, Emily M. Heckman,
Fahima Ouchen, and James G. Grote

Chapter 10 Nucleic Acids-Based Biosensors...291

Sara Tombelli, Ilaria Palchetti, and Marco Mascini

Chapter 11 Materials Science of DNA—Conclusions and Perspectives...........311

James G. Grote

Index...319

Preface

DNA has been at the center of bioscience and biotechnology for more than a half century since Watson and Crick described its double helical structure. While the biological roles of DNA were being explored vigorously, mainly by biologists, physicochemical properties and the materials applications of DNA have been investigated by chemists, physicists, and engineers, especially for the past 30 years or so.

Self-assembly of DNAs, transport of charge carriers along the helical axis of DNA, and sensing applications of DNA science (i.e., the concept of base-pairing) are representative examples of recent development. Utilization of modified DNAs as new functional materials in photonics and photoelectronics devices is an exciting recent advance in the practical applications of DNAs.

Converting water-soluble DNA by chemical modification to organic-soluble compositions enables us to work with high quality film-formers, which is rapidly gaining a significant importance in the area of bioelectronics or biotronics.

The materials science of DNA is an emerging research area of an inter- and multidisciplinary nature and is expected to broaden immensely the horizon of material science and technology in this century. In fact, the science hidden in DNA is teaching us so much as it is revealed that the end of materials science of DNA is not possible to see. It is no secret that much of the contemporary popular nanoscience and nanotechnology are being developed based on the structural characteristics of DNA.

The topics covered in this book highlight the most important subjects and perspectives of the materials science of DNA written by the pioneers of related researches. It is our hope that this volume will serve as a stimulus for the interdisciplinary community to intensify their research endeavor to bring maturation to this intensively interesting, emerging field of molecular-scale materials science. We would like to thank all scientists whose contributions have made the publication of this book possible.

Our special thanks go to Dr. Young-Wan Kwon for his meticulous proofreading of the final manuscripts.

Finally, the editors have not only been involved in the research of materials science of DNA for the past decade, but also leading the series of International Biotronics Workshop supported by the U.S. Air Force Research Laboratory (AFRL). We, the editors, are very thankful to the U.S. AFRL for the continued support of our activities and also to the enthusiastic participants in the workshops. These gatherings generated the seeds for this book.

Jung-Il Jin
James G. Grote

The Editors

JUNG-IL JIN

Professor Emeritus, Chemistry Department, Korea University, Jung-Il Jin is the immediate past president of the International Union of Pure and Applied Chemistry, IUPAC. He is the founding president of the Federation of the Asian Polymer Societies (FAPS).

He obtained his PhD in 1969 from the City University of New York, where his advisor was Richard H. Wiley, and was a Visitor at the University of Massachusetts and Cambridge University. His research areas have been liquid crystalline and poly-conjugated polymers and the materials science of DNA. He has published about 400 original research articles and contributed many chapters to various monographs related to functional polymers.

JAMES G. GROTE

James G. Grote is a Principal Electronics Research Engineer with the Air Force Research Laboratory, Materials and Manufacturing Directorate at Wright-Patterson Air Force Base, Ohio, where he conducts research in polymer and biopolymer-based opto-electronics. He is also an adjunct professor at the University of Dayton and University of Cincinnati. Dr. Grote received his BS degree in Electrical Engineering for Ohio University and both his MS and PhD degrees in Electrical Engineering from the University of Dayton, with partial study at the University of California, San Diego. He was a visiting scholar at the Institut d'Optique, Universite de Paris, Sud in the summer of 1995 and a visiting scholar at the University of Southern California, the University of California in Los Angeles and the University of Washington in 2001. He received Doctor Honoris Causa from the Politehnica University of Bucharest in 2010. Dr. Grote is an Air Force Research Laboratory Fellow, a Fellow of the International Society for Optics and Photonics (SPIE), a Fellow of the Optical Society of America (OSA), a Fellow of the European Optical Society (EOS) and a Senior Member of the Institute of Electrical and Electronics Engineers (IEEE). He has co-authored more than 130 journal and conference papers, including two book chapters, and has served as editor for more than 25 conference proceedings and journal publications.

Contributors

Carrie M. Bartsch
Air Force Research Laboratory
Sensors Directorate
Wright-Patterson Air Force Base, Ohio

Zhaoxiang Deng
CAS Key Laboratory of Soft Matter
    Chemistry
Department of Chemistry
University of Science and Technology
    of China
Anhui, China

Kurt E. Geckeler
Department of Materials Science and
    Engineering
Department of Nanobio Materials and
    Electronics
World-Class University (WCU)
Institute of Medical System
    Engineering
Gwangju Institute of Science and
    Technology (GIST)
Gwangju, South Korea

James G. Grote
Air Force Research Laboratory
Materials and Manufacturing
    Directorate
AFRL/RXPS
Wright-Patterson Air Force Base, Ohio

Joshua A. Hagen
Air Force Research Labs
711th Human Performance Wing
Human Effectiveness Directorate
    RHXB
Wright-Patterson Air Force Base, Ohio

Emily M. Heckman
Air Force Research Laboratory
Sensors Directorate
Wright-Patterson Air Force Base, Ohio

Heiko Ihmels
Organic Chemistry II
University of Siegen
Siegen, Germany

Jung-Il Jin
Department of Chemistry
Korea University
Seoul, South Korea

Young-Wan Kwon
Department of Chemistry
Korea University
Seoul, South Korea

Chang Hoon Lee
Department of Polymer Science &
    Engineering
Chosun University
Gwangju Institute of Science and
    Technology (GIST)
Gwangju, South Korea

Marco Mascini
Dipartimento di Chimica
Università di Firenze
Sesto Fiorentino (Fi), Italy

Naomi Nishimura
Department of Biotechnology
Tokyo University of Agriculture &
    Technology
Tokyo, Japan

Naoya Ogata
Ogata Research Laboratory, Ltd.
Hokkaido, Japan

Hiroyuki Ohno
Ogata Research Laboratory, Ltd.
Hokkaido, Japan

Fahima Ouchen
University of Dayton Research Institute
Dayton, Ohio

Ilaria Palchetti
Dipartimento di Chimica
Università di Firenze
Sesto Fiorentino (Fi), Italy

Thathan Premkumar
Department of Materials Science and
    Engineering
Department of Nanobio Materials and
    Electronics
World-Class University (WCU)
Institute of Medical System
    Engineering
Gwangju Institute of Science and
    Technology (GIST)
Gwangju, South Korea

Guru Subramanyam
Department of Electrical and Computer
    Engineering
University of Dayton
Dayton, Ohio

Laura Thomas
Organic Chemistry II
University of Siegen
Siegen, Germany

Sara Tombelli
Dipartimento di Chimica
Università di Firenze
Sesto Fiorentino (Fi), Italy

Perry P. Yaney
Department of Physics
University of Dayton
Dayton, Ohio
and
Materials and Manufacturing
    Directorate
Air Force Research Laboratory
    AFRL/RXPS
Wright-Patterson Air Force Base, Ohio

Yuanqin Zheng
CAS Key Laboratory of Soft Matter
    Chemistry
Department of Chemistry
University of Science and Technology
    of China
Anhui, China

1 Materials Science of DNA
An Introduction

Jung-Il Jin

CONTENTS

1.1 Naturally Occurring Organic Polymers 1
1.2 Structures of Nucleic Acids 2
1.3 Materials Science of DNA... 7
References.. 9

1.1 NATURALLY OCCURRING ORGANIC POLYMERS

The field of materials science and technology is currently undergoing revolutionary advances due to a recent trend in which various materials are coupled with different disciplines: for example, biomolecules with electronics (Willner and Katz 2005; Hoffmann 2002), nanomaterials with medical science (Wang, Katz, and Willner 2005; Alcamo 1996), atom-scale materials with the development of novel advanced analytical tools (Fukui and Tatsuo 2008; Salmeron et al. 1990; Knapp et al. 1995; Samorì 2008), new functional materials with multidisciplinary optical science and engineering (Wise et al. 1998), etc. Additionally, theoretical predictions rendered with increasingly improved models and computational capabilities (Grant and Richards 1995; Leach 2001) are also making impressive contributions to the progress of materials science and technology. The eagerness of scientists to embrace multidisciplinary approaches is another important driving force in the recent rapid changes in materials science. On the other hand, it has to be pointed out that since human beings came into being on this planet, their lives have been largely dependent on the use of naturally occurring polymers, namely, biopolymers. Proteins, polynucleotides, and polysaccharides are three major examples of biopolymers. All three of these biopolymer groups reveal very specific physiological activities. Additionally, their versatile applications as materials have already been, or are currently being, developed. The chemical structures and uses of these polymers are summarized in Table 1.1.

Natural rubbers, shellacs, lignins, and bitumens are other examples of naturally occurring organic polymers. Coals and ambers also belong to the class of natural organic polymers.

TABLE 1.1

Three Major Biopolymers

Biopolymer	Chemical Structure	Use
Protein	Repeated peptides bonds of α-amino acids	Fibers (silk, wool), foods (vegetable and animal proteins), adhesive
Polynucleotides (nucleic acids)	Phosphate ionic polyesters consisting of phosphoric acid and nucleosides residues	Biotechnology (genetic engineering), forensic detection (DNA sequencing), new functional materials (biosensing)
Polysaccharides	Polyacetals and/or polyketals of monosaccharides	Food fibers, plastics, papers, adhesives, fuel (biofuel)

1.2   STRUCTURES OF NUCLEIC ACIDS

A single strand of deoxyribonucleic acid (DNA) consists of three fundamental chemical units: heterocyclic bases, sugar deoxyribose, and the linking phosphate groups (Figure 1.1).

The molecular structure of a DNA segment is shown in Figure 1.2. The phosphate groups connect deoxyribose sugar groups, thereby forming a long single strand of DNA. The contents of the four bases in DNA vary among various species (Table 1.2). For example, the DNA of *Homo sapiens* contains 31.5% thymine, 31.0% adenine, 19.1% guanine, and 18.4% cytosine. By way of contrast, as can be observed in Table 1.2, the contents of the four bases in corn are approximately the same. Not only the base compositions but also their sequences in the DNA strand are related closely to the inherited characteristics of a specific species.

The famous research paper by Watson and Crick (Watson and Crick 1953), "Molecular structure of nucleic acids: A structure for deoxyribose nucleic acid" was published in *Nature* on April 24, 1953. Their proposed double helix structure (Figure 1.3) of DNA is based on the complementary base pairing of adenine (A) with thymine (T) and of cytosine (C) with guanine (G) (Figure 1.4). Base pairing between the bases occurs via hydrogen bonds between the carbonyl oxygens and the sterically best-fit amino hydrogens in the pairs. Watson, Crick, and Wilkinson shared the 1962 Nobel Prize in physiology for their contribution in establishing the structure of DNA by x-ray analysis.

The stereochemical structure of the DNA double helix depends heavily on the quantity of water molecules, that is, water of hydrations: in the so-called wet B-form, the average distance between the basal planes is 3.4 Å, and the helical diameter is 20 Å (Dickerson et al. 1982). The basal planes are aligned perpendicular to the helical axis. Ten base pairs make one helical turn. By way of contrast, in the so-called dry A-form, the distance between the basal planes is substantially shorter, 2.55 Å, but the diameter is significantly larger, at 26 Å (Dickerson et al. 1982). Additionally, the basal planes are tilted relative to the helical axis (Figure 1.5).

It is important to note that the majority of undergraduate textbooks depict the structure of B-form DNA without explaining the dependence of its stereostructure on the water content. The science of DNA currently constitutes the center of all

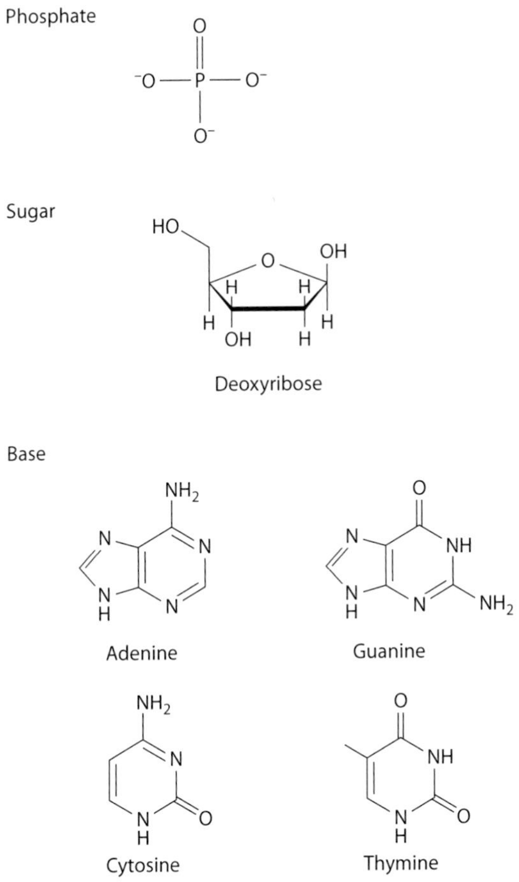

Phosphate

Sugar

Deoxyribose

Base

Adenine

Guanine

Cytosine

Thymine

FIGURE 1.1 The components of DNA.

TABLE 1.2
The Base Compositions of DNA for Various Species

Species	Adenine	Thymine	Guanine	Cytosine
Homo sapiens (human)	31.0	31.5	19.1	18.4
Drosophila melanogaster (fruit fly)	27.3	27.6	22.5	22.5
Zea mays (corn)	25.6	25.3	24.5	24.6
Neurospora crasa (mold)	23.0	23.3	27.1	26.6
Escherichia coli (bacterium)	24.6	24.3	25.5	25.6
Bacillus subtilis (bacterium)	28.4	29.0	21.0	21.6

Source: Alcamo, I. E. 1996. *DNA Technology: The Awesome Skill.* Dubuque, IA: Wm. C. Brown Publishers.

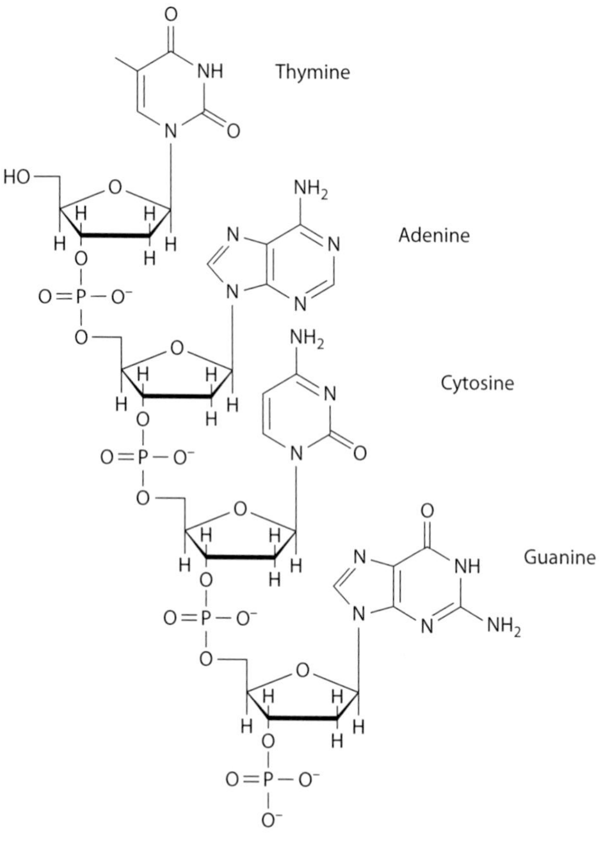

FIGURE 1.2 The molecular structure of a segment of DNA.

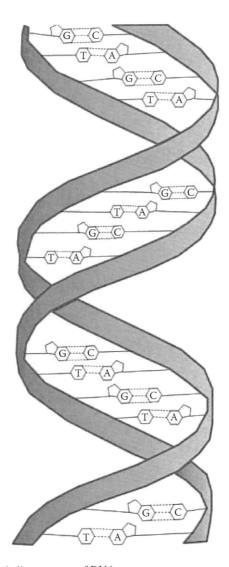

FIGURE 1.3   Double helix structure of DNA.

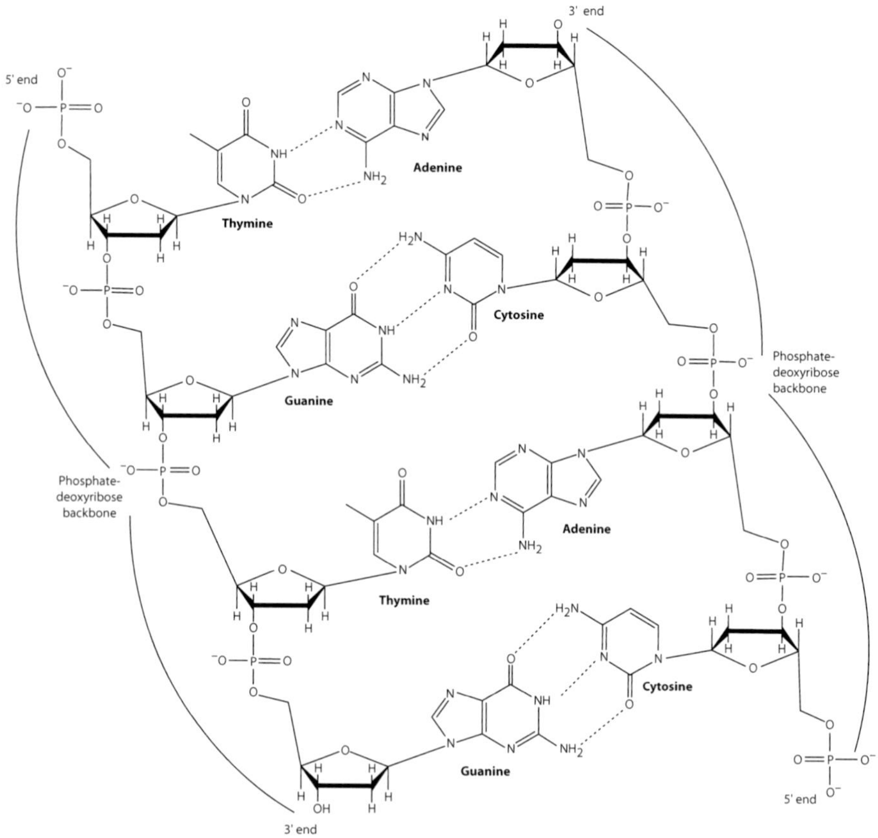

FIGURE 1.4 Base pairing of adenine with thymine and of cytosine with guanine in DNA.

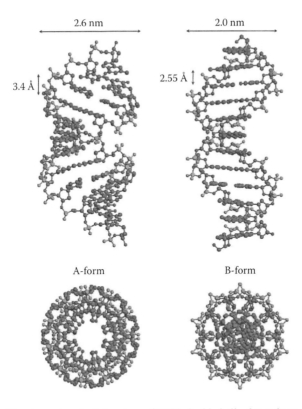

FIGURE 1.5 The stereochemical structure of DNA double helix depends on the amount of water molecules (A-form, B-form). (From Ng, H. L. et al. 2000. *Proc. Natl. Acad. Sci. USA.* 97 (5):2035–2039. Copyright 2000, National Academy of Science, USA.)

life sciences, including genetic engineering and many other new emerging areas of research, including proteomics, metabolomics, and chemical biology.

The chemistry relevant to the hydrogen bonds between the DNA base pairs has been extended to current popular approaches to self-assembly (Lehn 2002, 2004), molecular recognition, sensors, molecular computation, medical diagnosis, and forensic science, to list a few examples. The unveiling of the basic physicochemical and biological principles hidden within the structure of DNA has led to the opening up of new horizons in materials science, which is the subject of this chapter.

1.3 MATERIALS SCIENCE OF DNA

As mentioned earlier, human beings utilize a number of natural polymers and biopolymers as indispensable materials for life. DNA, however, has only recently begun to attract the attention of scientists and engineers, and its possible applications as a new class of material are the subject of extensive study. As a matter of fact, DNA is the most abundant of the natural polymers. All living creatures, including bacteria, not only contain DNA molecules, but their genetics also depend completely on those molecules. Despite the ubiquitous nature of DNA molecules, skillful separation

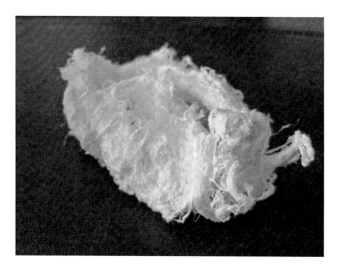

FIGURE 1.6　A DNA sample obtained from salmon sperm.

techniques are required for the acquisition of pure DNA. DNA almost always exists as a complex mixture with proteins, and thus it is very difficult to selectively remove or isolate it (Figure 1.6).

Secondly, due to its ionic chemical structure, DNA exhibits an extremely high affinity toward water molecules that can readily form channels or layers of hydration between the double helices. Some of the water molecules can also enter the core of helix-forming hydrogen bonds with nuclear bases. Therefore, one should exercise caution whenever attempting to measure any physical or mechanical property of DNA. The properties being measured must necessarily be affected by the presence of possible impurities such as water and oxygen (Kwon et al. 2008).

Prior to the rapid growth of research interest in the electrical, electrooptical, and optical properties of DNA and modified DNAs, scientists began to focus on the possible use of DNA science in the construction of a broad variety of self-assembled structures. Seeman (Seeman 1982, 2003) was a pioneer in advocating the spontaneous formation of self-assembled supramolecules using specifically designed DNA molecules. Nonlinear, even three-dimensional, structures can be constructed (Seeman 1998). This progress in the science of self-assembly is making possible the design of new materials at nanoscale, and even at larger scales. This general bottom-up approach is expected to revolutionize the miniaturization of future electronic or optoelectronic devices. The most important molecular parameters employed in these approaches are hydrogen bonding and stereochemical fitting learned from DNA science. If one adds electrical interactions between the counterparts, much more complex structures can be envisaged. Many molecular recognition examples are based heavily on the concept of DNA base pairing; many electrochemical sensors (Palecek and Fojta 2005) being developed using DNA segments result from the application of these same principles.

The electrical properties of DNA itself have been the subject of increased attention from researchers of electronic structures, particularly along the helical axis. Hole and electron transport properties, in particular, have been popular subjects among many research groups (Schuster 2004a, 2004b). Although debates continue on the subject, DNA appears to be a better positive hole transporter than a negative electron transporter. Despite the existence of a variety of contradictory reports regarding its electrical conductivity, DNA appears to function as a wide bandgap semiconductor. In order to prevent problematic experimental artifacts, my group (Lee et al. 2006; Kwon et al. 2008, 2009) has been attempting to evaluate the magnetic properties of natural and modified DNA; these attempts require no direct contact between the samples and probes.

Chemical modification of natural DNAs can be conducted in a variety of different directions: degradation to reduce the size, that is, molecular weight, of the parent DNA; chemical modification of terminal groups to render specific reactivities such as the formation of mercaptan (-SH) or amino (-NHR) groups; the replacement of sodium cations with organic cations converting water-soluble compositions to organic-soluble compositions; and the introduction of cross-linking reactions to attenuate solubility in water or organic solvents. Complexation with transition metals (Keren et al. 2002; Richter et al. 2000, 2001; Ciacchi et al. 2001; Ford et al. 2001; Seidel et al. 2004; Monson and Woolley 2003; Kwon and Jin 2004), wherein DNA molecules function as template-providing nuclear bases at ligand sites, has been a long-term subject of many investigations. This research has led to the formation of interesting self-assembled gold nanoparticles, as reported recently by Mirkin et al. (Mirkin et al. 1996; Park et al. 2008). The formation of both nanowires and two-dimensional metal planes has been reported previously (Koplin, Niemeyer, and Simon 2006; Park, Taton, and Mirkin 2002; Niemeyer et al. 2003; Alivisatos et al. 1996). Intercalation of DNA with a wide variety of flat molecules has been achieved: this approach is another attractive method to modify the properties of DNA (Ihmels and Otto 2005).

Thus far, no examples of practical applications of DNA—whether natural, modified, or synthetic—have been definitively demonstrated. There are, however, abundant examples to demonstrate the application of scientific principles relevant to DNA in a broad variety of devices. DNAs may, ultimately, turn out to be less than satisfactory for certain applications. However, new concepts and insights gained from DNA research are expected to prove genuinely useful in a variety of devices, in nano, micro, and macro dimensions, in the future.

REFERENCES

Alcamo, I. E. 1996. *DNA Technology: The Awesome Skill.* Dubuque, IA: Wm. C. Brown Publishers.

Alivisatos, A. P., K. P. Johnsson, X. G. Peng, T. E. Wilson, C. J. Loweth, M. P. Bruchez Jr, and P. G. Schultz. 1996. Organization of "nanocrystal molecules" using DNA. *Nature* 382 (6592):609–611.

Ciacchi, L. C., W. Pompe, and A. De Vita. 2001. Initial nucleation of platinum clusters after reduction of K_2PtCl_4 in aqueous solution: A first principles study. *J. Am. Chem. Soc.* 123 (30):7371–7380.

Dickerson, R. E., H. R. Drew, B. N. Conner, R. M. Wing, A. V. Fratini, and M. L. Kopka. 1982. The anatomy of a-DNA, B-DNA, and Z-DNA. *Science* 216 (4545):475–485.

Ford, W. E., O. Harnack, A. Yasuda, and J. M. Wessels. 2001. Platinated DNA as precursors to templated chains of metal nanoparticles. *Adv. Mater.* 13 (23):1793–1797.

Fukui, K., and U. Tatsuo. 2008. *Chromosome Nanoscience and Technology.* Boca Raton, FL: CRC Press.

Grant, G. H., and W. G. Richards. 1995. *Computational Chemistry.* 1st ed., *Oxford Chemistry Primers.* Oxford, England: Oxford University Press.

Hoffmann, K. H. 2002. *Coupling of Biological and Electronic Systems: Proceedings of the 2nd Caesarium,* Bonn, November 1–3, 2000. Berlin: Springer.

Ihmels, H. and D. Otto, 2005. Intercalation of Organic Dye Molecules into Double Stranded DNA-General principles and Recent Developments in *Supramolecular Dye Chemistry,* edited by F. Wűrthner, Berlin: Springer-Verlag, pp. 161–204.

Keren, K., M. Krueger, R. Gilad, G. Ben-Yoseph, U. Sivan, and E. Braun. 2002. Sequence-specific molecular lithography on single DNA molecules. *Science* 297 (5578):72–75.

Knapp, H. F., R. Wyss, R. Haring, C. Henn, R. Guckenberger, and A. Engel. 1995. Hybrid scanning-transmission electron scanning tunneling microscope system for the preparation and investigation of biomolecules. *J. Microsc.(Oxf)* 177:31–42.

Koplin, E., C. M. Niemeyer, and U. Simon. 2006. Formation of electrically conducting DNA-assembled gold nanoparticle monolayers. *J. Mater. Chem.* 16 (14):1338–1344.

Kwon, Y.-W., E. D. Do, D. H. Choi, J.-I. Jin, C. H. Lee, J. S. Kang, and E.-K. Koh. 2008. Hydration effect on the intrinsic magnetism of natural deoxyribonucleic acid as studied by EMR spectroscopy and SQUID measurements. *Bull. Korean Chem. Soc.* 29 (6):1233–1242.

Kwon, Y.-W., and J.-I. Jin. 2004. DNA mediated gold nanoparticles formation. Presented in part at IUMRS-ICA 2004, November, at Hsinchu Taiwan.

Kwon, Y.-W., C. H. Lee, D. H. Choi, and J.-I. Jin. 2009. Materials science of DNA. *J. Mater. Chem.* 19 (10):1353–1380.

Leach, A. R. 2001. *Molecular Modelling: Principles and Applications.* 2nd ed. New York: Prentice Hall.

Lee, C. H., Y.-W. Kwon, E.-D. Do, D. H. Choi, J.-I. Jin, D.-K. Oh, and J. Kim. 2006. Electron magnetic resonance and SQUID measurement study of natural A-DNA in dry state. *Phys. Rev. B* 73:224417.

Lehn, J.-M. 2002. Toward complex matter: Supramolecular chemistry and self-organization. *Proc. Natl. Acad. Sci. USA.* 99:4763–4768.

Lehn, J.-M. 2004. Supramolecular chemistry: From molecular information toward self-organization and complex matter. *Rep. Prog. Phys.* 167:249–265

Mirkin, C. A., R. L. Letsinger, R. C. Mucic, and J. J. Storhoff. 1996. A DNA-based method for rationally assembling nanoparticles into macroscopic materials. *Nature* 382 (6592):607–609.

Monson, C. F., and A. T. Woolley. 2003. DNA-templated construction of copper nanowires. *Nano Lett.* 3 (3):359–363.

Ng, H. L., M. L. Kopka, and R. E. Dickerson. 2000. The structure of a stable intermediate in the A ↔ B DNA helix transition. *Proc. Natl. Acad. Sci. USA.* 97 (5):2035–2039.

Niemeyer, C. M., B. Ceyhan, M. Noyong, and U. Simon. 2003. Bifunctional DNA-gold nanoparticle conjugates as building blocks for the self-assembly of cross-linked particle layers. *Biochem. Biophys. Res. Commun.* 311 (4):995–999.

Palecek, E., and M. Fojta. 2005. Electrochemical DNA Sensors. In *Bioelectronics,* edited by I. Willner and E. Katz. Weinheim, Germany: Wiley-VCH.

Park, S. J., T. A. Taton, and C. A. Mirkin. 2002. Array-based electrical detection of DNA with nanoparticle probes. *Science* 295 (5559):1503–1506.

Park, S. Y., A. K. R. Lytton-Jean, B. Lee, S. Weigand, G. C. Schatz, and C. A. Mirkin. 2008. DNA-programmable nanoparticle crystallization. *Nature* 451 (7178):553–556.

Richter, J., M. Mertig, W. Pompe, I. Monch, and H. K. Schackert. 2001. Construction of highly conductive nanowires on a DNA template. *Appl. Phys. Lett.* 78 (4):536–538.

Richter, J., R. Seidel, R. Kirsch, M. Mertig, W. Pompe, J. Plaschke, and H. K. Schackert. 2000. Nanoscale palladium metallization of DNA. *Adv. Mater.* 12 (7):507–510.

Salmeron, M., T. Beebe, J. Odriozola, T. Wilson, D. F. Ogletree, and W. Siekhaus. 1990. Imaging of biomolecules with the scanning tunneling microscope—problems and prospects. *J. Vac. Sci. Technol., A* 8 (1):635–641.

Samorì, P. 2008. STM and AFM Studies on (Bio)molecular Systems: Unravelling the Nanoworld in *Topics in Current Chemistry*. Heidelberg: Springer-Verlag. http://dx.doi.org/10.1007/978-3-540-78395-4.

Schuster, G. B., ed. 2004a. *Long-Range Charge Transfer in DNA I*. Vol. 236, *Topics In Current Chemistry*. Heildelberg: Springer.

Schuster, G. B., ed. 2004b. *Long-Range Charge Transfer In DNA II*. Vol. 236, *Topics In Current Chemistry*. Heidelberg: Springer.

Seeman, N. C. 1982. Nucleic-acid junctions and lattices. *J. Theoret. Biol.* 99 (2):237–247.

Seeman, N. C. 1998. DNA nanotechnology: Novel DNA constructions. *Annu. Rev. Biophys. Biomol. Struct.* 27:225–48.

Seeman, N. C. 2003. DNA nanotechnology. *Materials Today* 6 (1):24–29.

Seidel, R., L. C. Ciacchi, M. Weigel, W. Pompe, and M. Mertig. 2004. Synthesis of platinum cluster chains on DNA templates: Conditions for a template-controlled cluster growth. *J. Phys. Chem. B* 108 (30):10801–10811.

Wang, J., E. Katz, and I. Willner. 2005. Biomaterial-nanoparticle Hybrid Systems for Sensing and Electronic Devices in *Bioelectronics-From Theory to Applications*, edited by I. Willner and E. Katz. Weinheim, Germany: Wiley-VCH.

Watson, J. D., and F. H. Crick. 1953. Molecular structure of nucleic acids; a structure for deoxyribose nucleic acid. *Nature* 171 (4356):737–738.

Willner, I., and E. Katz. 2005. *Bioelectronics: From Theory to Applications*. Weinheim, Germany: Wiley-VCH.

Wise, D. L., G. E. Wnek, D. J. Trantolo, T. M. Cooper, and J. D. Gresser, eds. 1998. *Photonic Polymer Systems-Fundamentals, Methods, and Applicaitons, Plastics Engineering*. New York: Marcel Dekker.

2 Nanostructures and Nanomaterials via DNA-Based Self-Assembly

Yuanqin Zheng and Zhaoxiang Deng

CONTENTS

2.1   Introduction 13
2.2   Self-Assembly of DNA Nanostructures... 15
    2.2.1   One-Dimensional DNA Nanostructures.. 16
    2.2.2   Two-Dimensional DNA Self-Assembly... 18
    2.2.3   3D Self-Assembly of DNA Polyhedra 22
    2.2.4   3D Self-Assembly of Periodical DNA Crystals.................................... 25
    2.2.5   DNA Origami 27
2.3   DNA-Based Self-Assembly of Nanomaterials.. 30
    2.3.1   DNA-Based Self-Assembly of Gold Nanoparticles in One and
        Two Dimensions 30
    2.3.2   3D Ordering of Gold Nanoparticles with DNA.................................... 32
    2.3.3   DNA-Based Self-Assembly of Carbon Nanotubes 34
    2.3.4   DNA-Based Self-Assembly of Gold–Carbon Hybrid
        Nanostructures... 37
    2.3.5   Molecular Lithography 41
2.4   Summary and Outlook 41
Acknowledgments... 43
References... 43

2.1   INTRODUCTION

Nature builds biological materials and develops their functional pathways on nano (biomacromolecular) or micro (cellular) levels with superfine spatial (structural) and temporal (process) controls, securing their roles in a living system. Humans often find it hard to duplicate the structures and functions of such biomaterials even after the underlying chemical, physical, and biological "languages" that govern the formation of these materials have been gradually "cracked." On the other hand, though lithography-based top-down techniques have laid a solid foundation for microelectronics, they are facing unprecedented difficulties in nanoelectronics. Self-assembly is believed to be the most promising "key" to deal with some of these challenges that will be able to provide new methodologies and generate materials with novel functions.

DNA, life's programmer from billions of years of evolution, has been taken as an excellent nanoscale building block that enables probably the most-controllable self-assembly processes [Aldaye et al. 2003; Deng et al. 2005, 2006, 2009; He et al. 2007; Kata and Willner 2004; Lin et al. 2006; Niemeyer 2001; Seeman 2003], based on Prof. Nadrian C. Seeman's pioneering work. The past decades have witnessed a rapid and fruitful development in DNA nanotechnology, evidencing the critical role of DNA in the programmed self-assembly of various DNA nanostructures and the use of DNA to direct material syntheses.

Genome DNA adopts a highly wrapped conformation under in vivo conditions assisted by histone proteins in order to fit in a cell nucleolus and to facilitate its biological functions. Also, short DNA duplexes can crystallize into macrosized crystals under favorable conditions. However, these DNA molecules, though existing in a hierarchically or periodically ordered superstructure, exclusively take a simple linear shape. In order to boost the complexities of DNA-based nanostructures, branched DNA tiles are needed. As a matter of fact, Seeman's original dream was to employ a DNA module branched in six unique directions (i.e., ±x, ±y, ±z) so that it would self-associate into three-dimensional (3D) crystals through programmable sticky-end cohesions [Seeman 2010].

Holliday junction, named after Robin Holliday, is a four-branched DNA formed via strand exchange between two juxtaposed duplexes [Holliday 1964], which is a central intermediate in various genetic processes. Branch migration is a key feature of such a junction accounting for its genetic roles. In 1983, Kallenbach et al. [1983] constructed an artificial four-way DNA junction (J1 junction) composed of four synthesized 16-base oligonucleotides designed on the basis of minimized sequence symmetry (MSS) criteria [Seeman 1990]. This artificial mimic of Holliday junction achieved a stable crossover structure against branch point migration (Figure 2.1). Ideally, by simply appending single-stranded DNA overhangs (sticky ends) on the four arms of this junction, a planar tetragonal DNA lattice may be formed via sticky end-directed assembly. Unfortunately, successes in constructing a well-defined DNA lattice directly based on a Holliday junction–like module was very rare [Malo et al. 2005]. This is due to the fact that a branched DNA duplex still lacks a suitable mechanical rigidness in order to minimize connectivity errors between the assembly tiles. In 1993, Fu and Seeman reported DNA double-crossover molecules (DX) with much more enhanced rigidity [Fu and Seeman 1993]. A DX molecule contains two strand-exchanging points (Holliday junction like) that tightly hold two juxtaposed helical domains (Figure 2.1). Further experimental investigations indicated that the antiparallel (DNA strands reverse their directionalities after passing the junction point and entering another helix domain) DX molecules were much more stable than their parallel counterparts. Following the building logic of the DX molecules, various rigid DNA structural motifs have been investigated for their suitability in building DNA nanostructures with different lattice types and dimensionalities, which includes TX and multiple crossover modules [LaBean et al. 2000; Liu et al. 2004; Mathieu et al. 2005; Reishus et al. 2005], DX triangles [Zheng et al. 2009], 4×4 four-way junctions

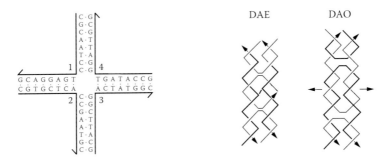

FIGURE 2.1 An immobile four-way DNA branched junction (reprinted by permission from Macmillan Publishers Ltd: [*Nature*] Kallenbach. N. R. et al. 1983. *Nature* 305: 829–831. Copyright 1983) and two antiparallel double crossover molecules (reprinted with permission from Fu, T. J. and Seeman, N. C. 1993. *Biochemistry* 32: 3211–3220, Copyright 1993 American Chemical Society) designed based on a sequence symmetry minimization (SSM) rule. Letter "D" depicts double crossover; "A" means antiparallel; "E" or "O" represents even or odd number of helical half-turns separating adjacent crossovers.

[He, Chen et al. 2005; Yan et al. 2003], and three, four, five, and six-point-star motifs [He, Tian et al. 2005; He et al. 2005; Zhang et al. 2008, 2009].

In 2006, Paul Rothemund introduced DNA origami [Rothemund 2006], a DNA self-assembly technique by which various complicated DNA structures and patterns can be easily and accurately assembled by folding, other than sticky-end cohesions, a long single-stranded M13 virus genome with more than 200 short "helper" strands. Besides the increased structural complexities, DNA origami has achieved a full addressability at each pixel on its lattice surface, which is important for its nanotechnological applications [Sanderson 2010].

The almost three decades of innovations and developments have demonstrated the superb ability of DNA in controlling the structure of a nanoscale object by simply altering DNA's base sequences. DNA probably has no rivals regarding the ease, accuracy, flexibility, and programmability in self-assembly. The next challenge is to realize the transition from structural DNA nanotechnology to functional DNA nanotechnology. Nanomaterials such as inorganic nanoparticles often lack an ability to self-organize into desired structures, which is, however, critically important for the realizations of some of their important functions such as building blocks in nanoelectronics and nanosensors. A possibly wise choice is to use DNA as a structural scaffold or smart glue to direct the assembly of functional nanomaterials, thanks to the fact that the DNA sequence can be easily synthesized and functionalized with various chemical groups. This chapter will concentrate our discussions on the young and rapidly evolving research field of DNA nanotechnology with topics covering DNA-based self-assembly and its applications in organizing inorganic nanomaterials.

2.2 SELF-ASSEMBLY OF DNA NANOSTRUCTURES

The nanostructures built from DNA can be conveniently classified according to their different dimensionalities. Because two- and three-dimensional nanostructures require more structural controls and thus are more challenging to build this chapter starts off with relatively simple one-dimensional (1D) DNA nanostructures. (*Note*: As one part of the content in this section, DNA nanotubes came out to the world much later due to their relatively high technological requirements that were not really "simple" to deal with.) The area of DNA nanotechnology started its efforts from finding suitable building blocks for DNA-based self-assembly and reached a phase of rapid development successfully worked out about one decade ago, and are now quickly evolving to include a third-dimensional control. Another important clue in the evolution of DNA nanotechnology is the increased addressability of the assemblies, which has now achieved an unprecedented success with the advent of DNA origami.

2.2.1 ONE-DIMENSIONAL DNA NANOSTRUCTURES

One of the early examples of self-assembled 1D DNA structures is a DAE-E array [Deng and Mao], which is well suited to demonstrate how structural rigidity (greater persistence length) of an assembly tile helps the formation of a well-defined DNA nanostructure. In the case of the DAE-E module, micrometer long 1D arrays could be easily generated by sticky-end cohesions (Figure 2.2). The enhanced rigidity of the DAE-E tile was responsible for the formation of a 1D DNA lattice reaching such a length. Atomic force microscope (AFM) images readily resolved some bumpy morphological features of the assembled structures with a repeating distance of 17.5 nm, consistent with the 53-base pair separation between adjacent DAE-E motifs.

DNA nanotubes represent another important class of 1D DNA nanoassemblies characteristic of a hollow interior and structural chirality similar to single-walled carbon nanotubes (SWNTs) that require more careful sequence designs. Starting from the initially undesired formations of double-layered tubular "by-products" during the assembly of a tetragonal 2D DNA lattice [Yan et al. 2003], the concept of

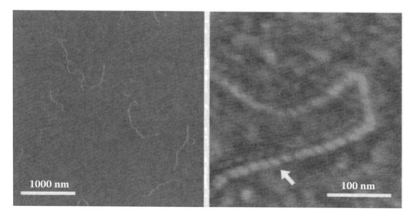

FIGURE 2.2 Assembly of one-dimensional DAE-E array based on sticky-end cohesions. The clearly resolvable bumps on a high-resolution AFM image (right panel) were consistent with the 53-base long repeating unit of a DAE-E tile. An arrow on the right panel indicates such a bumpy feature on a DAE-E array, probably corresponding to a unit tile. (From Z. Deng and C. Mao, unpublished data, 2002.)

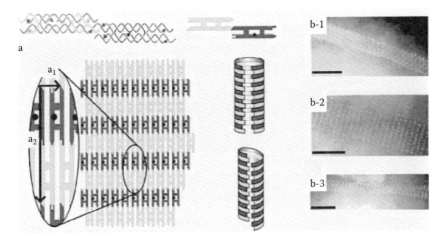

FIGURE 2.3 (**See color insert.**) Assembly of chiral DNA nanotubes from two DAE-O tiles. Samples were protein labeled and negatively stained for an easy observation under TEM. (Reprinted with permission from Mitchell, J. C. et al. 2004. *J. Am. Chem. Soc.* 126: 16342–16343. Copyright 2004 American Chemical Society.)

DNA nanotubes has attracted a great deal of research interest due to their potential applications in nanofabrications, helping the understanding of a 3D nanoassembly process and targeted cargo deliveries. In 2004, Rothemund and coworkers reported the assembly of DNA nanotubes from DX molecules (DAE-E tiles) [Rothemund et al. 2004]. Each DNA tile was composed of a central core (two double helices held together by two crossovers) and four single-stranded sticky ends. These authors employed a DAE-E tile to generate an appropriate curvature of a patch of DNA tiles to induce the formation of DNA nanotubes. The as-formed DNA tubules reached a length up to 50 μm (persistence length close to 4 μm) with diameters ranging from 7 to 20 nm.

The chirality of a SWNT is well known to determinative of its electronic properties. DNA nanotube, as a structural mimic of carbon nanotubes, could have a similar definition of chirality according to its different wrapping vectors of a scenario 2D lattice. Mitchell et al. [2004] revealed the existence of different chiral forms of DAE-O DNA nanotubes as observed by transmission electron microscope (TEM) facilitated by protein labeling (Figure 2.3). Different helical patterns of surface-bound protein molecules were clearly resolved under TEM, corresponding to different chiralities of as-formed DNA nanotubes.

Interestingly, Sharma et al. [2009] found that gold nanoparticles could efficiently control the conformation of a DNA nanotube based on the strong and tunable steric hindrance between adjacent gold nanoparticles attached to the outer surface of the nanotube. It was found that large gold nanoparticles almost exclusively favored the formation of stacked-ring nanotubes, while small gold nanoparticles resulted in either helical or staked-ring nanotubules. The formation of stacked-ring structures could be related to the strong steric hindrance between large nanoparticles that significantly increased the energy penalty associated with an in-plane twist of the DNA lattice, bearing in mind that such a twist was necessary for the formation of a helical nanotube. Besides, the strong short-range steric repulsion between large gold nanoparticles also resulted in a greater curvature of a DNA lattice and, therefore, thinner tubes.

By maximizing the use of a sequence symmetry rule, Liu et al. [2006] realized the assembly of 1D DNA nanotubes from a single DNA strand composed of four self-complementary segments. Such a strand would dimerize into a quasi-DX tile by base pairing of the middle segments in two strands. The as-formed "DX" motif could further self-associate into an extended 2D lattice by sticky-end cohesions. Some unidentified strain forces within the DNA lattices were believed to be the driving force for the formation of a screwed DNA nanotube via scrolling up a 2D lattice. In this design, merely a single DNA strand was needed, which largely reduced the load in sequence design and eliminated commonly existing stoichiometry problem in a multicomponent self-assembly system.

Ke et al. [2006] investigated the formation mechanisms of DNA nanotubes using rectangular DNA tiles with different aspect ratios containing 4, 8, and 12 juxtaposed duplexes. After a systematic experimental survey, they finally came to the conclusion that tile anisotropy, stronger sticky-end cohesions, and uncorrugated design all favor the formation of a tubular geometry during a 2D assembly.

2.2.2 Two-Dimensional DNA Self-Assembly

It has been well recognized that the rigidity of a DNA tile is a key factor that determines the formation of well-defined DNA lattices. Strategies to enhance the rigidity of a DNA tile include the use of double or multiple crossover DNA molecules that have achieved great successes [LaBean et al. 2000; Liu et al. 2004; Mathieu et al. 2005; Reishus et al. 2005]. Besides that, the use of a rhombus-shaped building block with its duplex edges jointed at each corner in the form of a Holliday junction also allowed for a successful assembly of 1D and 2D DNA nanostructures [Mao et al. 1999].

In 1998, Eric Winfree et al. brought the first series of 2D DNA "crystals" into the world [Winfree et al. 1998]. As shown in Figure 2.4, the authors used two sets of DX

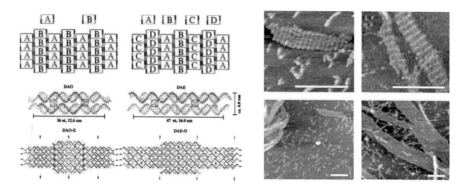

FIGURE 2.4 Programmed assembly of two-dimensional DNA lattices with DX tiles. (Reprinted by permission from Macmillan Publishers Ltd: [*Nature*] Winfree, E. et al. 1998. *Nature* 394: 539–544. Copyright 1998.)

tiles, DAO and DAE, to assemble the 2D lattices. Each of the DX tiles had four properly chosen sticky ends to guide the self-assembly pathway. Ligation of as-formed 2D lattices provided a convenient way to verify the correct connectivity of the DX tiles. The basic process of this ligation assay was to resolve a ladder of DNA bands (isotropically labeled by ^{32}P) on a denaturing PGAE (polyacrylamide gel electrophoresis). These visible bands indicated the polymerization of some specially chosen report strands that should run continuously and infinitely within a correctly assembled DNA lattice. Directly observing the lattice "fringes" of the structures by AFM was possible, which was further facilitated by labeling one tile with a hairpin structure or a protein–gold conjugate. These molecular labels appeared as "highlighted" (greater in heights) stripes on the lattices with clearly resolvable periodicities consistent with the designs. The labeling of the DNA lattice with a streptavidin-coated nanogold might represent the first-time realization of a DNA-guided nanoparticle assembly, although the order and identity of the particles within the "stripe" was not further characterized in order to address the organization power of the DNA lattice.

In 1999, Mao et al. designed a rhombus DNA tile composed of four Holliday junctions [Mao et al. 1999]. Although the Holliday junction itself was not rigid, the fusion of four junctions into a rhombus tile made it possible to further assemble into 1D and 2D lattices (Figure 2.5). Such an ordered 2D lattice also revealed a 63.5 degree equilibrium conformational angle between two stacked double helices constituting the Holliday junction, in good agreement with previous reports (about 60 degrees).

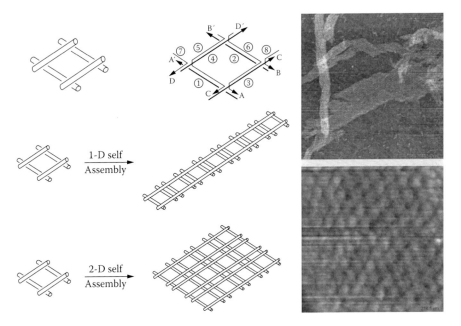

FIGURE 2.5 A rhombus-like DNA motif constructed from Holliday junctions was assembled into one- and two-dimensional DNA lattices. (Reprinted with permission from Mao, C. D. et al. 1999. *J. Am. Chem. Soc.* 121: 5437–5443. Copyright 1999 American Chemical Society.)

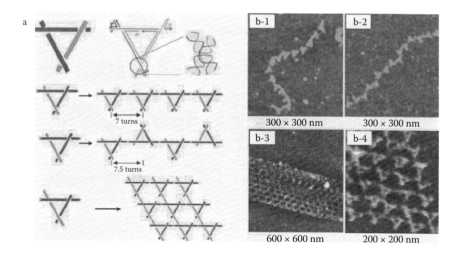

FIGURE 2.6 Construction of DNA lattices based on a tensegrity triangle. Left: schematic representations of a rigid DNA triangle that was further assembled into one- and two-dimensional structures. Right: AFM images of as-formed one- and two-dimensional DNA lattices. (Reprinted with permission from Liu, D. G. et al. 2004b. *J. Am. Chem. Soc.* 126: 2324–2325. Copyright 2004 American Chemical Society.)

In 2004, Liu et al. built a tensegrity DNA triangle containing three identical Holliday junction corners [Liu et al. 2004]. In this structural module, rigidity stemmed from the balance between the push force of rigid struts (double helix) and the pull force of flexible tendons (junctions). The shape of the triangle was fully determined by the lengths of its three edges. By properly introducing sticky ends at one or two of the edges, the DNA triangle would readily assemble into 1D or 2D structures in desired fashions (Figure 2.6). It is noteworthy that the three duplex edges of such a triangle pointed to six distinct directions in a real space that gradually allowed the assembly of millimeter-sized 3D DNA crystals as will be discussed later in Section 2.2.4 [Boer et al. 2010; Zheng et al. 2009].

Another very important assembly unit was developed by Yan and coworkers [Yan et al. 2003], which had a cross shape and contained four four-arm (4×4) junctions (Figure 2.7). The advantage from the increased rigidity of the 4×4 tile, as compared to a J1 junction, was demonstrated by the successful assembly of a 2D DNA lattice with a pseudotetragonal symmetry (considering the heterogeneous sequences). It was also found that some strain forces. *Note:* these inherent forces were reasonable but hard to be predicted and introduced to the lattice by design could facilitate the scrolling of a lattice into tubular forms. In order to obtain an extended monolayer lattice, a "corrugated" design was adopted, which alternated the face of a 4×4 tile up and down during an assembly process. This could be easily achieved by setting the four-stranded DNA "bridge" between two tiles to have an odd number of helical half-turns. In this way, any up and down curlings within the lattice would be immediately canceled out between adjacent tiles (one facing up, the other facing down, or vice versa).

The tubular by-products observed earlier were soon realized to be important and now could be more rationally prepared by introducing some controllable

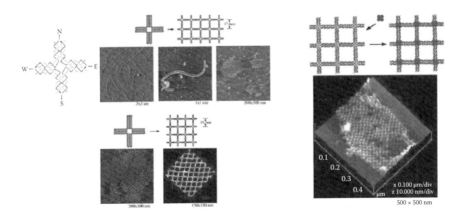

FIGURE 2.7 Self-assembly of a 4×4 tile into DNA nanogrids and the use of a DNA lattice to guide the 2D assembly of a streptavidin molecule. (From Yan, H. et al. 2003. *Science* 301: 1882–1884. Reprinted with permission from AAAS.)

curvatures into the lattice. This work also demonstrated that a DNA lattice could be used to organize nanosized objects such as a streptavidin molecule in a well-controlled order.

He, Tian et al. [He, Chen et al. 2005; He et al. 2005] introduced the rule of sequence symmetry to minimize the number of required DNA sequences to form a symmetric DNA tile, as well as to achieve a symmetry-related homogeneous distribution of the DNA sequences in a DNA lattice. In these cases, self-complementary sticky ends had to be utilized to enable single tile-based self-assemblies. As shown in Figure 2.8, the success of this concept was demonstrated by a tetragonal and a hexagonal array, with the size of a lattice reaching up to 1 mm, which could be attributed to more accurate interstrand stoichiometries and improved sequence homogeneities with the formed lattices.

So far, the 1D and 2D structures we have discussed mostly contained one or two unique tiles that associated into "infinite" periodical arrays. In order to obtain a DNA lattice with a definite size and improved addressability for each tile, a 2D DNA array containing a limited number of DNA tiles was desirable. A primitive idea to achieve a finite-size array would need unique sequences for all component strands (at least sticky ends) that might not be realistic for very large systems. Therefore, it is smart to find a suitable way to reuse the same DNA sequences as more times as possible, sometimes at the price of losing some addressability.

Liu et al. [2005] took advantage of the geometric symmetry of a finite-size DNA array to minimize the number of unique DNA strands. For a 2D array with C_m rotational symmetry containing N DNA tiles, the number of required unique tiles was N/m or Int(N/m)+1. All unique tiles shared the same core strands, different only in their sticky ends. By this means, two types of finite-size arrays containing up to 25 (5×5) assembly tiles were successfully constructed. Park et al. [2006] adopted a different strategy to save the use of unique DNA sequences by a hierarchical multistep assembly that required 18 unique sticky ends (plus their complements) to assemble a 16 (4×4) tile array in 4 steps. Later in this chapter, we will introduce a universal

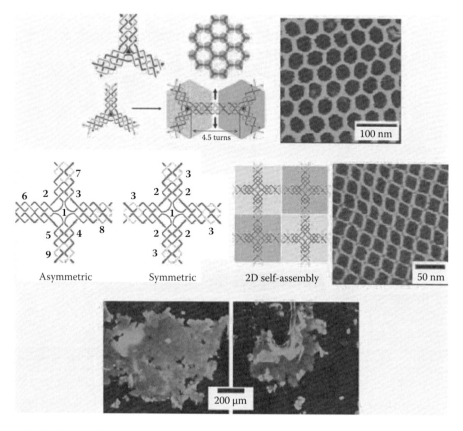

FIGURE 2.8 Self-assembly of three-point-star and four-point-star DNA motifs with three-and fourfold rotational symmetries into two-dimensional honeycomb (reprinted with permission from He, Y., Chen, Y. et al. 2005a. *J. Am. Chem. Soc.* 127: 12202–12203. Copyright 2005 American Chemical Society) and tetragonal (reproduced with permission from He, Y., Tian, Y. et al. 2005b. *Angew. Chem. Int. Ed.* 44: 6694–6696. Copyright Wiley-VCH Verlag GmbH & Co, KGaA) networks. The bottom row shows low-magnification fluorescent microscope images of some super large DNA lattices that was hard to locate under AFM.

method called *DNA origami* that is able to realize a finite-size DNA lattice with arbitrary shapes and full addressability to each pixel of the lattice.

2.2.3   3D SELF-ASSEMBLY OF DNA POLYHEDRA

Compared to a planar array, a DNA polyhedron is even more challenging to build due to the requirement of an efficient structural control in three dimensions. Initial work in making DNA-based wireframe polyhedra emphasized on the correct topographical connectivities of a polyhedron. The first 3D DNA object was a DNA nanocube fabricated by Chen and Seeman in 1991 [Chen and Seeman 1991]. The central task in fabricating such an object was to synthesize six catenated DNA circles representing the six square facets (Figure 2.10) of a cubic geometry. This could be done by

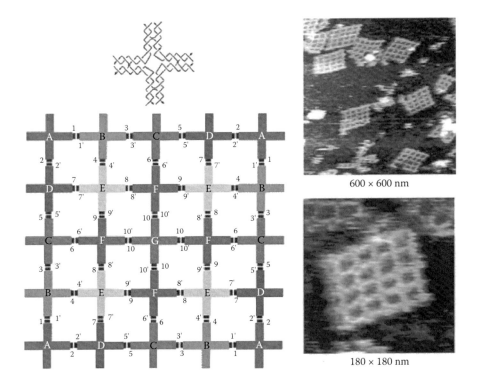

600 × 600 nm

180 × 180 nm

FIGURE 2.9 **(See color insert.)** A 5×5 DNA array assembled from seven unique tiles (only sticky ends were unique) based on a fourfold rotational symmetry of the lattice. (Reprinted with permission from Liu, Y. et al. 2005. *J. Am. Chem. Soc.* 127: 17140–17141. Copyright 2005 American Chemical Society.)

four separate enzymatic ligation steps to covalently circularizing selected (by selective phosphorylation) DNA strands under the help of some short "sealing" strands (Figure 2.10). The first two parallel steps generated two individual circles, the third ligation produced a triclcyclic belt, and the final ligation formed all the six circles with each circle catenated with four neighboring circles (faces) in a DNA cube. The whole set of the six circular DNA strands determined the correct connectivity of a DNA cube. Each edge of the cube was a double-helical DNA, while the eight vertices were three-way DNA junctions (Figure 2.10). The correct connectivity of the cube could be verified by enzymatic cuttings of selected one, two, and four edges of the cube to degenerate the product into four-, three-, two-, and single-ring structures.

In 2004, Shih et al. reported the self-folding of a 1,669 nucleotide DNA single strand into an octahedron with the help of five synthetic 40-mer DNA oligos [Shih et al. 2004]. As shown in Figure 2.11, 5 double crossover (DX) struts and 7 paranemic crossover (PX) struts together constituted the 12 edges of the DNA octahedron. The vertices of the octahedron were four-way junctions that joined the inner-layer double helices to define eight equilateral triangular facets as shown in Figure 2.11. The structure was formed in the absence of knottings or catenations simply by a heat denaturation and following up cooling renaturation, which was different from the

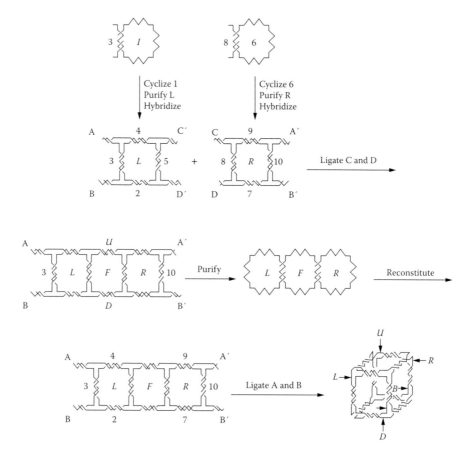

FIGURE 2.10 Schematic synthetic process of a DNA cube composed of six catenated DNA circles. (Reprinted by permission from Macmillan Publishers Ltd: [*Nature*] Chen, J. H. and Seeman, N. C. 1991. *Nature* 350: 631–633. Copyright 1991.)

DNA cube. Under cryo-conditions, 3D reconstructed TEM images clearly revealed the octahedron geometry of the DNA assembly. This work could be viewed as a primitive version of the later-on developed technology of DNA origami for the building of almost arbitrarily complicated 2D and 3D patterns, restricted only by imagination and the length of available single-stranded DNA scaffolds.

Another important work by Goodman et al. [2004] demonstrated the use of short oligonucleotides to quickly assemble DNA tetrahedra in one step with impressively high yields. In a following work, these authors further demonstrated the stereoselective assembly of the tetrahedron by an improved design [Goodman et al. 2005]. The resulting tetrahedron could withstand significant compressive force up to 0.1 N applied by an AFM tip.

To construct a finite-size 3D polyhedral object often requires the use of large numbers of unique DNA strands, which, though resulting in full addressability for each strand, will inevitably reach a limit with the increased size and complexity of a DNA

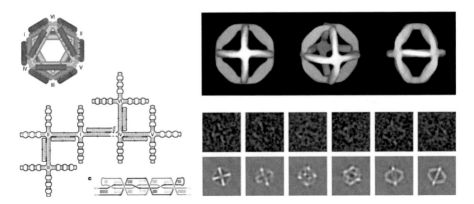

FIGURE 2.11 A hollow wireframe DNA octahedron folded from a long single-strand DNA. Left: schematic structures; Right: representative raw cryo-TEM micrographs accompanied by corresponding projections of a reconstructed octahedron. (Reprinted by permission from Macmillan Publishers Ltd: [*Nature*] Shih, W. M. et al. 2004. Nature 427: 618–621. Copyright 2004.)

object. In response to these challenges, He et al. [2008] employed three-point-star motifs with a threefold sequence symmetry as three-way junction vertices that were further fused into the edges of a specific polyhedron. Very impressively, by simply controlling the flexibility (varying the lengths of three central T-loop bulges of the motif) and the concentration of the three-point-star tile as well as the relative face orientation between adjacent tiles the authors demonstrated the highly selective syntheses of a DNA tetrahedron, a dodecahedron, and even a buckyball structure without the need of any product purifications. It is noteworthy that the syntheses of all these complicated 3D structures only required three unique strands. Reconstructed 3D images based on Cryo-TEM data provided convincing evidence for the formation of desired polyhedral structures (Figure 2.12). This high-fidelity one-step assembly of polyhedral structures makes it possible to encapsulate nanosized molecules such as an enzyme or a catalyst particle for the purpose of protection against deactivations, with the flux of small reactant and product molecules in and out of the cages unaffected. Besides, nanobiomedical applications with these polyhedra as nanosized drug carriers also have a great promise.

2.2.4 3D SELF-ASSEMBLY OF PERIODICAL DNA CRYSTALS

Seeman's initial motivation to launch the field of structural DNA nanotechnology was to grow 3D DNA crystals with fully designable and predictable crystalline phases. These crystals are expected to find important applications in crystallizing proteins (or nanocrystals) for diffraction-based structural determinations [Seeman 2003, 2010]. The following decades of development, albeit fruitful in overcoming many scientific and technical challenges, did not really achieve this goal. One possible reason is that sticky-end cohesion is useful in controlling the connectivity of a DNA lattice but lacks the control in determining the coordinates of every atom of

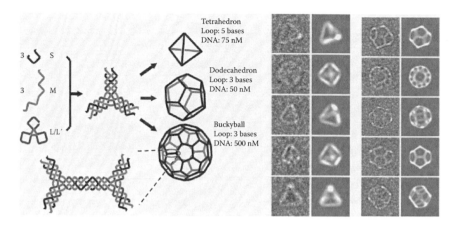

FIGURE 2.12 Design of threefold symmetric three-point-star motifs and their assembly into various supramolecular polyhedra. The right panel showed raw cryo-TEM images of individual particles along with corresponding projections of reconstructed three-dimensional objects. (Reprinted by permission from Macmillan Publishers Ltd: [*Nature*] He, Y. et al. 2008. *Nature* 452: 198–202. Copyright 2008.)

a DNA strand. This means a special building block that has a well-defined shape, high mechanical rigidity, reasonable molecular symmetry, and suitable cohesive interaction strength has to be considered. Benefiting from Chengde Mao's tensegrity triangle [Liu et al. 2004] and the sequence symmetry methodology [He, Tian et al. 2005], a milestone breakthrough happened in the year 2009, when Seeman and Mao's research teams collaboratively addressed this challenge [Zheng et al. 2009]. As shown in Figure 2.13, the authors utilized a small (21 base pairs for each edge with an inter-junction distance of 7 base pairs) DNA tensegrity triangle with a three-fold rotational symmetry as a single robust module for the assembly of macrosized DNA crystals, and achieved an X-ray diffraction resolution down to 4 Å. The three Holliday junction–like corners of the triangle determined that the three duplex edges were stacked one on another and pointed to six distinctive directions in a real space. By appending six very short (2-base long) sticky ends to the triangle, the authors successfully obtained DNA crystals with a typical size reaching 0.1 mm. This long-

FIGURE 2.13 Designed assembly of a rhombohedral DNA crystal from a tailor-made tensegrity triangle. (Reprinted by permission from Macmillan Publishers Ltd: [*Nature*] Zheng, J. P. et al. 2009. *Nature* 461: 74–77. Copyright 2009.)

dreamed success is expected to expedite the research in DNA crystal engineering toward various applications. For example, the DNA crystal can be taken as a host to put in various guest molecules and/or nanoclusters for X-ray–based structural and functional elucidations.

2.2.5 DNA Origami

The strategies we have mentioned so far are good at making periodically extended or finite-size DNA lattices based on the programmable interactions between DNA sticky ends. However, with these strategies, it was difficult to construct arbitrarily shaped DNA nanostructures with addressability for each componential strand. This well-recognized challenge was not satisfactorily tackled until the advent of DNA origami. In Shih and coworkers' masterpiece, a DNA octahedron, a 1.7 k base long single DNA strand folded on itself to form a desired structure with the assistance of five additional short oligos. Formation of PX and DX junctions drove the folding process to give a DNA octahedron. The concept of building complicated DNA nanostructures by folding a long single strand was further developed into a new technique called DNA origami. In 2006, Rothemund [2006] described the details of his DNA origami and employed this powerful tool to build various planar shapes and patterns with finite sizes around 100 nm (Figure 2.14). The core idea was to make use of a 7249nt DNA single strand of the virus M13 genome (also used as a cloning vector) as a scaffold that could be folded into desired geometries with the help of over 200 short "staple" strands. The final structure had a spatial resolution (pixel size) of about 5.4 × 6 nm. DNA origami provides a universal approach to construct finite-size structures with high complexity, has vastly expanded the library of fully addressable DNA structures. This excellent addressability could allow for many possible applications [Sanderson 2010], including a nucleic acid nanoarray [Ke et al. 2008] and a plug-and-play DNA "breadboard" to harbor a nanoscale circuitry and other functional units. Some of these applications have been demonstrated to some extent [Maune et al. 2010].

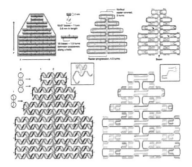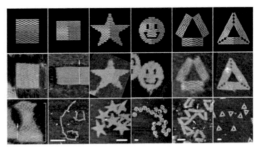

FIGURE 2.14   (See color insert.) Scaffolded DNA origami capable of making accurate geometric shapes in two dimensions. Left panel: schematic drawings showing how a long DNA strand is folded into a defined shape with the help of short-fixing strands. Right panel: theoretical models and experimentally achieved geometrical shapes. (Reprinted by permission from Macmillan Publishers Ltd: [*Nature*] Rothemund, P. W. K. 2006. *Nature* 440: 297–302. Copyright 2006.)

The discovery of DNA origami has opened up new horizons in DNA technology, based on which many sophisticated experiments can now be carried out. For example, the excellent addressability of DNA origami made it possible to build a soluble nanosized-hybridization array (namely, a nanoarray following the nomenclature of DNA microarray) for RNA detection [Ke et al. 2008]. In this work, AFM was used to detect the hybridization events based on the increased heights of some "nanospots" on the origami after hybridizing with nucleic acid targets. Such a homogeneous nanoarray has potential in ultimately reducing the spot size of a DNA array, improving hybridization efficiency, and minimizing surface adsorptions, all challenging for a regular heterogeneous microarray.

A majority of the following-up pursuits in DNA origami were directed toward the building of 3D objects. Typical examples include the assembly of nanocontainers and various complicated and reality-mimic nanoarchitectures. In 2009, Ke et al. made a DNA tetrahedron with closely packed mesh-structure faces using DNA origami [Ke et al. 2009]. Since previously reported DNA polyhedra had wireframe structures and open faces (a more suitable name is "cage"), the closed structure of the tetragonal DNA origami better matched the name "nanocontainer" and might be filled with even smaller molecules determined by mesh hole size. Therefore, such a molecular container could be used to encapsulate special cargos so that the highly addressable surfaces of the container might facilitate target-specific deliveries. Andersen et al. [2009] demonstrated the concept of a DNA origami-based nanobox that had a controllable lid, which could be conveniently opened and closed by externally supplied DNA "keys." The next step is to put some real things, such as a drug molecule, a catalyst particle, or an enzyme, inside these containers and see what happens next.

The Shih group further demonstrated the power of DNA origami in building various more complicated 3D nanoobjects with excellent controls on their shapes, curvatures, and helicities [Dietz et al. 2009; Douglas et al. 2009]. Most recently, Liedl et al. [2010] made progress in constructing prestressed tensegrity structures such as a prism, normal, and mechanically distorted kites, with rigid helical bundles as "wood" and soft DNA single strands as "cord" (Figure 2.15). These results show that DNA origami has an unprecedented ability to increase the complexities of nanosized objects and place excellent controls on almost every one of their structure details. Therefore, it is reasonable to expect some more important applications of DNA origami in the near future. For example, DNA origami might be able to help fabricating nanoscale electric circuits that are currently hindered by the challenging requirements for the locating and wiring of nanoelectronic units.

One realistic application of DNA nanotechnology was demonstrated by the use of a DNA origami nanotube as an alignment media for a surfactant-solubilized membrane protein that would facilitate its NMR (nuclear magnetic resonance) structural determination [Douglas et al. 2007]. A 0.8 µm long DNA nanotube with a single molecular weight was built by scaffolded DNA origami. This nanotube was a six double-helix bundle bound together by strand crossovers to give a hollow interior (Figure 2.16). This tubular form of DNA was able to form a liquid crystalline phase at a DNA concentration of 28 mg/mL (Figure 2.16c). One advantage of the DNA nanotube as compared to other alignment media was that the liquid crystalline phase it formed was almost unaffected by the coexistence of a large amount of surfactant

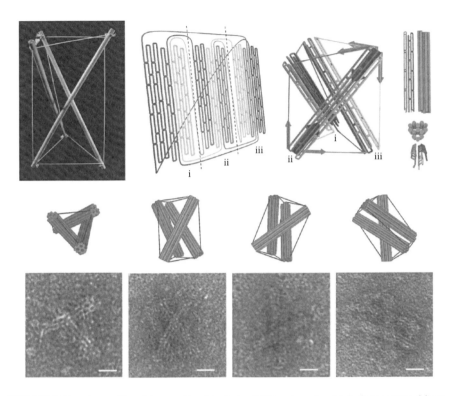

FIGURE 2.15 Assembly of tensegrity objects by DNA origami, which demonstrated how a complicated three-dimensional architecture can be constructed with rigid wood (duplex bundle) and soft cord (single-stranded DNA). (Reprinted by permission from Macmillan Publishers Ltd: [*Nature Nanotechnol.*] Liedl, T. et al. 2010. *Nat. Nanotechnol.* 5: 520–524. Copyright 2010.)

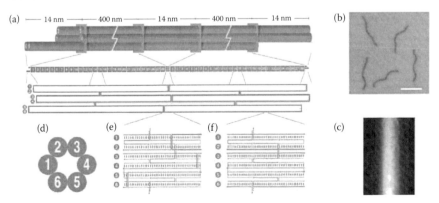

FIGURE 2.16 A DNA nanotube built by DNA origami. (a) Schematic of the assembly; (b) TEM images of a negatively stained sample; (c) birefringence exhibited between crossed polarizers by a DNA nanotube liquid crystalline phase. (Reprinted with permission from Douglas, S. M. et al. 2007. *Proc. Natl. Acad. Sci. USA.* 104: 6644–6648. Copyright (2007) National Academy of Sciences, USA.)

molecules for the solubilization of the membrane protein. Residual dipolar couplings (RSCs) of detergent-solubilized membrane protein were measured, which provided important NMR structural information.

2.3 DNA-BASED SELF-ASSEMBLY OF NANOMATERIALS

We have clearly seen that DNA is a very smart molecule that can form various specified structures by self-assembly. This ability of DNA nicely matches the requirements in nanomaterials science toward the organization of individual inorganic nanounits into functional nanoworks. The resulting superstructures of various single- or multiple-component nanomaterials are very promising for applications such as nanoelectronics, sensors, novel catalysts, and biological materials. Section 2.3 of this chapter will cover some recent methodological successes of using DNA to organize nanostructured materials. In most of the cases, DNA acted as a programmable structural module, smart glue, or a noncovalent molecular linker to build some important inorganic nanostructures mostly composed of metal nanoparticles and carbon nanotubes and nanosheets.

2.3.1 DNA-BASED SELF-ASSEMBLY OF GOLD NANOPARTICLES IN ONE AND TWO DIMENSIONS

Owing to the attractive optical, electronic, catalytic, and other properties of metal nanoparticles, research interest in their assembly into desired superstructures continued to grow during the past decade. The idea of using DNA to organize gold nanoparticles could be traced back to the pioneering work by Mirkin et al. [1996] and Alivisatos et al. [1996]. Two different strategies were developed with an emphasis on different levels of control on the assembly behaviors of gold nanoparticles, resulting in a diversity of applications. The method developed by Alivisatos et al. was able to produce monofunctionalized DNA–gold conjugates assisted by gel electrophoretic purifications which nicely met the purpose of using the nucleobase information within a DNA strand to guide the organization of gold nanoparticles. For example, various discrete gold nanoparticle structures have been achieved based on DNA-templated assemblies [Fu et al. 2004; Loweth et al. 1999].

Ordered assemblies of nanoparticles in the form of extended 1D, 2D, or even 3D arrays not only create possible couplings and synergies in their properties and functions but also make it possible to manipulate these nanosized objects on a scale readily accessible under an optical microscope (micrometer scale). Taking advantage of a rolling circle amplification (RCA) technique [Fire and Xu 1995; Liu et al. 1996], Deng et al. [2005] achieved micrometers-long single-stranded DNA composed of hundreds of tandemly linked oligo repeats complementary to a circular DNA template responsible for guiding the enzymatic DNA elongation. This long single-stranded DNA was well suited for the organization of gold nanoparticles into extended 1D arrays. Following Alivisatos' protocol, a 5 nm gold nanoparticle was functionalized with a single DNA strand that was complementary to the repeating unit of the RCA product. Simply mixing the DNA-conjugated gold nanoparticles

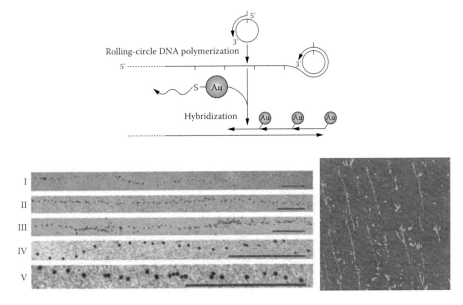

FIGURE 2.17   Self-assembly of a mono-functionalized DNA–gold conjugate along a rolling-circle polymerized DNA template to form a micrometer long linear array of gold nanoparticles. (Reproduced with permission from Deng, Z. X. et al. 2005b. *Angew. Chem. Int. Ed.* 44: 3582–3585. Copyright Wiley-VCH Verlag GmbH & Co, KGaA.)

with the linear template at an appropriate molar ratio resulted in 1D gold nanoparticle arrays up to 4 µm long. As shown in Figure 2.17, the as-formed linear gold nanoparticle arrays could be easily stretched and aligned either on a carbon-coated TEM grid or on a silicon wafer by molecular combing [Deng and Mao 2003].

The preceding work successfully demonstrated the possibility of assembling gold-tagged DNA into extended 1D arrays. However, several bottlenecks had to be properly dealt with before attempting 2D nanoparticle assemblies. For example, gold nanoparticles are not stable under the ionic conditions necessary for the stabilization of a DNA 2D lattice since as demonstrated by the fact that the presence of Mg^{2+} will turn the nanoparticles into irreversible random aggregates. Another well-known challenge is to find a suitable way to reduce DNA adsorptions on a gold surface, a major factor that significantly lowers the hybridizability of a gold nanoparticle-tagged DNA.

In 2006, the Yan group [Sharma et al. 2006] employed 4×4 DNA tiles to direct the assembly of gold nanoparticles into 2D periodical arrays (Figure 2.18a). One component strand of the cross-shaped assembly tile was conjugated with a 5 nm gold nanoparticle. During this experiment, the gold–DNA conjugate had to tolerate high salt buffers (in this work, high concentration of Na^+ was used instead of Mg^{2+}) that were essential for the formation of the DNA structures. To face this situation, a single-DNA-functionalized gold nanoparticle [Zanchet et al. 2001], after being produced, was backfilled with a monolayer of thiolated T5 oligomers on its surface. The coating of a layer of T5 strands greatly enhanced the resistivity of the gold nanoparticles against a strongly ionic environment. In the meantime, DNA adsorption was efficiently alleviated due to the existence of a high density of the T5 strands that blocked

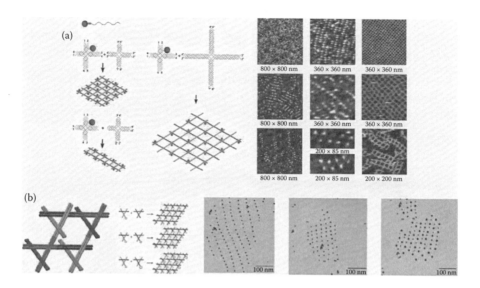

FIGURE 2.18 **(See color insert.)** (a,b) DNA-programmed two-dimensional assembly of gold nanoparticles into various predetermined arrays; (b) enhanced structural control on a nanoparticle array was achieved by the use of a rigid triangular DNA motif with DX edges for the assembly. ([a] Reproduced with permission from Sharma, J. et al. 2006. *Angew. Chem. Int. Ed.* 45: 730–735. Copyright Wiley-VCH Verlag GmbH & Co, KGaA. [b] Reprinted with permission from Zheng, J. W. et al. 2006. *Nano Lett.* 6: 1502–1504. Copyright 2006 American Chemical Society.)

the active adsorption sites on the gold nanoparticles. In this way, the Yan group successfully demonstrated the assembly of gold nanoarticles into well-defined 2D tetragonal arrays with easily tunable periodicities based on a two-tile assembly system.

As the in-plane rigidity of a DNA 20 lattice was a critical factor that will determine the structural regularity of a nanoparticle array, Seeman and coworkers [Zheng et al. 2006] designed a triangular motif (can be viewed as an altered version of Chengde Mao's tensegrity triangle) with rigid DX edges to guide the nanoparticle assembly. Two edges of the triangular tile were involved in the formation of a DNA lattice, with the third edge attached with a gold nanoparticle. TEM micrographs strongly supported the notion that such a rigid motif had an impressive control on the interparticle spacings and the positions of nanoparticles on the DNA lattices (Figure 2.18b).

2.3.2　3D Ordering of Gold Nanoparticles with DNA

In addition to the attempts to organize gold nanoparticles in one and two dimensions, it is equally important to achieve a 3D-ordered array of nanoparticles that will further enrich the structural and functional varieties of nanoparticle assemblies. Previous work in controlling the aggregation of gold nanoparticles with DNA did not achieve a relatively long-range order of nanoparticles [Mirkin et al. 1996]. More recently, the Mirkin group realized DNA-assisted crystallizations of gold nanoparticles into both close-packed and non-close-packed crystalline phases by fine-tuning

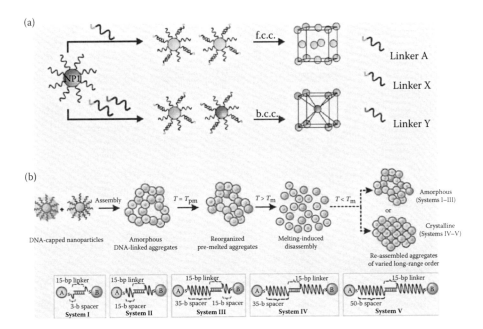

FIGURE 2.19 (**See color insert.**) Schematic drawings showing the formation of three-dimensional gold nanoparticle supra-crystals in two different crystalline phases based on a single- and a binary-component system. (Copyright request pending, credit line to be added) ([a] Reprinted by permission from Macmillan Publishers Ltd: [*Nature*] Park, S. Y. et al. 2008. *Nature* 451: 553–556. Copyright 2008. [b] Reprinted by permission from Macmillan Publishers Ltd: [*Nature*] Nykypanchuk, D. et al. 2008. *Nature* 451: 549–552. Copyright 2008.)

the DNA linkers [Park et al. 2008] (see Figure 2.19a) on the gold nanoparticles. The authors designed a single-component system with a self-complementary DNA linker to produce a close-packed face-centered-cubic (f.c.c) colloidal crystal as evidenced by small-angle X-ray diffractions. Alternatively, a binary system with complementary DNA linkers located on two different nanoparticle units resulted in non-close-packed body-centered-cubic (b.c.c) supracrystals of gold nanoparticles. There were various factors affecting the orders of the assembled gold colloids including thermal annealing scheme, a molecular spacer located between DNA and gold, and a nonpairing flexor base bridging the linking sequence and a double-stranded DNA domain. The formation of a special crystalline phase could be rationalized from an energy minimization standpoint that favored maximized DNA linkages around a gold nanoparticle. The maximized number of hybridized DNA linkages around each particle is 12 in the single component f.c.c. close-packed system, and 8 for the binary component b.c.c. (non-close-packed) assembly. On the other hand, an entropic contribution (favoring a close-packed structure) was dominant when samples were thermally annealed from a temperature above the melting point of the system that voided the enthalpic contribution reflective of interparticle attractions. As a result, a substitutionally disordered f.c.c. phase was formed for the binary system when the nanoparticles were heated above the melting temperature of the system followed by

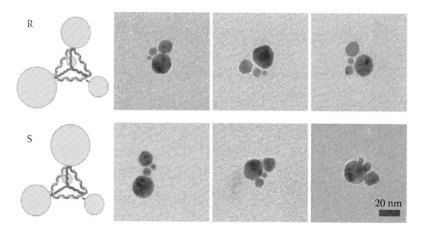

FIGURE 2.20 Self-assembly of differently sized gold nanoparticles into chiral pyramids. (Reprinted with permission from Mastroianni, A. J. et al. 2009. *J. Am. Chem. Soc.* 131: 8455–8459. Copyright 2009 American Chemical Society.)

a slow cooling. At the same time, Nykypanchuk et al. [2008] reported the formation of 3D crystals of gold nanoparticles through controlling the length of a flexible single-stranded DNA spacer between the gold particle and a hybridizing sequence (Figure 2.19b). It was found that more flexible spacers favored the formation of a better-crystallized structure. This method used thermal treatment of the sample across the melting point to achieve a desired crystallinity (structural order).

Apart from packing nanoparticles into a periodically ordered array, the Alivisatos group continued their efforts to build discrete nanoparticle assemblies [Mastroianni et al. 2009]. The authors successfully appended gold nanoparticles on the four vertices of a DNA tetrahedron. In addition, as shown in Figure 2.20, the authors achieved chiral assemblies based on a tetrahedral arrangement of four different-sized gold nanoparticles. Since TEM images only reflected various projections of a flattened nanoparticle assembly, cryo-TEM or optical characterizations of the samples were necessary to evidence the native structures and the properties of the 3D pyramidal assemblies.

Another important work regarding the 3D assembly of nanoparticles was reported by Sharma et al. [2009]. As shown in Figure 2.21, gold nanoparticles were assembled on a nanotubular DNA template to form a helical or stacked-ring structure as confirmed by TEM imagings. Most interestingly, it was found that the conformation of a DNA nanotube could be dictated by the use of large gold nanoparticles due to enhanced steric hindrances. Detailed explanations of the underlying mechanism for such a control have been presented in Section 2.1.1.

2.3.3 DNA-Based Self-Assembly of Carbon Nanotubes

Carbon nanotubes have been playing a key role in materials science since the findings by Iijima and Bethune et al. [Bethune et al. 1993; Iijima 1991; Iijima and Ichihashi 1993], thanks to its unique physical and chemical properties and various potential applications. Although significant breakthroughs have been achieved regarding

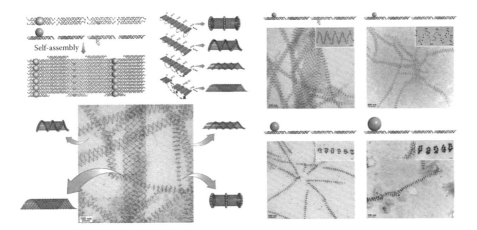

FIGURE 2.21   DNA nanotube-guided assembly of gold nanoparticles. The right panel shows that the gold nanoparticles exhibited an efficient determination of the nanotube's conformations by interparticle steric hindrance: larger particles meant a stronger repulsion and thus favored the formation of a thinner nanotube with a stacked-ring conformation. (From Sharma, J. et al. 2009. *Science* 323: 112–116. Reprinted with permission from AAAS.)

the DNA-programmed assembly of gold nanoparticles, more functional building blocks are definitely required to build a multifunctional and integrated nanosystem. Obviously, single-walled carbon nanotubes (SWNTs) would be the next candidate following the great success of gold nanoparticles on the basis of the following considerations: (1) a SWNT can be metallic or semiconducting depending on its structural chirality; (2) Carbon nanotube has some unique optical properties such as the chirality-related aborbance and fluorescence spectra in the visible to near infrared (vis-NIR) domain as well as Raman scattering properties, which give it high potential in various realistic applications; (3) carbon nanotubes promise to increase the structural and functional diversities of self-assembled nanomaterials, and allow for easy tailoring and measurement of their electronic/optoelectronic properties.

To develop carbon nanotube-based building blocks for DNA-programmed assembly, two problems need to be tackled. First, sidewall DNA functionalization of SWNTs is necessary in order to achieve orientation control of SWNTs on a DNA-patterned landscape, important for a self-assembly based device fabrication. However, high-density decoration of DNA on the sidewall of a SWNT is difficult due to the lack of suitable reactive groups. Second, in the case of DNA-programmed material assembly, it will never be inappropriate to overemphasize the importance of the high hybridization activity of DNA. Unfortunately, hybridizations of DNA on an SWNT may be hindered due to strong pi-staking interactions between DNA bases and the electron-conjugated surface of the SWNT. Therefore, a suitable way to prevent this adsorption is highly demanded.

The Deng group [Li et al. 2007] achieved some important successes in solving the aforementioned problems toward the sidewall DNA decorations of SWNTs based on a DNA wrapping strategy, which was originally employed by Zheng et al. [Zheng et al. 2003; Zheng et al. 2003] to realize chromatographic length sorting of carbon

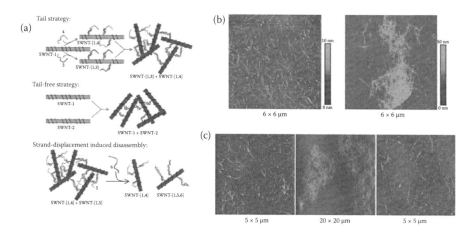

FIGURE 2.22 A "tail" strategy achieved high hybridization efficiency of DNA-conjugated carbon nanotubes as demonstrated by a hybridization-controlled assembly and disassembly of carbon nanotubes. (Reproduced with permission from Li, Y. L. et al. 2007. *Angew. Chem. Int. Ed.* 46: 7481–7484. Copyright Wiley-VCH Verlag GmbH & Co, KGaA.)

nanotubes. Unfortunately, the DNA sequences tightly wrapped on carbon nanotubes were not expected to have a satisfactory hybridization activity [Li et al. 2007]. To improve this situation, Li and Deng et al. grafted another DNA sequence as a hybridization tail. Wrapping DNA around the carbon nanotubes was achieved by reacting an excess amount of the secondary strand with the DNA–nanotube conjugate for a prolonged time followed by a purification of the product by centrifugation. In this way, a DNA duplex was formed between the wrapping and the hybridizing DNA segments. The rigidity of such a duplex insert was responsible for disabling a tight wrapping of the grafted tail on the SWNT. In addition, the DNA strands wrapping around the vicinity tubular areas also helped to prohibit a strong adsorption of the DNA graft on the SWNTs through electrostatic repulsions. Actually, the DNA tails were found to have greatly improved hybridization kinetics as compared to a directly wrapped hybridizing sequence (a tail-free strategy; see Figure 2.22), as revealed by monitoring the DNA-hybridization-induced aggregation reaction between two parts of carbon nanotubes grafted with complementary DNA tails. The carbon nanotube aggregates could reach a size above 10 microns, providing further evidence for the high hybridization efficiencies of the grafted DNA tails. Redispersion of the carbon nanotube aggregates was possible based on a DNA-strand-displacement reaction. The authors believed that such a DNA–SWNT conjugate would find important applications in DNA-programmed self-assembly and DNA-directed nanofabrication of carbon nanotube-based hybrid nanomaterials.

Recently, DNA origami has been employed for nanoparticle organizations [Ding et al. 2010; Pal et al. 2010]. Especially, DNA origami could be used as a template to realize an orientation control of SWNTs [Maune et al. 2010]. As shown in Figure 2.23, the highly addressable surface of DNA origami made it possible to pattern two lines (arranged in a cross shape) of a capture DNA that could be recognized by the DNA decorated on the carbon nanotubes. The two lines of the single-stranded

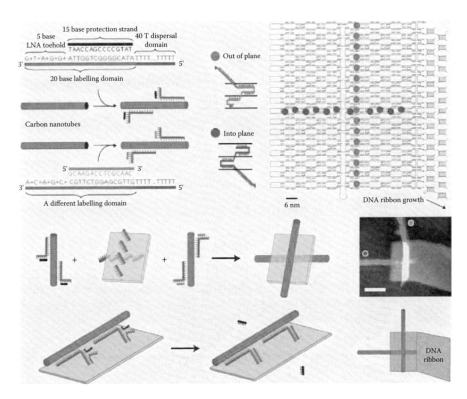

FIGURE 2.23 (**See color insert.**) DNA origami-based self-assembly of two single-walled carbon nanotubes into a cross junction with a 90° orientation related to each other. Note that the two nanotubes were assembled on opposite faces of a DNA origami. (Reprinted by permission from Macmillan Publishers Ltd: [*Nature Nanotechnology*], Maune, H. T. et al. 2010. *Nat. Nanotechnol.* 5: 61–66. Copyright 2010.)

DNA "hooks" protruded out of the two opposite surfaces of the DNA origami. The authors overcame the DNA hybridizability problem by using a duplex-protection strategy. As a result, a strand-displacement-based deprotection of the hybridizing sequence was needed during the assembly process of carbon nanotubes. In this way, two carbon nanotubes could be attached on the upper and lower surfaces of the DNA origami, respectively, with a 90° orientation related to each other (Figure 2.23). Such a cross-junction formed by single-walled carbon nanotubes could be viewed as a prototype field-effect transistor with the DNA origami acting as a "gate oxide" (insulating layer) sandwiched between the two carbon nanotubes (Figure 2.23).

2.3.4 DNA-BASED SELF-ASSEMBLY OF GOLD–CARBON HYBRID NANOSTRUCTURES

Forming heteronanostructures between nanoparticles and single-walled carbon nanotubes provides a useful way to tailor the electric properties of carbon nanotubes. In addition, hybrid nanosystems provide novel interfaces to control electron and energy

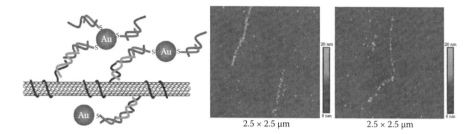

FIGURE 2.24 Sequence-specific one-dimensional assembly of gold nanoparticles on single-walled carbon nanotubes. (Reproduced with permission from Li, Y. L. et al. 2007. *Angew. Chem. Int. Ed.* 46: 7481–7484. Copyright Wiley-VCH Verlag GmbH & Co, KGaA.)

transfers between different materials. Therefore, applications of these hybrid nanostructures in catalysis, sensing, and electronic optoelectronic devices are desirable. However, suitable methodologies toward the constructions of these nanostructures have to be developed.

Li et al. [2007] successfully realized the grafting of highly hybridizable DNA sequences on the sidewalls of SWNTs, which made it possible to further assemble nanoparticles on the carbon nanotubes by sequence-specific DNA hybridizations. To achieve this goal, gold nanoparticles bearing complementary DNA sequences were allowed to locate and then hybridize with the DNA "glues" on the carbon nanotubes (Figure 2.24). The resulting nanoparticle–nanotube adducts were stable enough to allow for an electrophoretic purification of the products. The realization of DNA-programmable assembly of heteronanostructured materials offered a new dimension to harness the interactions between different materials toward applications such as biosensing, nanocatalysis, and electronic optoelectronic devices.

Besides the DNA hybridization-driven self-assembly, a simpler gold–thiol bonding interaction could also be used to guide the 1D assembly of gold nanoparticles on SWNTs to produce linear metal–carbon hybrid nanostructures [Han et al. 2007], and one such strategy was shown in Figure 2.25. The researchers took a thiolated DNA, $d(GT)_{29}SH$, to wrap around the SWNTs and serve as a molecular linker between carbon nanotubes and gold nanoparticles. During the experiment, a large excess of gold nanoparticles were added to interact with the DNA-wrapped carbon nanotubes in order to minimize the formation of cross-linked structures. The excessive gold nanoparticles were removed by agarose gel electrophoresis-based purifications. The successful assembly of the gold–carbon hybrid nanomaterials was initially judged electrophoretic analyses, and was further confirmed by AFM imaging (Figure 2.25).

As a new member of the carbon materials family, graphene has recently attracted considerable attentions compared to nanotubes a significant part of these interests stem from some attractive features of graphene including low synthetic cost, scalable production, and the absence of catalyst impurities. Moreover, the existence of a dissolvable oxidized form of graphene, ultimately a large specific surface area and high electron mobility makes graphene and graphene-based hybrid nanomaterials very promising as catalytic, electronic, and battery materials. Liu et al. [2010] found that single-stranded DNA could tightly adsorb on the surface of a graphene oxide (GO)

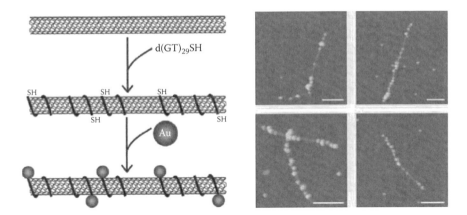

FIGURE 2.25   Gold-thiol bonding driven self-assembly of gold nanoparticles on DNA-wrapped carbon nanotubes. (Reproduced with permission from Han, X. G. et al. 2007. *Adv. Mater.* 19: 1518–1522. Copyright Wiley-VCH Verlag GmbH & Co, KGaA.)

nanosheet by noncovalent pi-pi stacking interactions (Figure 2.26), similar to the case of carbon nanotubes, resulting in a GO–DNA conjugate. By introducing hydrazine as a chemical reductant, the GO–DNA conjugate was further converted into DNA-coated reduced graphene oxide (RGO–DNA). When thiolated DNA oligomers were used to decorate GO and RGO, hybrid nanomaterials between gold nanoparticles and the GO/RGO nanosheets could be obtained through gold–thiol bondings, a strategy similar to the nanoparticle functionalization of carbon nanotubes. Water-soluble metal–graphene hybrid nanostructures were obtained accordingly, which could be employed as novel building blocks for various possible applications as mentioned earlier.

The self-assembly based methodology has potential in assembling different nano-objects on a single GO/RGO nanosheet or a SWNT due to the successful elimination of a compatibility problem between nanoparticle synthesis and assembly. In this regard, the assembly-based strategy is superior to in-situ chemical reduction based techniques for the preparation of metal–carbon nanostructures. In addition, the non-covalent DNA modifications of the sp^2 carbon materials prevented any permanent alterations of their pristine chemical structures, a common drawback of covalent chemical decorations.

Instead of DNA, recent research [Liu et al. 2010] has demonstrated that protein could serve well as a multi-functional molecular linker to realize a graphene-based universal self-assembly platform for the assembly of various metallic and nonmetallic nanoparticles (Figure 2.27). By this strategy, nanoparticles with different sizes, compositions, and surface properties could be assembled either individually or combinatorially on the carbon nanosheets. The special role of protein in such a process was closely related to its surface-exposed amino acid residues that led to different interactions with a solid surface. Possible interactions responsible for the self-assembly process included but were not limited to hydrogen and chemical bondings, hydrophobic force, and electrostatic interaction. Such a universal method would enable a self-assembly based combinatorial material synthesis with easily achieved

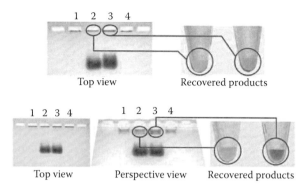

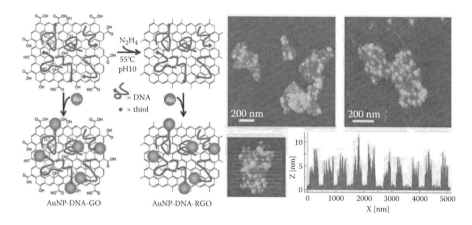

FIGURE 2.26 DNA-mediated assembly of gold nanoparticles on graphene oxide and reduced graphene oxide. (From Liu, J. B. et al. 2010b. *J. Mater. Chem.* 20: 900–906. Reproduced by permission of The Royal Society of Chemistry (RSC).)

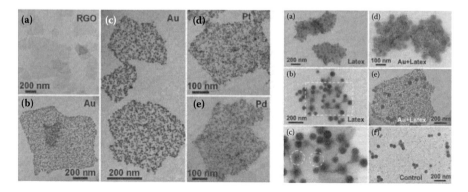

FIGURE 2.27 Protein-directed assembly of various nanoparticle materials (see information in the figure) on RGO (left) and GO (right). (Reprinted with permission from Liu, J. B. et al. 2010a. *J. Am. Chem. Soc.* 132: 7279–7281. Copyright 2010, American Chemical Society.)

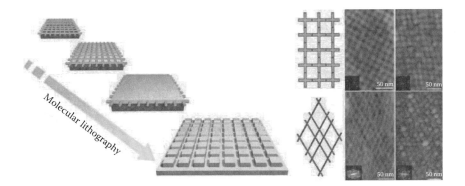

FIGURE 2.28 Molecular lithograghy was utilized to replicate a DNA nanostructure on a gold substrate. The left panel illustrates a typical process of molecular lithography, and the right panel presents AFM images of the DNA structures and their gold replicas. (Reproduced with permission from Deng, Z. X. and Mao, C. D. 2004. *Angew. Chem. Int. Ed.* 43: 4068–4070. Copyright Wiley-VCH Verlag GmbH & Co, KGaA.)

structural, compositional, and functional tailorability facilitating explorations of the structure–function relationship of multicomponent nanomaterials.

2.3.5 MOLECULAR LITHOGRAPHY

The term *molecular lithography* was introduced by Keren et al. in 2002 [Keren et al. 2002] to name a RecA filament-masked DNA metallization process that resulted in a small insulating gap flanked by two metallized parts on a linearized lambda phage DNA molecule. Deng et al. developed a different meaning of the term molecular lithography aiming at transferring the pattern of a DNA lattice to the surface of a solid substrate. The initial motivation was to develop a DNA-based nanofabrication technique to surpass the diffraction limit of optical lithography-based surface patternings. As shown in Figure 2.28, a typical molecular lithography process comprises several key steps. First, self-assembled DNA nanostructures were prepared and deposited onto a freshly cleaved mica surface. Choosing an atomically smooth mica surface was critical for a successful replication of a DNA nanostructure that had a typical vertical feature height around 1 nm (in air) after being deposited on mica. Second, a thin layer of gold was thermally evaporated to cover the DNA lattice (molecular mask) supported on mica. Finally, the gold film was separated from the DNA mold, and the negative replica of the DNA lattice was exposed. AFM imaging clearly evidenced DNA's replica on the gold film with a lateral resolution <12 nm. In the future, DNA origami could be employed as a supramolecular mask for the molecular lithography such that complexity requirements for applications such as nanocircuitry might be gradually satisfied.

2.4 SUMMARY AND OUTLOOK

It will be interesting and inspiring to look back at some earlier works in this area in order to see what challenges have been solved and what breakthroughs have been achieved since. By looking at a review article published five years ago [Deng et al. 2005], we are excited to see that some of the technical challenges have been solved. For example, the assembly of a randomly shaped DNA array is now an easy job thanks to Rothemund's DNA origami. At the same time, DNA origami shows unprecedented addressability of a 2D DNA lattice. In the previous review, it was stressed that incompatibility problems of some nanoscale building blocks with DNA nanotechnology had to be solved for DNA-programmed material assemblies. Today, gold nanoparticles, carbon nanotubes, and graphene, three important nanomaterials, have been successfully interfaced with DNA. In the year 2005, it was expected that algorithmic DNA assemblies could be executed in higher dimensions to achieve a highly programmable and accurately controlled DNA self-assembly. Although the algorithmic assembly of DNA did not move in this direction, 3D DNA origami, DNA nanotubes, DNA polyhedra and, especially, sticky end-directed macroscopic DNA crystals have constituted meaningful progress in 3D nanoassembly. In the case of nanocircuitry, DNA origami has allowed carbon nanotubes to be assembled into functional cross junctions that can be viewed as prototype field-effect transistors.

With the ever-continuing efforts in DNA nanotechnology, new challenges will be faced. DNA origami has clearly demonstrated the great advantages of fully address-able 2D or 3D DNA arrays. It is still difficult to build a micrometer-sized DNA origami, limited by the availability of a much longer single-stranded DNA scaffold. The combination of sticky-end cohesion-based assembly and top-down fabrications may be able to bring the DNA origami to such a scale at the price of decreasing some addressability. Therefore, a better solution is definitely needed to this problem [Zhao et al. 2010]. The 3D structure of a DNA polyhedron, which can now be easily assem-bled, is an ideal host to encapsulate guest molecules and nanoparticles [Erben et al. 2006; Lo et al. 2010], which may find important applications in catalytic systems with restricted flux for undesirable substrates or products, targeted cargo delivery, and in vivo bioimaging.

With a marriage between DNA-based nanoassembly and the principles of bio-technology and bioengineering, plus the dynamic control realized by DNA-based molecular machines [Bath and Tuberfield 2007; Han et al. 2008; Gu et al. 2010], the area of DNA nanotechnology might be able to make important contributions to the growing discipline of synthetic biology. One important mission of synthetic biology is to seek artificial syntheses of previously nonexistent biological parts and pathways and the redesign of natural biosystems to enable new or enhanced functions and applications. So far, some functioning DNA nanomechanical devices have also been realized and their further integration to achieve a more sophisticated function in vivo or in vitro might be possible [Bath et al. 2007; Chen and Mao 2004; Han et al. 2008; Liao and Seeman 2004]. The rapid evolution of this area seems to surpass the chal-lenges mentioned at the beginning of this chapter.

The use of DNA structures to build materials will keep developing to enter a new phase that may bring the most promising applications for DNA nanotechnol-ogy. DNA-directed assembly is believed to play an important role in fabricating nanostructured materials for the exploration of new properties and functions. For example, DNA wrapping now allows both length and chirality sortings of SWNTs [Tu et al. 2009] that may be further utilized in creating novel structured SWNT-based nanomaterials. On the other hand, we also expect that well-controlled DNA nanostructures can be incorporated with functional nucleic acids for various applications [Lu et al. 2008]. Regarding the methodological developments in DNA-based self-assembly, dynamic [Gu et al. 2010], sequential [Maye et al. 2009], and surface-directed [Hamada and Murata 2009; Sun et al. 2009] assemblies may be further investigated toward more complex hierarchical nanostructures and nanomaterials that may not be achievable through one-pot processes. Also, we can start thinking about the use of reengineered proteins or other synthesized small or macromolecular materials instead of a pure DNA system for functional extensions of DNA-based nanostructures and nanomaterials [Boer et al. 2010; Lo et al. 2010; Mitra et al. 2004]. Apart from ordered structures, recent years' developments also showed great promise for some relatively "disordered" DNA materials such as DNA hydrogels that could be used as protein manufacturing factories [Cheng et al. 2009; Li et al. 2004; Park et al. 2009; Um et al. 2006]. More interesting applications for these materials, especially biomedical applications, are expected.

Another aspect of DNA nanotechnology we did not mention in this chapter is the DNA-templated syntheses of inorganic nanomaterials [Deng and Mao 2003; Deng et al. 2009; Keren et al. 2002; Liu et al. 2004; Yan et al. 2003], including DNA metallization and DNA-controlled nucleation and growth of functional nanocrystals. When combined with DNA-programmed nanoassembly, new developments are expected to flourish in this field. Finally, we would like to add that building an integrated nanomechanical system with various DNA nanomachines acting cooperatively will be important to deliver a true "working" system for highly controlled material synthesis and assembly. This endeavor has already progressed well, an example being a recently realized nanoparticle assembly line [Gu et al. 2010]. We are full of confidence that the next decade will keep changing the face of DNA nanomaterials science both in methodological developments and in application explorations.

ACKNOWLEDGMENTS

Z. Deng is grateful to financial supports from the National Natural Science Foundation of China (Grant No. 20873134, 91023005, 20605019), the Chinese Academy of Sciences (a Bairen start-up support), the University of Science and Technology of China, and PCSIRT (IRT0756).

REFERENCES

Aldaye, F. A., Palmer, A. L., and Sleiman, H. F. 2003. Assembling materials with DNA as the guide. *Science* 321: 1795–1799.

Alivisatos, A. P., Johnsson, K. P., Peng, X. G., Wilson, T. E., Loweth, C. J., Bruchez Jr. M. P., and Schultz, P. G. 1996. Organization of nanocrastal molecules using DNA. *Nature* 382: 609–611.

Andersen, E. S., Dong, M. D., Nielsen, M. M., Jahn, K., Subramani, R., Mamdouh, W., Golas, M. M., Sander, B., Stark, H., Oliveira, C. L. P., Pedersen, J. S., Birkedal, V., Besenbacher, F., Gothelf, K. V., and Kjems, J. 2009. Self-assembly of a nanoscale DNA box with a controllable lid. *Nature* 459: 73–77.

Bath, J., and Turberfield, A. J. 2007. DNA nanomachines. *Nat. Nanotechnol.* 2: 275–284.

Bethune, D. S., Klang, C. H., de Vries, M. S., Gorman, G., Savoy, R., Vazquez, J., and Beyers, R. 1993. Cobalt-catalysed growth of carbon nanotubes with single-atomic-layer walls. *Nature* 363: 605–607.

Boer, R., Kerckhoffs, J. M. C. A., Parajo, Y., Pascu, M., Usón, I., Lincoln, P., Hannon, M. J., and Coll, M. 2010. Self-assembly of functionalizable two-component 3D DNA arrays through the induced formation of DNA three-way-junction branch points by supramolecular cylinders. *Angew. Chem. Int. Ed.* 49: 2336–2339.

Chen, J. H., and Seeman, N. C. 1991. Synthesis from DNA of a molecule with the connectivity of a cube. *Nature* 350: 631–633.

Chen, Y., and Mao, C. D. 2004. Reprogramming DNA-directed reactions on the basis of a DNA conformational change. *J. Am. Chem. Soc.* 126: 13240–13241.

Cheng, E. J., Xing, Y. Z., Chen, P., Yang, Y., Sun, Y. W., Zhou, D. J., Xu, L. J., Fan, Q. H., and Liu, D. S. 2009. A pH-triggered, fast-responding DNA hydrogel. *Angew. Chem. Int. Ed.* 48: 7660–7663.

Deng, Z. X., Chen, Y., Tian, Y., and Mao, C. D. 2006. A fresh look at DNA nanotechnology. In *Nanotechnology: Science and Computation*, ed. J. Chen, N. Jonoska, and G. Rozenberg, 2324. Berlin: Springer-Verlag.

Deng, Z. X., Han, X. G., Li, Y. L., and Mao, C. D. 2009. Bio meets nano: DNA-based synthesis and assembly toward one-dimensional nanostructures. In *One-dimensional nanostructures: Concepts, Applications and Perspectives*, ed. Y. Zhou, 194–219. Hefei: Univ. of Science and Technology of China Press.

Deng, Z. X., Lee, S. H., and Mao, C. D. 2005a. DNA as nanoscale building blocks. *J. Nanosci. Nanotechnol.* 5: 1954–1963.

Deng, Z. X. and Mao, C. D. 2003. DNA-templated fabrication of 1D parallel and 2D crossed metallic nanowire arrays. *Nano. Lett.* 3: 1545–1548.

Deng, Z. X. and Mao, C. D. 2004. Molecular lithography with DNA nanostructures. *Angew. Chem. Int. Ed.* 43: 4068–4070.

Deng, Z. X., Tian, Y., Lee, S. H., Ribbe, A. E., and Mao, C. D. 2005b. DNA-encoded self-assembly of gold nanoparticles into one-dimensional arrays. *Angew. Chem. Int. Ed.* 44: 3582–3585.

Dietz, H., Douglas, S. M., and Shih, W. M. 2009. Folding DNA into twisted and curved nanoscale shapes. *Science* 325: 725–730.

Ding, B., Deng, Z., Yan, H., Cabrini, S., Zuckermann, R., and Bokor, J. 2010. Gold nanoparticles self-similar chain structure organized by DNA origami. *J. Am. Chem. Soc.* 132: 3248–3249.

Douglas, S. M., Chou, J. J., and Shih, W. M. 2007. DNA-nanotube-induced alignment of membrane proteins for NMR structure determination. *Proc. Natl. Acad. Sci. USA.* 104: 6644–6648.

Douglas, S. M., Dietz, H., Liedl, T., Högberg, B., Graf, F., and Shih W. M. 2009. Self-assembly of DNA into nanoscale three-dimensional shapes. *Nature* 459: 414–418.

Erben, C. M., Goodman, R. P., and Turberfield, A. J. 2006. Single-molecule protein encapsulation in a rigid DNA cage. *Angew. Chem. Int. Ed.* 45: 7414–7417.

Fire, A. and Xu, S. Q. 1995. Rolling replication of short DNA circles. *Proc. Natl. Acad. Sci. USA* 92: 4641–4645.

Fu, A. H., Micheel, C. M., Cha, J., Chang, H., Yang, H., and Alivisatos, A. P. 2004. Discrete nanostructures of quantum dots/Au with DNA. *J. Am. Chem. Soc.* 126: 10832–10833.

Fu, T. J. and Seeman, N. C. 1993. DNA Double-crossover molecules. *Biochemistry* 32: 3211–3220.

Goodman, R. P., Berry, R. M., and Turberfield, A. J. 2004. The single-step synthesis of a DNA tetrahedron. *Chem. Comm.* 1372–1373.

Goodman, R. P., Schaap, I. A. T., Tardin, C. F., Erben, C. M., Berry, R. M., Schmidt, C. F., and Turberfield, A. J. 2005. Rapid chiral assembly of rigid DNA building blocks for molecular nanofabrication. *Science* 310: 1661–1665.

Gu, H. Z., Chao, J., Xiao, S. J., and Seeman, N. C. 2010. A proximity-based programmable DNA nanoscale assembly line. *Nature* 465: 202–205.

Hamada, S. and Murata, S. 2009. Substrate-assisted assembly of interconnected single-duplex DNA nanostructures. *Angew. Chem. Int. Ed.* 48: 6820–6823.

Han, X. G., Li, Y. L., and Deng, Z. X. 2007. DNA-wrapped single-walled carbon nanotubes as rigid templates for assembling linear gold nanoparticle arrays. *Adv. Mater.* 19: 1518–1522.

Han, X. G., Zhou, Z. H., Yang, F., and Deng, Z. X. 2008. Catch and release: DNA tweezers that can capture, hold and release an object under control. *J. Am. Chem. Soc.* 130: 14414–14415.

He, Y., Chen, Y., Liu, H. P., Ribbe, A. E., and Mao, C. D. 2005a. Self-assembly of hexagonal DNA two-dimensional (2D) arrays. *J. Am. Chem. Soc.* 127: 12202–12203.

He, Y., Liu, H. P., Chen, Y., Tian, Y., Deng, Z. X., Ko, S. H., Ye, T., and Mao, C. D. 2007. DNA-based nanofabrications. *Microscopy Res. & Tech.* 20: 522–529.

He, Y., Tian, Y., Chen, Y., Deng, Z. X., Ribbe, A. E., and Mao, C. D. 2005b. Sequence symmetry as a tool for designing DNA nanostructures. *Angew. Chem. Int. Ed.* 44: 6694–6696.

He, Y., Ye, T., Su, M., Zhang, C., Ribbe, A. E., Jiang, W., and Mao, C. D. 2008. Hierarchical self-assembly of DNA into symmetric supramolecular polyhedra. *Nature* 452: 198–202.

Holliday, R. 1964. A mechanism for gene conversion in fungi. *Genet. Res.* 5: 282–304.

Iijima, S. 1991. Helical microtubules of graphitic carbon. *Nature* 354: 56–58.

Iijima, S. and Ichihashi, T. 1993. Single-shell carbon nanotubes of 1-nm diameter. *Nature* 363: 603–605.

Kallenbach. N. R., Ma, R. I., and Seeman, N. C. 1983. An immobile nucleic acid junction constructed from oligonucleotides. *Nature* 305: 829–831.

Kata, E. and Willner, I. 2004. Integrated nanoparticle-biomolecule hybrid systems: synthesis, properties, and applications. *Angew. Chem. Int. Ed.* 43: 6042–6108.

Ke, Y. G., Lindsay, S., Chang, Y., Liu, Y., and Yan, H. 2008. Self-assembled water-soluble nucleic acid probe tiles for label free RNA detection. *Science* 319: 180–183.

Ke, Y. G., Liu, Y., Zhang, J. P., and Yan, H. 2006. A study of DNA tube formation mechanisms using 4-, 8-, and 12-helix DNA nanostructures. *J. Am. Chem. Soc.* 128: 4414–4421.

Ke, Y. G., Sharma, J., Liu, M. H., Jahn, K., Liu, Y., and Yan, H. 2009. Scaffold DNA origami of a DNA tetrahedron molecular container. *Nano Lett.* 9: 2445–2447.

Keren, K., Krueger, M., Gilad, R., Ben-Yoseph, G., Sivan, U., and Braun, E. 2002. Sequence-specific molecular lithography on single DNA molecules. *Science* 297: 72–75.

LaBean, T. H., Yan, H., Kopatsch, J., Liu, F. R., Winfree, E., Reif, J. H., and Seeman, N. C. 2000. Construction, analysis, ligation, and self-assembly of DNA triple crossover complexes. *J. Am. Chem. Soc.* 122: 1848–1860.

Li, Y. G., Tseng, Y. D., Kwon, S. Y., D'Espaux, L., Bunch, J. S., Mceuen, P. L., and Luo, D. 2004. Controlled assembly of dendrimer-like DNA. *Nature Materials* 3: 38–42.

Li, Y. L., Han, X. G., and Deng, Z. X. 2007. Grafting single-walled carbon nanotubes with highly hybridizable DNA sequences: potential building blocks for DNA-programmed materials assembly. *Angew. Chem. Int. Ed.* 46: 7481–7484.

Liao, S. and Seeman, N. C. 2004. Translation of DNA signals into polymer assembly instructions. *Science* 306: 2072–2074.

Liedl, T., Högberg, B., Tytell, J., Ingber, D. E., and Shih, W. M. 2010. Self-assembly of 3D prestressed tensegrity structures from DNA. *Nat. Nanotechnol.* 5: 520–524.

Lin, C. X., Liu, Y., Rinker, S., and Yan, H. 2006. DNA tile based self-assembly: building complex nanoarchitectures. *ChemPhysChem* 7: 1641–1647.

Liu, D. G., Park, S. H., Reif, J. H., and Labean, T. H. 2004a. DNA nanotubes self-assembled from triple-crossover tiles as templates for conductive nanowires. *Proc. Natl. Acad. Sci. USA*. 101: 717–722.

Liu, D. G., Wang, M. S., Deng, Z. X., Walulu, R., and Mao, C. D. 2004b. Tensegrity: construction of rigid DNA triangles with flexible four-arm DNA junctions. *J. Am. Chem. Soc.* 126: 2324–2325.

Liu, D. Y., Daubendiek, S. L., Zillman, M. A., Ryan, K., and Kool, E. T. 1996. Rolling circle DNA synthesis: small circular oligonucleotides as efficient templates for DNA polymerases. *J. Am. Chem. Soc.* 118: 1587–1594.

Liu, H. P., Chen, Y., He, Y., Ribbe, A. E., and Mao, C. D. 2006. Approaching the limit: can one DNA oligonucleotide assemble into large nanostructures. *Angew. Chem. Int. Ed.* 45: 1942–1945.

Liu, J. B., Fu, S. H., Yuan, B., Li, Y. L., and Deng, Z. X. 2010a. Toward a universal "adhesive nanosheet" for the assembly of multiple nanoparticles based on a protein-induced reduction/decoration of graphene oxide. *J. Am. Chem. Soc.* 132: 7279–7281.

Liu, J. B., Li, Y. L., Li, Y. M., Li, J. H., and Deng, Z. X. 2010b. Noncovalent DNA decorations of graphene oxide and reduced graphene oxide toward water-soluble metal-carbon hybrid nanostructures via self-assembly. *J. Mater. Chem.* 20: 900–906.

Liu, Y., Ke, Y. G., and Yan, H. 2005. Self-assembly of symmetric finite-size DNA nanoarrays. *J. Am. Chem. Soc.* 127: 17140–17141.

Lo, P., Karam, P., Aldaye, F., Hamblin, G., Cosa, G., and Sleiman, H. F. 2010. Loading and selective release of cargo in DNA nanotubes with longitudinal variation. *Nat. Chem.* 2: 319–328.

Loweth, C. J., Caldwell, W. B., Peng, X. G., Alivisatos, A. P., and Schultz, P. G. 1999. DNA-based assembly of gold nanocrystals. *Angew. Chem. Int. Ed.* 38: 1808–1812.

Lu, N., Shao, C. Y., and Deng, Z. X. 2008. Rational design of an optical adenosine sensor by conjugating a DNA aptamer with split DNAzyme halves. *Chem. Commun.* 6161–6163.

Lund, K., Liu, Y., Lindsay, S., and Yan, H. 2005. Self-assembling a molecular pegboard. *J. Am. Chem. Soc.* 127: 17606–17607.

Malo, J., Mitchell, J. C., Vénien-Bryan, C., Harris, J. R., Wille, H., Sherratt, D. J., and Turberfield, A. J. 2005. Engineering a 2D protein-DNA crystal. *Angew. Chem. Int. Ed.* 44: 3057–3061.

Mao, C. D., Sun, W. Q., and Seeman, N. C. 1999. Designed two-dimensional DNA Holliday junction arrays visualized by atomic force microscopy. *J. Am. Chem. Soc.* 121: 5437–5443.

Mastroianni, A. J., Claridge, S. A., and Alivisatos, A. P. 2009. Pyramidal and chiral groupings of gold nanocrystals assembled using DNA scaffolds. *J. Am. Chem. Soc.* 131: 8455–8459.

Mathieu, F., Liao, S. P., Kopatsch, J., Wang, T., Mao, C. D., and Seeman, N. C. 2005. Six-helix bundles designed from DNA. *Nano Lett.* 5: 661–665.

Maune, H. T., Han, S. P., Barish, R. D., Bockrath, M., Goddard III, W. A., Rothermund, P. W. K., and Winfree, E. 2010. Self-assembly of carbon nanotubes into two-dimensional geometries using DNA origami templates. *Nat. Nanotechnol.* 5: 61–66.

Maye, M. M., Nykypanchuk, D., Cuisinier, M., van der Lelie, D., and Gang, O. 2009. Stepwise surface encoding for high-throughput assembly of nanoclusters. *Nat. Mater.* 8: 388–391.

Mirkin, C. A., Letsinger, R. L., Mucic, R. C., and Storhoff, J. J. 1996. A DNA-based method for rationally assembling nanoparticles into macroscopic materials. *Nature* 382: 607–609.

Mitchell, J. C., Harris, J. R., Malo, J., Bath, J., and Turberfield, A. J. 2004. Self-assembly of chiral DNA nanotubes. *J. Am. Chem. Soc.* 126: 16342–16343.

Mitra, D., Di Cesare, N., and Sleiman, H. F. 2004. Self-assembly of cyclic metal-DNA nanostructures using ruthenium bipyridine branched oligonucleotides. *Angew. Chem. Int. Ed.* 43: 5804–5808.

Niemeyer, C. M. 2001. Nanoparticles, proteins, and nucleic acids: biotechnology meets materials science. *Angew. Chem. Int. Ed.* 40: 4128–4158.

Nykypanchuk, D., Maye, M. M., van der Lelie, D., and Gang, O. 2008. DNA-guided crystallization of colloidal nanoparticles. *Nature* 451: 549–552.

Pal, S., Deng, Z., Ding, B., Yan, H., and Liu, Y. 2010. DNA Origami directed self-assembly of discrete silver nanoparticle architectures. *Angew. Chem. Int. Ed.* 49: 2700–2704.

Park, N. Y., Um, S. H., Funabashi, H., Xu, J., and Luo, D. 2009. A cell-free protein-producing gel. *Nat. Mater.* 8: 432–437.

Park, S. H., Pistol, C., Ahn, S. J., Reif, J. H., Lebeck, A., Dwyer, C., and LaBean, T. H. 2006. Finite-size, fully-addressable DNA tile lattices formed by hierarchical assembly procedures. *Angew. Chem, Int. Ed.* 45: 735–739.

Park, S. Y., Lytton-Jean, A. K. R., Lee, B., Weigand, S., Schatz, G. C., and Mirkin, C. A. 2008. DNA-programmable nanoparticle crystallization. *Nature* 451: 553–556.

Reishus, D., Shaw, B., Brun, Y., Chelyapov, N., and Adleman, L. 2005. Self-assembly of DNA double-double crossover complexes into high-density, doubly connected, planar structures. *J. Am. Chem. Soc.* 127: 17590–17591.

Rothemund, P. W. K. 2006. Folding DNA to create nanoscale shapes and patterns. *Nature* 440: 297–302.

Rothemund, P. W. K., Ekani-Nkodo, A., Papadakis, N., Kumar, A., Fygenson, D. K., and Winfree, E. 2004. Design and characterization of programmable DNA nanotubes. *J. Am. Chem. Soc.* 126: 16344–16352.

Sanderson, K. 2010. What to make with DNA origami. *Nature* 464: 158–159.

Seeman, N. C. 1990. De Novo design of sequences for nucleic acid structure engineering. *J. Biomol. Struct. Dyn.* 8: 573–581.

Seeman, N. C. 2003. DNA in a material world. *Nature* 421: 427–431.

Seeman, N. C. 2010. Structural DNA nanotechnology: growing along with Nano Letters. *Nano Lett.* 10: 1971–1978.

Sharma, J., Chhabra, R., Cheng, A., Brownell, J., Liu, Y., and Yan, H. 2009. Control of self-assembly of DNA tubules through integration of Gold nanoparticles. *Science* 323: 112–116.

Sharma, J., Chhabra, R., Liu, Y., Ke, Y. G., and Yan, H. 2006. DNA-templated self-assembly of two-dimensional and periodical gold nanoparticle arrays. *Angew. Chem. Int. Ed.* 45: 730–735.

Shih, W. M., Quispe, J. D., and Joyce, G. F. 2004. A 1.7-kilobase single-stranded DNA that folds into a nanoscale octahedron. *Nature* 427: 618–621.

Sun, X. P., Ko, S. H., Zhang, C., Ribbe, A. E., and Mao, C. D. 2009. Surface-mediated DNA self-assembly. *J. Am. Chem. Soc.* 131: 13248–13249.

Tu, X. M., Manohar, S., Jagota, A., and Zheng, M. 2009. DNA sequence motifs for structure-specific recognition and separation of carbon nanotubes. *Nature* 460: 250–253.

Um, S. H., Lee, J. B., Park, N., Kwon, S. Y., Umbach, C. C., and Luo, D. 2006. Enzyme-catalysed assembly of DNA hydrogel. *Nat. Mater.* 5: 797–801.

Winfree, E., Liu, F. R., Wenzler, L. A., and Seeman, N. C. 1998. Design and self-assembly of two-dimensional DNA crystals. *Nature* 394: 539–544.

Yan, H., Park, S. H., Finkelstein, G., Reif, J. H., and Labean, T. H. 2003. DNA-templated self-assembly of protein arrays and highly conductive nanowires. *Science* 301: 1882–1884.

Zanchet, D., Micheel, C. M., Parak, W. J., Gerion, D., and Alivisatos, A. P. 2001. Electrophoretic isolation of discrete Au nanocrystal/DNA conjugates. *Nano Lett.* 1: 32–35.

Zhang, C., Ko, S. H., Su, M., Leng, Y., Ribbe, A. E., Jiang, W., and Mao, C. 2009. Symmetry controls the face geometry of DNA polyhedra. *J. Am. Chem. Soc.* 131: 1413–1415.

Zhang, C., Su, M., He, Y., Zhao, X., Fang, P., Ribbe, A. E., Jiang, W. and Mao, C. 2008. Conformational flexibility facilitates self-assembly of complex DNA nanostructures. *Proc. Natl. Acad. Sci. USA.* 105: 10665–10669.

Zhao, Z., Yan, H., and Liu, Y. 2010. A route to scale up DNA origami using DNA tiles as folding staples. *Angew. Chem. Int. Ed.* 49: 1414–1417.

Zheng, J. P., Birktoft, J., Chen, Y., Wang, T., Sha, R. J., Constantinou, P. E., Ginell, S. L., Mao, C. D., and Seeman, N. C. 2009. From molecular to macroscopic via the rational design of a self-assembled 3D DNA crystal. *Nature* 461: 74–77.

Zheng, J. W., Constantinou, P. E., Micheel, C., Alivisatos, A. P., Kiehl, R. A., and Seeman, N. C. 2006. Two-dimensional nanoparticle arrays show the organization power of robust DNA motifs. *Nano Lett.* 6: 1502–1504.

Zheng, M., Jagota, A., Semke, E. D., Diner, B. A., McLean, R. S., Lustig, S. R., Richardson, R. E., Tassi, N. G. 2003a. DNA-assisted dispersion and separation of carbon nanotubes. *Nat. Mater.* 2: 338–342.

Zheng, M., Jagota, A., Strano, M. S., Santos, A. P., Barone, P., Chou, S. G., Diner, B. A., Dresselhaus, M. S., McLean, R. S., Onoa, G. B., Samsonidze, G. G., Semke, E. D., Usrey, M., Walls, D. J. 2003b. Structure-based carbon nanotube sorting by sequence-dependent DNA assembly. *Science* 302: 1545–1548.

3 Intercalation of Organic Ligands as a Tool to Modify the Properties of DNA

Heiko Ihmels and Laura Thomas

CONTENTS

3.1 Introduction 49
3.2 Intercalation—General Principles...54
    3.2.1 Determination of the Intercalator–DNA Association.........................55
    3.2.2 Thermodynamics...57
    3.2.3 Dynamic Aspects of Intercalation...60
3.3 Structural Changes of DNA upon Intercalation 62
References..69

3.1 INTRODUCTION

Nucleic acids represent remarkable host systems that may associate with a large variety of different guest molecules (Haq 2006; Hannon 2007; Strekowski and Wilson 2007; Xie et al. 2010). In principle, external ligands may bind to DNA by three significantly different binding modes, namely, electrostatic association to the phosphate backbone (also referred to as outside-edge binding), groove binding, and intercalation. The outside binding is mainly governed by the electrostatic interactions, whereas groove binding and intercalation are based on several supramolecular interactions simultaneously operating between the ligand and particular regions of the DNA through, such as π stacking or hydrogen bonding and van der Waals or hydrophobic interactions. In the process of groove binding, an appropriately substituted ligand with crescent-type shape is accommodated in the minor or major groove of double-stranded DNA (Reddy et al. 1999; Nelson et al. 2007). Large molecules such as proteins or oligonucleotides bind to the major groove of DNA by the formation of hydrogen bonds with the functionalities of the DNA bases that point inside the groove. This complex formation is physiologically relevant as it is the basis of the highly selective recognition of DNA sequences by proteins (Rice and Correll 2008) or the specific formation of triplex DNA (Escudé and Sun 2005). On the other hand, the minor groove is the preferred binding site for smaller ligands (Neidle 2001;

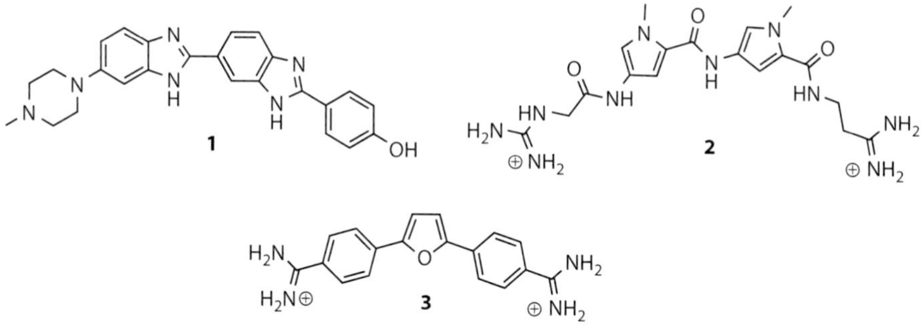

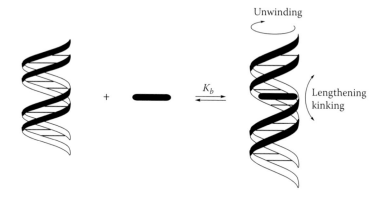

FIGURE 3.1   Typical minor groove binders.

Baraldi et al. 2004; Wilson et al. 2005; Haq 2006). The binding pocket of a DNA groove may be defined by two different regions, namely, the bottom, formed by the edges of the nucleic bases that face into the groove, and the walls, which are formed from the deoxyribose-phosphate backbone of the DNA. Groove binders usually consist of at least two aromatic or heteroaromatic rings whose connection allows conformational flexibility such that the molecule fits perfectly into the groove. In addition, functional groups are required to form hydrogen bonds with the nucleic bases at the bottom of the groove. Typical minor-groove binders are Hoechst 33258 (**1**), netropsin (**2**), and furamidines such as **3** (Wemmer 2001). It was demonstrated that polyamides, specifically hairpin-forming polyamides, are highly sequence-specific groove binders (Dervan 2001).

   In the double-stranded DNA helix, the nucleic bases are located in an almost coplanar arrangement. Upon unwinding of the DNA, the two neighboring base pairs may be separated to such an extent that external small molecules can intercalate into the DNA interior (Scheme 3.1) (Lerman 1961, 1963). Apparently, this binding mode involves significant π stacking between the ligand and the base pairs, so that an intercalator usually consists of a flat aromatic system, ideally with a fused arene unit, that enables a maximal overlap of π surfaces (Li et al. 2009). As a consequence,

SCHEME 3.1   Intercalation of a ligand into DNA and the accompanying structural perturbations.

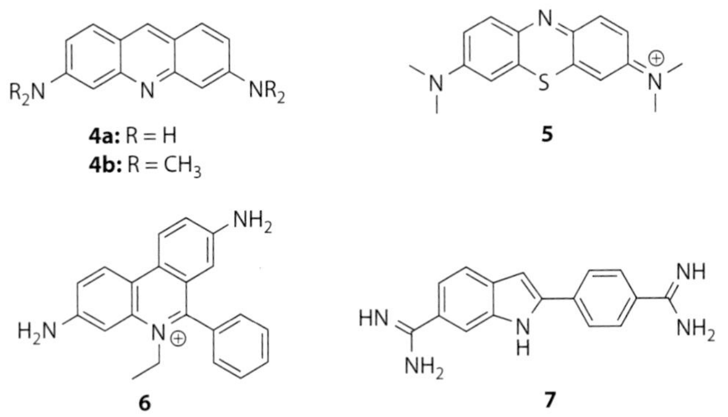

FIGURE 3.2 Structure of proflavine (4a), acridine orange (4b), ethidium bromide (6), and DAPI (7).

polycyclic aromatic and heteroaromatic compounds exhibit a high propensity to intercalate into DNA. Representative examples of intercalators are acridine derivatives, for example, proflavine (**4a**) and acridine orange (**4b**) (Lerman 1961, 1963), methylene blue (**5**) (Hossain and Kumar 2009), phenanthridinium derivatives, with ethidium bromide (**6**) being the most prominent example (Luedtke et al. 2005; Kubar et al. 2006; Tsuboi et al. 2007), monomethine dyes (Armitage 2005; Deligeorgiev and Vasilev 2006), metallorganic complexes (Richards and Rodger 2007; Zeglis et al. 2007), and annelated azinium and quinolizinium derivatives (Ihmels et al. 2005a).

It should be emphasized that groove binding and intercalation do not exclude each other. In other words, a ligand may be an intercalator and a groove binder at the same time and, mostly depending on the conditions, one of the two binding modes is chosen preferentially. Thus, it was demonstrated that 4′,6-diamidino-2-phenylindole (**7**), that is, a compound with characteristic features of a groove binder, indeed associates in the minor groove of AT-rich DNA. However, in GC-rich DNA sequences, this ligand actually intercalates preferentially (Kim et al. 1993; Trotta et al. 1996). A similar behavior was observed for **1** (Bailly et al. 1993). Also, even the "classical" intercalator ethidium bromide (**6**) was proposed to bind in the major groove of B-form DNA at very low occupancy ratios (Monaco 2007). In another case, it was demonstrated that the cyanine dye **8** binds to triplex DNA by intercalation, whereas it binds to the nonalternating duplex poly(dA)-poly(dT) through minor groove binding. This tricationic ligand intercalates also into the alternating duplex [poly(dA-dT)]₂ at low concentration and dimerizes within the minor groove at higher concentration (Cao et al. 2001).

Also, a half-intercalation model was proposed and studied in detail. According to this model, one part of the ligand is accommodated in the intercalation pocket, whereas a significant portion of the molecule is also placed in the minor groove (Ogul'chansky et al. 2001; Yarmoluk et al. 2001). In this model, the orientation of a half-intercalated ligand is mainly controlled by steric interactions and by the distribution of the electron density within the ligand. In detailed studies with a series of monomethine cyanine dyes such as **9**, it was demonstrated that the part of the ligand with the higher positive charge density is placed in the DNA groove due to

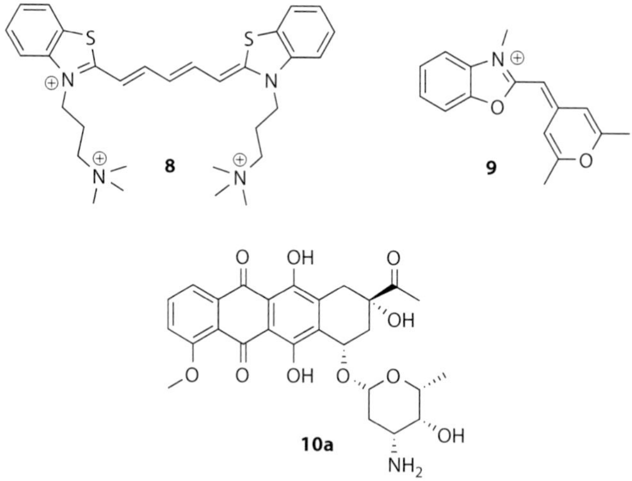

FIGURE 3.3 Structure of DNA intercalators 8–10a.

attractive interactions with the negative charge of the phosphate backbone, whereas the part of the ligand with the less positive polarization is intercalated between the base pairs. A DNA-binding ligand may also consist of one distinct intercalating part and one groove-binding unit such as in the case of daunorubicin (**10a**), that is, a typical anthracycline antibiotic (Chaires 1995). In this case, the anthraquinone part intercalates between two DNA base pairs, whereas the protonated aminopyranose unit is bound within the minor groove, where it forms a series of noncovalent bonds with the DNA.

Recently, an interesting system was presented in which the intercalating properties of the photochromic ligand **11**closed are controlled by irradiation (Scheme 3.2). Thus, the nonintercalating spiropyran **11**closed is transformed to the intercalating merocyanine **11**open upon irradiation with UV light (Andersson et al. 2008). Notably, the photochromic system is reversible, so that the intercalator **11**open may be reverted back to the closed nonintercalating form **11**closed by irradiation with visible light (>475 nm), that is, the ring-opening and ring-closing photoreactions are induced at significantly different wavelengths. It was also demonstrated that the intercalating properties of a photochromic azobenzene unit may be controlled by irradiation, because

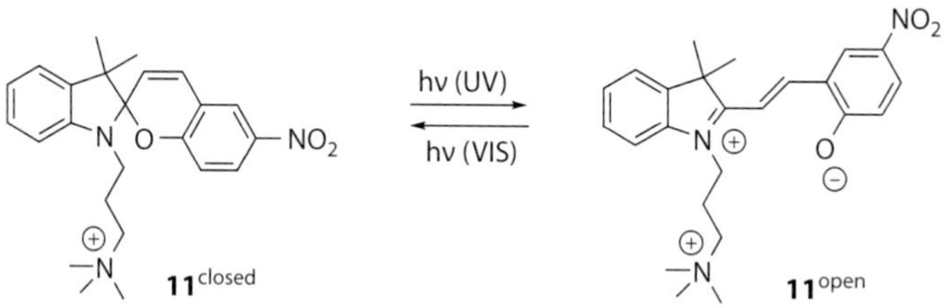

SCHEME 3.2 Photoinduced switching of the DNA-intercalating properties of the photochromic spiropyran 11.

only the *E*-configured azobenzene intercalates into double-stranded DNA whereas the *Z*-configured does not due to its increased steric demand (Liang et al. 2008). In this case, however, the azobenzene unit is connected directly to the DNA as part of the oligonucleotide backbone of the single strands, such that the intercalation is accomplished as an intramolecular process and not as an intermolecular complex formation.

It should be noted that several other variations of the intercalative binding mode are known, for example bis- and tris-intercalation (Tumir and Piantanida 2010), threading intercalation (Chu et al. 2009), and specific DNA forms such as triplex or quadruplex DNA may be the host for DNA-binding ligands (Waring 2006; Du et al. 2010), but these variations will not be discussed in this review.

In living systems, the association of a groove-binding or intercalating ligand with DNA may have a significant influence on the physiological function of the nucleic acid. A ligand may occupy binding sites of DNA such that enzymes are inhibited whose physiological activity requires the association with the DNA. Along these lines it was demonstrated that DNA intercalators are genotoxic (Ferguson and Denny 2007) or may act as efficient topoisomerase inhibitors (McClendon and Osheroff 2007, Pommier 2009). Also, the ligand–DNA interaction may induce a drastic change of the overall DNA structure so that specific recognition processes between DNA and enzymes are suppressed that are important for DNA-based cellular processes (Hurley 2002). As a consequence, DNA-binding reagents exhibit a promising potential as chemotherapeutic drugs that suppress the gene replication or transcription in tumor cells. For that reason, several classes of DNA-binding molecules have been established and investigated in detail with respect to their potential as DNA-targeting drugs (Baraldi et al. 2004; Hendry et al. 2007; Wheate et al. 2007).

In contrast to groove-binding, the intercalation of a ligand into DNA has a significant influence on the structure of the nucleic acid, because the double helix needs to unwind in order to accommodate the intercalator between two neighboring base pairs. This unwinding leads to the lengthening of the double helix for approximately 340 pm per intercalator molecule, which is often accompanied by a kinking and stiffening of the DNA around the intercalation site (Scheme 3.1). These changes of the DNA structure upon intercalation may have a notable impact on the material properties of the biomacromolecule, so that it may be tempting to employ intercalation processes to modify the material properties of DNA in an almost predictable way. Surprisingly, studies along these lines are rather rare, although the potential application of DNA-intercalator complexes, for example, as materials with relevant optical and electronic properties, was addressed recently in a review article (Kwon et al. 2009). Also, DNA-based materials were employed as host systems that may intercalate organic ligands, but most of these systems aimed at the removal of intercalating contaminants from a solution, whereas the dependence of material properties on the intercalation process was not studied explicitly (Yamada et al. 2001, 2002; Liu et al. 2003). Therefore, the general principles of intercalation, specifically the changes of the DNA structure upon intercalation of a ligand, will be presented in this review and exemplarily demonstrated on the basis of representative examples in order to motivate materials scientists to employ intercalation as a versatile tool to modify the material properties of nucleic acids. In this context it should be noted that a series of excellent reviews on DNA binders with different focus and emphasis have been published in the last few years (e.g., Graves and Velea 2000; Dervan 2001; Ihmels et al. 2005a; Haq 2006; Waring 2006; Hannon 2007; Hendry et al. 2007; Richards and Rodger 2007; Zeglis et al. 2007; Nakamoto et al. 2008; Xie et al. 2010), and the reader is advised to consult these reviews for an extensive coverage of all aspects of intercalation and DNA-binding ligands.

3.2   INTERCALATION—GENERAL PRINCIPLES

Several structural features are required for a molecule to act as an intercalator, as exemplified by the representative intercalators proflavine (**4a**), acridine orange (**4b**), methylene blue (**5**), ethidium bromide (**6**), anthraquinone **12**, coralyne (**13**), and the anthracene derivative **14**. First an intercalator usually consists of a flat aromatic system, ideally with fused arene units, that provides a maximal overlap of π surfaces. Thus, almost all intercalators exhibit a conjugated aromatic system (marked red in Figure 3.4) that enables the π-stacking interaction with the DNA base pairs (Li et al. 2009). A fused bicyclic system is considered to be the minimum size; however, most intercalators exhibit at least three fused aromatic units (Denny 2004). Theoretical studies indicated that the dispersion energy gained from the π stacking between ligand and base pairs represents the dominant energetic contribution to the complex formation (Reha et al. 2002). The stabilization energy of cationic intercalators with DNA is considerably larger than the one of uncharged intercalators, because charged ligands provide significantly larger energy contributions based on charge-transfer and, most dominantly, on electrostatic interactions (Kubar et al. 2006). Therefore, many simple intercalators carry a positive charge, which is established by a quaternary nitrogen atom, which is conjugated with the aromatic π system (marked blue in Figure 3.4). In contrast, the positive charge may also be introduced by aminoalkyl substituents, such as in the anthracene derivative **14**, which are protonated under

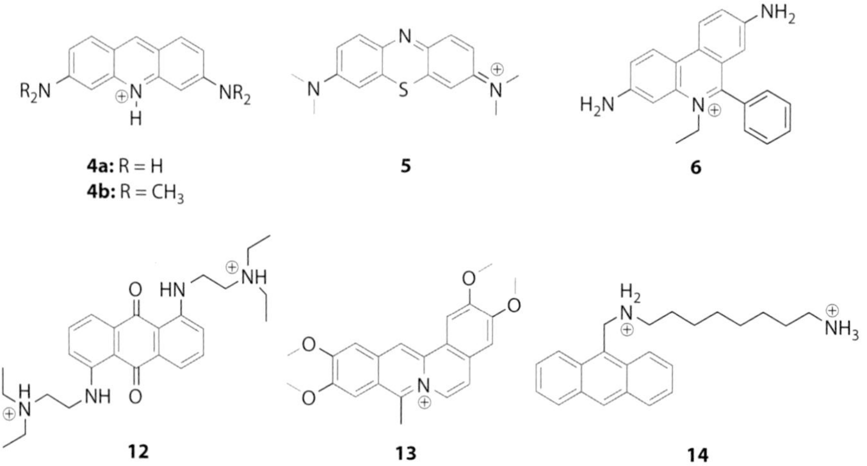

FIGURE 3.4   (**See color insert.**) Structure of representative DNA intercalators.

physiological conditions, but whose positive charge is not conjugated with the inter-calating aromatic system (Modukuru et al. 2005).

In these cases, the ammonium functionality provides attractive electrostatic inter-actions with the phosphate residues of the DNA backbone and it is, as in the case of spermine or spermidine, accommodated in the DNA grooves. Thus, most likely this positive charge does not directly contribute to the intercalating property of the ligand. Many typical cationic aromatic intercalators carry one or more additional donor substituents, such as amino functionalities or hydroxy/alkoxy substituents (marked green in Figure 3.4). Although it was suggested that these functionalities contribute to the overall binding energy by hydrogen-bonding interactions with the DNA grooves, it was calculated for an ethidium–DNA complex that the hydrogen bonding of the ethidium ligand **6** to DNA contributes less than 10% to the overall stabilization energy (Kubar et al. 2006); it may be proposed that the introduction of the donor substituent increases the affinity of the ligand towards DNA by the intro-duction of a more pronounced donor–acceptor interplay, which in turn increases the extent of delocalization of π electrons.

3.2.1 DETERMINATION OF THE INTERCALATOR–DNA ASSOCIATION

The association of an intercalator with DNA usually influences its absorption and emission properties, so that this interaction may be evaluated qualitatively and quan-titatively by spectrophotometric and spectrofluorimetric titrations (Graves 2001). In most cases, the absorption and emission of the DNA bases ($\lambda_{max} \approx 260$ nm) do not interfere with these methods because most intercalators usually absorb and emit at significantly longer wavelengths. The binding isotherms obtained from these titra-tions may be used to determine the binding constants K_b and the binding site size n, that is, the number of occupied binding sites. In most cases, the Scatchard plot is analyzed by the classical model of McGhee and von Hippel (1974), which also con-siders cooperative or anticooperative binding; however, improved methods for data analysis have been proposed recently (Kudrev 2001; Stootman et al. 2006).

The absorption of circularly or linearly polarized light may be applied in circular dichroism (CD) spectroscopy (Kypr et al. 2009) and linear dichroism (LD) spectros-copy (Nordén et al. 1992; Nordén and Kurucsev 1994) to obtain further information about the orientation of the ligand relative to the DNA and to determine the binding mode. Achiral molecules do not give a CD signal in homogeneous solution, but when they associate with the chiral DNA they exhibit an induced CD (ICD) signal because of the coupling between the transition dipole moment of the bound ligand and the one of the DNA bases. Thus, the ICD signal of a ligand indicates unambiguously the ligand-DNA interaction. Moreover, the orientation of an intercalator within the bind-ing site of DNA may be deduced from the ICD signal, if the intercalator is placed close to the helix axis and if the transition related to the ICD does not overlap with other independent transitions of the chromophore (Schipper et al. 1980; Schipper and Rodger 1983). According to this simplifying model, a positive ICD indicates a perpendicular orientation of the transition dipole moment of the ligand relative to the average dipole moment of the two nucleic bases that constitute the intercalation site, which is almost perpendicular relative to the long axis of the intercalation site.

In contrast, a negative ICD indicates a parallel orientation of the transition dipole moment of the ligand relative to the long axis of the binding pocket. Thus, with the knowledge of the transition dipole moment of the intercalator and the sign of the ICD it is possible to deduce the relative orientation of the ligand within the binding site.

Complementary to the CD-spectroscopic investigations, LD spectroscopy is a useful tool to determine the binding modes between ligands and nucleic acids (Nordén et al. 1992; Nordén and Kurucsev 1994), and it uses the fact that the absorption of polarized light by a chromophore depends on the geometrical orientation of its transition dipole moment relative to the polarization plane of the light. By definition, linear dichroism is the differential absorption of light that is polarized parallel and perpendicularly with respect to a given reference axis ($LD = A_{\parallel} - A_{\perp}$). In a homogeneous solution the LD will add up to zero because of the statistical distribution of molecules. Therefore, the DNA molecules need to be arranged in a fixed orientation relative to the planes of the linearly polarized light, which may be achieved by an external electric field (*electric linear dichroism*) or by a hydrodynamic field (*flow linear dichroism*). Usually, the direction of the hydrodynamic flow lines is used as the reference axis. In such an arrangement, the chromophores, whose transition dipole moment lies within the plane of the molecule and whose transition dipole moment is oriented at an angle $\alpha > 55°$ relative to the reference axis, result in a negative LD signal, while chromophores with a smaller angle relative to the reference axis, $\alpha < 55°$, give a positive LD signal. As a consequence the nucleic bases of the DNA, as well as intercalated ligands, show a negative LD signal, because they are positioned perpendicular to the reference axis ($\alpha \approx 90°$), whereas groove binders ($\alpha \approx 45°$) give a weak positive LD signal.

If a molecule exhibits pronounced fluorescence, that is, with a reasonable emission quantum yield, the interaction with DNA may result in shifts of the emission maxima and changes of the fluorescence quantum yield (Lakowicz 2006). Emission quenching is usually the result of electron- or energy-transfer reactions between the excited ligand and the DNA; however, this effect should be used carefully as a specific detection signal for DNA binding, because the emission quenching may also be induced by other external sources. Thus, it is more desirable that the emission intensity of the ligand increases upon intercalation into DNA. Such an effect requires that in a homogeneous aqueous solution the excited ligand exhibits a radiationless deactivation, for example, due to rapid conformational changes or by hydrogen-, electron-, or energy-transfer reactions with the solvent (Neto and Lapis 2009). Within the binding site of DNA, these deactivating processes are suppressed, or at least are slower than fluorescence, due to the restricted conformational flexibility of the ligand and/or the shielding of the ligand from solvent molecules. As a consequence, the emission intensity increases significantly.

In the latter case, the increasing fluorescence intensity of the ligand upon binding to DNA may be employed to evaluate the binding process with spectrofluorimetric titrations. Even more important, this light-up effect enables the staining of DNA, as is well established with the intercalating DNA stain ethidium bromide (**6**). In addition, steady-state fluorescence polarization measurements as well as fluorescence energy transfer from the DNA bases to the bound ligand were employed as reliable criteria to determine the binding mode (Suh and Chaires 1995).

The absorption of the DNA bases decreases upon hybridization (hypochromic effect). As a consequence, the thermally induced dissociation of the double strand (DNA denaturation) may be monitored by the increasing absorption of the DNA bases. The detection of the temperature-dependent absorption of DNA at ca. 260 nm (thermal denaturation) enables the determination of the melting temperature T_m of the DNA double helix, that is, the temperature at which 50% of the DNA is in the double-helical state, and which is a characteristic physical property of each particular DNA. The resulting plot of the normalized absorbance of the DNA bases versus the temperature is commonly termed the *melting curve* of DNA, and the melting temperature T_m is the maximum of the first derivative thereof. It should be noted that the melting temperature T_m depends on the buffer composition (e.g., pH, ionic strength, type of cations), so that the latter must be given explicitly along with the T_m data. In general, the determination of the thermal denaturation of DNA-intercalator complexes may also be employed to study the association of a ligand with the double helix (Mergny and Lacroix 2003). Ligands that bind to the double-stranded DNA, but not to the corresponding single strands, stabilize the ds DNA against thermal denaturation as indicated by increasing T_m values. The induced shifts of the melting temperature, ΔT_m, are characteristic of the affinity of a given ligand towards the DNA double strand.

3.2.2 THERMODYNAMICS

The change of the DNA structure upon formation of the intercalation complex is energetically unfavorable because the nucleic acid gives up its thermodynamically most stable structure. In addition, the intercalator loses a significant degree of translational, rotational, and vibrational freedom when transferred from the homogeneous solution to the constrained intercalation binding site. These energetic penalties from the complex formation need to be compensated by favorable energetic terms, and it should be helpful for the understanding of the intercalation process to identify the latter as the actual driving forces for the complex formation. Along these lines, several theoretical studies aim at the assessment of the different energetic contributions to the overall binding process.

Although several attractive noncovalent interactions between the intercalator and DNA contribute simultaneously to the overall binding (see below), it is commonly accepted that the dispersion energy contributes predominantly to the stability of intercalator–DNA complexes (Reha et al. 2002). Nevertheless, for a complete understanding of the factors that determine the intercalation of a ligand into DNA the detailed partition of the Gibbs energy of ligand–DNA binding is desirable. For this purpose, the experimental free energy of the intercalation, ΔG^{bind}, of several ligands with DNA was dissected according to Equation 3.1.

$$\Delta G^{bind} = \Delta G_{conf} + \Delta G_{t+r} + \Delta G_{hyd} + \Delta G_{pe} + \Delta G_{mol} \qquad (3.1)$$

In this equation, the terms ΔG_{conf} and ΔG_{t+r} refer to the free energy involved in the change of DNA structure and in the loss of translational and rotational degree of freedom. ΔG_{hyd} is the free energy of the hydrophobic transfer, that is, the change of the microenvironment of a hydrophobic guest molecule from the hydrophilic aqueous solution to the hydrophobic binding sites of the nucleic acids. The term ΔG_{pe} refers to the polyelectrolyte contribution. ΔG_{mol} is the combined free energy of all noncovalent bonding interactions such as hydrogen bonding, π stacking, or Lewis-acid-base interactions. The contributions of the particular free energies to the intercalation of ethidium bromide (**6**) into DNA were deduced by a combination of kinetic and thermodynamic data, theoretical calculations, as well as reasonable estimations. Thus, it was shown that indeed ΔG_{conf} ($\approx$ +4 kcal mol^{-1}) and ΔG_{t+r} ($\approx$ + 15 kcal mol^{-1}) are energetically unfavorable energy terms. The value for ΔG_{pe} was obtained according to the polyelectrolyte theory and is mostly determined by the release of condensed counterions from the DNA helix upon binding of the charged ligand. Notably, these data reveal only a marginal contribution of ΔG_{pe} ($\approx$ –1 kcal/ mol). In contrast, the free energy of the hydrophobic transfer ΔG_{hyd} ($\approx$ –11 kcal mol^{-1}) and of the noncovalent interactions between the ligand and DNA ΔG_{mol} ($\approx$ –13 kcal mol^{-1}) were identified as the main thermodynamic factors that provide the energetic driving force for the intercalation.

More recently, a refined theoretical model was presented that divides the free energy ΔG^{bind} into selected components. First, the free energy of DNA unwinding, ΔG^{uw}, and the free energy of the insertion of the intercalator into the binding pocket, ΔG^{ins}, were treated separately, where ΔG^{uw} corresponds to the term ΔG_{conf} in Equation 3.1. Secondly, solvation effects on the total Gibbs energy were taken into account for all calculations, namely for the intermolecular and intramolecular interactions of DNA and the intercalator in the vacuum, ΔG_{im}, and in aqueous solution, ΔG_{solv}. As compared with the analysis of the free energy according to Equation 3.1, another significant improvement was employed by an additional dissection of the term ΔG_{mol}. Thus, the specific noncovalent interactions between the intercalator and the DNA were treated separately, namely van der Waals interactions ΔG^{bind}_{vdW}, hydrogen bonding ΔG^{bind}_{HB}, electrostatic interactions ΔG^{bind}_{el}, and charge-transfer interactions ΔG^{bind}_{ct}. Also, an entropic component of the complex formation was employed that is comparable ΔG_{t+r} in Equation 3.1; however, in the latter case, only changes in translational and rotational degrees of freedom were considered. In an improved model, the change in vibration of chemical bonds upon complexation was introduced as ΔG_{vibr} because the intercalation may also lead to new vibration modes, which were further differentiated into vibrations of chemical bonds and mechanical oscillations. In summary, the changes of translational, rotational, and vibrational degrees of freedom were calculated as ΔG_{entr}. Notably, a detailed analysis was performed to assess the ideal platform to calculate the hydrophobic energy for binding of aromatic ligands with DNA. As a result, it was demonstrated that data based on the changes of solvent-accessible surface areas (SASA) upon intercalation correspond better with the hydrophobic energy than the changes of heat capacity, ΔC_p, of the intercalation reaction. With these various components, the free energy of DNA unwinding, ΔG^{uw}, was calculated according to Equation 3.2. The additive relationship between ΔG^{uw} and the free energy of the insertion ΔG^{ins} leads to Equation 3.3, which was employed as the basis for the partitioning of the total binding energy into its various components.

$$\Delta G^{uw} = \Delta G^{uw}_{vdW} + \Delta G^{uw}_{el} + \Delta G^{uw}_{pe} + \Delta G^{uw}_{HB} + \Delta G^{uw}_{ct} + \Delta G^{uw}_{hyd} + \Delta G^{uw}_{entr} \quad (3.2)$$

$$\Delta G^{bind} = \Delta G^{uw} + \Delta G^{ins}_{vdW} + \Delta G^{ins}_{el} + \Delta G^{ins}_{pe} + \Delta G^{ins}_{HB} + \Delta G^{ins}_{ct}$$
$$+ \Delta G^{ins}_{hyd} + \Delta G^{ins}_{entr}$$

$$(3.3)$$

With these equations at hand, the different free energy terms were calculated for the intercalation of selected ligands, namely proflavine (**4a**), ethidium bromide (**6**), dauno-mycin (**10a**), nogalamycin (**10b**), and novantrone (**15**), into the self-complementary DNA dodecamer d(CGCTCGAGCG)$_2$. For example, the following free energies were obtained for the intercalation of **6** (note that the energies of hydrogen bonding and electrostatic interactions are given as combined value as the latter also influ-ences significantly to the hydrogen bonds): $\Delta G^{uw} = 12$ kcal/mol; $\Delta G^{ins}_{el} + \Delta G^{ins}_{HB} = 17$ kcal/mol; $\Delta G^{ins}_{entr} = 8$ kcal/mol; $\Delta G^{ins}_{pe} = -1$ kcal/mol; $\Delta G^{ins}_{vdW} - 11$ kcal/mol; $\Delta G^{ins}_{hyd} = -35$ kcal/mol. ΔG^{ins}_{ct} was assumed to be zero. The inspection of these data, as well as the ones obtained with the other ligands, reveals—as a general trend—a relatively small value of the total free energy of DNA binding (< 10 kcal/mol), which is the sum of large compensating energy contributions with opposite signs.

In summary, the calculated data show the general trend that the unwinding of the DNA (ΔG^{uw}), as well as the loss of translational, rotational, and vibrational degrees of freedom (ΔG^{ins}_{entr}), are thermodynamically unfavorable processes of the DNA intercalation and are in agreement with previous analyses (as previously shown). Additional factors were identified that bring about a destabilization of the interca-lation complex, namely electrostatic interactions and hydrogen bonding ($\Delta G^{ins}_{el} + \Delta G^{ins}_{HB}$). It should be emphasized, however, that the contribution of the electrostatic interactions deserves special attention as it was observed that the total sum of elec-trostatic interactions at the level of binding, ΔG^{bind}_{el}, and ligand insertion, ΔG^{ins}_{el}, is essentially the result of two large components with opposite signs, namely, the interaction of the dissolved complex with water, $\Delta G^{bind}_{solv} = 176$ kcal/mol, and the intermolecular Coulomb interactions in a vacuum, $\Delta G^{bind}_{im} = -174$ kcal/mol.

This balance between two large quantities that add up to a small overall energy led to the conclusion that both terms need to be fully considered for the detailed thermodynamic analysis of the electrostatic contribution to intercalation processes. At the same time, it was observed that hydrogen bonding of the unbound ligand with water destabilizes the DNA–intercalator complex because of the loss of these

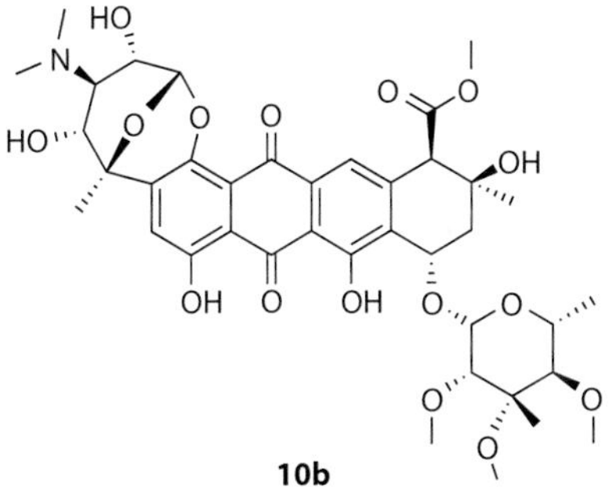

10b 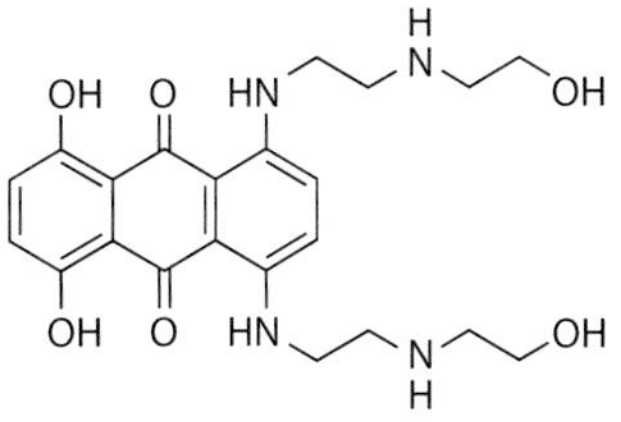   **15**

FIGURE 3.5   Structure of nogalamycin (10b) and novantrone 15.

hydrogen bonds upon intercalation of the ligand. Entropic factors, namely, the loss of translational, rotational, and vibrational degrees of freedom upon intercalation, destabilize the ligand-DNA complex. In contrast, the polyelectrolyte contribution, ΔG^{ins}_{pe}, the van der Waals interactions, ΔG^{ins}_{vdW}, and the hydrophobic forces provide favorable terms to the overall binding energy. Among these factors, the polyelectrolyte contribution on the intercalation is only nominal. Such as in previous studies, it was observed that the intercalation complexes of aromatic ligands are mainly stabilized by van der Waals interactions between the ligand and the base pairs and by hydrophobic forces, with the latter providing the most dominant impact on the overall energy. In a few exceptional cases, namely, proflavine (**4a**) and nogalamycin (**10b**), the van der Waals interaction destabilize the intercalation complex.

Overall, the studies described above enable a reasonable dissection and identification of the pertinent thermodynamic forces that govern the intercalation of an organic ligand into DNA. In addition, a principal conclusion was made regarding the scope and limits of the evaluation and discussion of calculated data on DNA intercalation. Specifically, based on a reasonable analysis of the errors in the calculations and comparison with the separate absolute values, it was concluded that it is not possible to obtain a sum of calculated binding energies with an absolute value that is smaller than the error limits of the experimentally determined energy ΔG^{exp}. The study also showed that a reasonable evaluation of the physicochemical properties of intercalator–DNA complexes should consider the individual components of the Gibbs free energy separately rather than the total Gibbs energy of binding.

3.2.3 Dynamic Aspects of Intercalation

Thermodynamic aspects of intercalation were assessed extensively by experimental and theoretical data, however, fewer studies are known on the kinetics of the intercalation process. The kinetics of the intercalation process were studied with different physicochemical techniques, such as temperature jump experiments, stopped flow, surface plasmon resonance (SPR), NMR spectroscopy, or time-resolved optical methods such as laser flash spectroscopy or fluorescence correlation spectroscopy, as discussed in detail in Pace and Bohne (2008). Notably, the reported data are not always consistent, mainly because of significantly different experimental conditions, such as composition of the DNA, temperature, buffer composition, pH, etc., which usually have a significant influence on the kinetic data.

For example, several studies were performed to elucidate the several steps during intercalation of ethidium bromide (**6**) into double-stranded DNA, but, unfortunately, the respective interpretations of the data sets, as well as mechanistic discussions, vary significantly and sometimes even lead to contradicting conclusions. Nevertheless, most studies show consistently that the association rate constants for intercalation are significantly slower (10^5–10^6 M^{-1} s^{-1}) than a diffusion-controlled reaction and that the dissociation rate constants are in the order of 10^1–10^2 s^{-1}, i.e., the intercalator has a residence time in the binding site of < 100 ms and > 10 ms. Almost all high-resolution techniques confirm the original proposal by Bresloff and Crothers of a ligand redistribution reaction between two separate DNA molecules at very low ligand-to-DNA ratios, according to Scheme 3.3 (Bresloff and Crothers 1975;

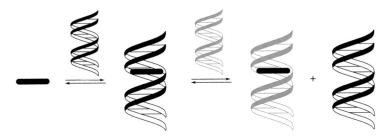

SCHEME 3.3   Ligand redistribution between two DNA molecules.

Wakelin and Waring 1980; Meyer-Almes and Pörschke 1993). Moreover, it was shown by the kinetic studies that the ethidium ligand binds to DNA in two different types of intercalation complexes, which differ in the orientation of the ligand relative to the intercalations site. Specifically, in one complex the phenyl substituent of **6** points inside the minor groove, whereas in the other type of complex it is pointing into the major groove of the double helix.

It should be stressed that so far, to the best of our knowledge, the discussion of the experimentally assessed kinetics of the intercalation processes of ethidium (**6**) neglects any dynamics involved in the formation of an outer complex (preequilibrium complex), because the latter complexes are very weak and are assumed to be formed with diffusion-control (Meyer-Almes and Pörschke 1993). In the case of daunomycin (**10a**), however, the outside binding was addressed, and explicitly included in the discussion of the intercalation mechanism. In one detailed study, the kinetics of the daunomycin–DNA interaction was examined by the stopped-flow and temperature-jump relaxation methods (Chaires et al. 1985). The kinetic data revealed three essential steps of the intercalation process, interpreted as an introducing rapid "outside" binding of **10a** to DNA with a subsequent intercalation of the ligand. It was proposed that in a final step the ligand reorganizes its orientation within the binding site or even changes the binding site of the DNA host molecule. In a later stopped-flow study, in which fluorescence instead of absorption was employed for detection, three different relaxation times were observed, also; however, the kinetic analysis led to the proposal of five different steps of the intercalation process (Rizzo et al. 1989). Thus, an additional bimolecular association step to a very weakly bound state was suggested, which does not lead directly to intercalation. Also, two simultaneous final steps were proposed instead of one for the reorganization of the ligand within the host system. In addition, more recent theoretical studies of daunomycin–DNA interactions revealed a sequence of DNA roll-and-rise distortions, as well as changes in water–hydrogen bonds during the intercalation process (Mukherjee et al. 2008). According to free energy calculations, the mechanism of the intercalation is introduced by a downhill transition in free energy from the completely separated ligand to a DNA minor groove-bound state, followed by an activated transition of the ligand from the groove-bound state to the intercalated state. These theoretical results are in agreement with the experimental data as they support at least the two essential steps, namely the primary bimolecular "outside binding" with a subsequent monomolecular penetration into an intercalation site. Nevertheless, it should be noted that the analysis of the kinetic data considers labile states during intercalation, which are not explicitly considered in the theoretical calculations.

3.3 STRUCTURAL CHANGES OF DNA UPON INTERCALATION

The assembly of the DNA double helix depends on a delicate balance between several, often interactive electronic/stereoelectronic and steric parameters of either the nucleic bases, the ribose residues, or the phosphate units; however, it is well accepted that the stacking interactions between the neighboring hydrogen-bonded base pairs are among the fundamental features that determine the structure of the DNA double helix. Specifically, variables have been defined that may be used to characterize the relative orientation of base pairs towards each other, most notably the three angular variables tilt (τ), roll (ρ), twist (Ω), and the three variables shift (D_x), slide (D_y), and rise (D_z) that describe the displacement of a base pair within the helix (Figure 3.6). Although each DNA structure represents a thermodynamic minimum of energy with an optimal combination of the above-mentioned structural parameters, this ideal structure may be disrupted in the presence of an appropriately sized ligand. For example, planar polycyclic aromatic molecules may be accommodated—depending on size and shape, either completely or partially—between two base pairs in an intercalative binding mode (Scheme 3.1). Since the intercalation has been first described as a possible DNA binding mode (Lerman 1961; Luzzati et al. 1961), the structural changes in the DNA double helix that accompany this binding process were observed and examined in several studies (Brana et al. 2001; McCauley and Williams 2006; Wheate et al. 2007; Richards and Rodger 2007).

In order to accommodate a planar molecule between adjacent base pairs of the DNA, the torsional angles of the proximate deoxyribose residues in the DNA backbone have to change, which decreases the helical twist (unwinding) (i.e., a change of twist Ω of the base pairs). Simultaneously, the separation of the base pairs at the intercalation site leads to a lengthening of the double helix (Scheme 3.1), that is, a change of rise of the base pairs D_z. Classically, the induced elongation and stiffening of DNA is detected by hydrodynamic methods such as viscometry (Waring and Henley 1975) and centrifugation (Waring 1970), whereas the unwinding of the

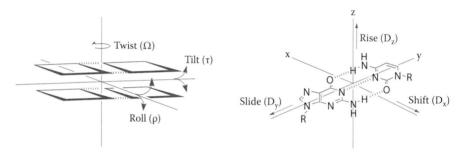

FIGURE 3.6 Schematic representation of the parameters tilt (τ), roll (ρ), twist (Ω), shift (D_x), slide (D_y), and rise (D_z) that describe the relative arrangement of DNA base pairs in a double helix.

helix is monitored by the relaxation of supercoiled circular DNA (Wang 1974). The increase of the viscosity of a DNA solution upon addition of a ligand is often used to confirm an intercalating binding mode, for example, in the case of proflavine (4a) (Aslanoglu 2006), because the association of a groove binder to DNA usually has no significant influence on the viscosity. The double helix is theoretically extended by 340 pm per intercalated molecule, which corresponds to the thickness of a typical unsubstituted aromatic compound, but electric dichroism experiments showed that the increase of the DNA length varies in a range from 200–370 pm (Hogan et al. 1979). The interactions of organoiridium intercalators **16a** and **16b** with ct DNA were investigated recently by viscometric analysis (Kokoschka et al. 2010). Compound **16a** contains a flexible linker chain and acts as bis-intercalator, whereas **16b** can only act as a monointercalator due to the more rigid cyclic linker. As a consequence, the measurements of the viscosity indicate that the increase of the DNA contour length induced by **16a** is approximately twice as large as the one caused by the association of **16b**. An increase of the viscosity of ct DNA solutions is also induced by intercalating ruthenium–complexes (Waywell et al. 2010a).

The helical twist Ω (i.e., the rotation of one base pair relative to the neighboring pair) is approximately 36° for B-DNA and decreases upon intercalation due to the unwinding process. Also, a significant twisting of the DNA axis may be induced upon intercalation, as demonstrated with the dipyridophenazine derivative **17**, which intercalates into the hexanucleotide d(GAGCTC) at the terminal base pair with a "side-on" geometry (Waywell et al. 2010b). In order to compensate the unfavorable energetic effects of unwinding (see above), helical steps that are located more distant to the intercalation site may be overwound (Lisgarten et al. 2002). It should be noted that in general, the amount of unwinding depends strongly on the nature of the intercalating ligand. Notably, an exceptionally high degree of unwinding was observed with the intercalation of the acridine-4-carboxamide **18** that leads to an overwinding of 19°, that is, the unwinding followed by subsequent rewinding with opposite handedness of the helix, as indicated by an x-ray structure analysis (Adams et al. 2000). Also, the DNA sequence around the binding site has an influence on the structural changes during intercalation. For example, it was shown that the intercalated triazoloacridone **19** causes exceptionally strong structural perturbations in DNA regions with three adjacent guanine residues, namely, the widening of both DNA grooves to which it binds preferentially (Lemke et al. 2005). In this case, it was suggested that the structural changes are mainly determined by a prototropic equilibrium of the ligand.

The trioxatriangulenium cation **20** intercalates into DNA with a preference for GC base pairs (Reynisson et al. 2003). The x-ray crystal structure of the complex provides evidence that binding of 20 induces an unusually large extension of the helical rise of the DNA. In contrast, the increase of the viscosity of DNA solutions due to intercalation of **20** is smaller as compared with the viscosity changes induced by other intercalators (Waring et al. 1979). This observation may be explained by the extended helical rise, which decreases stacking contacts and thus provides more flexibility.

Most of the above-mentioned structural changes of DNA upon intercalation of a ligand are directly related to conformational changes of the deoxyribose units

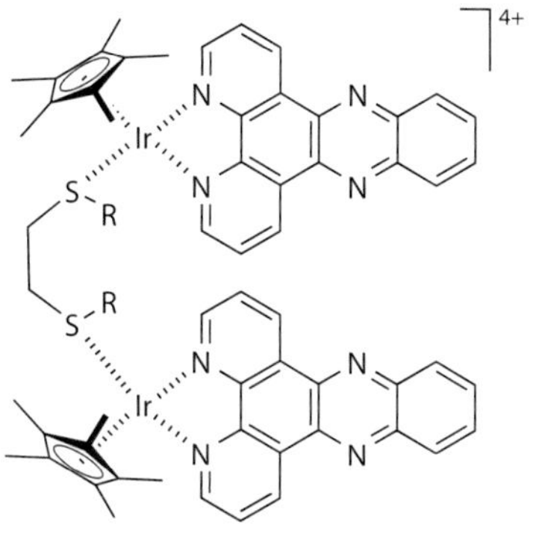

16a: R = CH₃
16b: R–R = –(CH₂)₂–

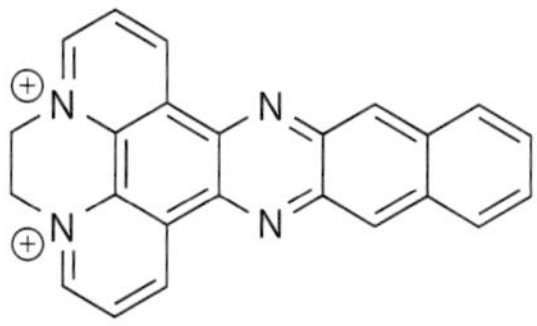

17

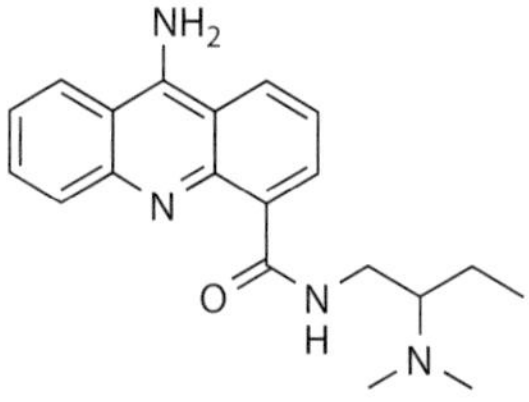

18 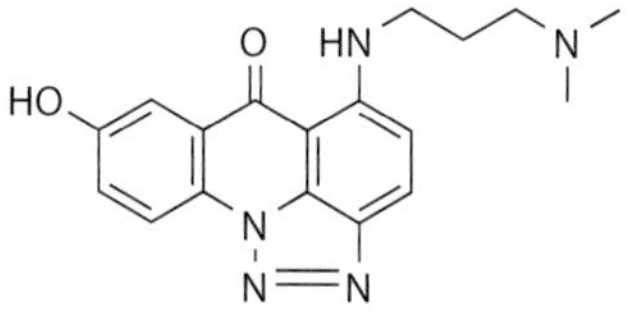   19 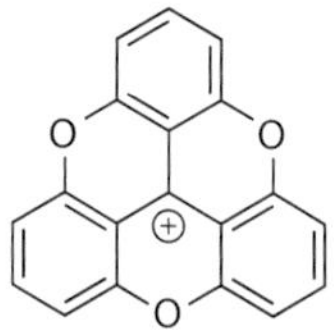   20

FIGURE 3.7 Structure of DNA intercalators 16–20.

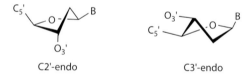

C2'-endo   C3'-endo

FIGURE 3.8 Conformations of the deoxyribose units ("sugar pucker") at the DNA intercalation site.

("sugar pucker") at the intercalation site (Figure 3.8). In regular B-DNA all deoxyribose units are in the C2'-endo conformation, but the association with an intercalator leads to a change of the conformation of the deoxyribose residues at the intercalation site: The sugar units at the strand with 5'→3' direction remain in the C2'-endo conformation, whereas the deoxyribose units at the strand with 3'→5' direction are in the C3'-endo conformation (Tsai et al. 1975; Sakore et al. 1977).

In summary, the intercalation of ligands into DNA leads to significant changes of the structure of the biopolymer. Exemplarily, the most relevant parameters for the structural changes of B-DNA that are induced upon intercalation of ethidium bromide (6) are presented in Table 3.1.

Notably, the association of a ligand into a DNA intercalation site limits the access of another ligand to the next binding site because the resulting conformational

TABLE 3.1
Structural Changes of B-DNA upon Intercalation of Ethidium Bromide (6)

Structural Parameter	Change upon Intercalation
Unwinding angle	12° (Waring 1970)
	26° (Wang 1974)
	28.7° (Zeman 1998)
	29° (Tsai et al. 1975)
Conformation of deoxyribose	5′→3′: C2′-endo; 3′→5′: C3′-endo (Tsai et al. 1975, Sakore et al. 1977, Benevides and Thomas 2005)
Increase of length per ligand	2.7 Å (Hogan et al. 1979)
Increase of contour length	40% (Smith 1992)
	11% (Tessmer 2003)
	27% (Sischka et al. 2005)
	39% (Rocha et al. 2007)
Decrease of persistence length	54% (Tessmer 2003)
	48% (Sischka et al. 2005)

changes induce a steric hindrance to the neighboring site. Furthermore, the intercalation process changes the electrostatic potential at the intercalation site, so that the attractive electrostatic interactions between another intercalator and the base pairs are slightly reduced. Both effects are assumed to contribute to the limitation of DNA intercalation to alternate occupation of the binding sites, commonly known as the *neighbor exclusion principle* (Rocha 2010). At the same time, theoretical studies have shown that intercalation affects the stretching flexibility of a DNA double strand, such that the base-pair step that is second-next to the intercalation site exhibits a higher deformability. It was proposed that this increased flexibility facilitates the preferential association of another intercalator into this particular binding site (Kubar et al. 2006).

Recently, a method was developed to observe the reversible unwinding of a single duplex-DNA molecule upon intercalation of ethidium bromide (6) (Hayashi and Harada 2007). Two associated streptavidin beads, named doublet beads, were linked to a short double-stranded DNA sequence, whose other end was bound to the glass surface of a flow chamber (Figure 3.9). The rotation of the bead after addition and removal of the intercalator 6 was monitored with an optical microscope. It was observed that upon binding of the intercalator to the DNA, the bead twisted counterclockwise because of the unwinding of the right-handed DNA helix.

Changes of the elastic properties of a DNA molecule in the presence of intercalators such as ethidium bromide (6) or daunomycin (10a) were examined using AFM-based single molecule force spectroscopy (Eckel et al. 2003; McCauley and Williams 2006). This technique employs DNA that is attached to a gold surface at one end and to an AFM tip at the other end, such that the molecular extension of the double helix may be measured as a function of the applied force (Figure 3.10). In a similar approach, a λ-DNA single strand, which is biotin-labeled at the 3' terminus,

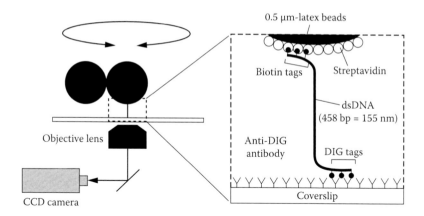

FIGURE 3.9 Setup to observe unwinding (not to scale). (From M. Hayashi and Y. Harada, *Nucl. Acids Res.*, 2007, 35, e125, Copyright 2007 Oxford University Press. With permission.)

is attached to a fixed streptavidin-coated bead in the focus of two laser beams. The complementary single strand of the DNA is linked to a bead that is attached to a micropipette tip and which may be moved to stretch the DNA with nanometer precision (Figure 3.10; Smith et al. 1996). This method was further developed to a single beam system (Sischka et al. 2003).

The foregoing experimental setups are usually analyzed with so-called force-extension or DNA-stretching curves, that is, a plot of the applied force versus the DNA extension per base pair as shown qualitatively in Figure 3.11. Thus, with double-stranded DNA a plateau emerges at forces around 65–75 pN due to overstretching of the double helix. Further stretching leads to the separation of the oppositely labeled strands at forces near 150 pN. Upon intercalation, a different plot is detected that may be used to follow the structural changes of the double strand. At small intercalator-to-DNA ratios, a rise in the overstretching force is observed, and the plateau is shifted to longer extensions (Vladescu et al. 2005). In contrast, the overstretching

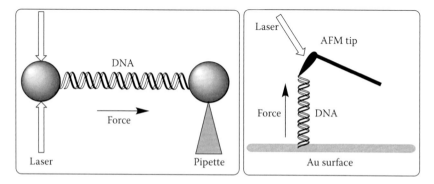

FIGURE 3.10 Force spectroscopy of single DNA molecules. Left: Optical-tweezers setup with DNA held between two beads. Right: Force-microscopy experiment with surface-immobilized DNA which may be stretched by a cantilever (according to McCauley and Williams 2006).

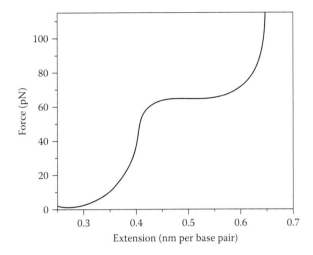

FIGURE 3.11   Qualitative presentation of a typical force extension curve for double-stranded DNA.

plateau vanishes in the force-extension curves at relatively larger intercalator-to-DNA ratios. It was concluded that intercalation of 6 stabilizes the double helix, and the contour length of the complex increases with increasing ligand concentration.

In principle, two theories were proposed to explain the nature of the overstretching plateau, where the DNA is extended to approximately double its contour length without a strong increase in force (Figure 3.12). The first model suggests a transition from B-DNA to a stretched form of DNA termed "S-DNA" in which the base pairs are tilted but remain intact (Cluzel et al. 1996; Smith et al. 1996). In an alternative model it is suggested that the hydrogen bonds between the base pairs break to yield

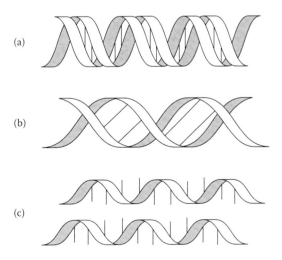

(a)

(b)

(c)

FIGURE 3.12   Schematic representation of B-DNA (A), S-DNA B) and fully dissociated DNA (C).

two separate single strands, which is referred to as force-induced melting (Rouzina and Bloomfield 2001a, 2001b). Studies on the dependence of the overstretching properties of DNA on different solution parameters, such as pH, ionic strength, and temperature, supported the latter model (Williams et al. 2001a, 2001b; Wenner et al. 2002). Additionally, results of experiments with so-called optical tweezers on DNA complexes with ethidium bromide (6) (Vladescu et al. 2005), metallointercalators (Mihailovic et al. 2006) and a bis-intercalator (Murade et al. 2010) are consistent with the proposed force-induced melting. The increase in overstretching force indicates that with increasing concentration of an intercalator additional energy is required to induce the dissociation of the DNA double strand by force. The overstretching plateau in force-extension curves of double-stranded DNA disappears when a critical concentration of the intercalator is reached, which means that the dissociation of the ligand-DNA complex to DNA single strands is not induced by force anymore.

In contrast to bulk thermal melting experiments (see above), intercalators such as 6 may bind to every base stack in violation of the *neighbor exclusion principle* when the DNA is stretched at high forces (Vladescu et al. 2005). In order to examine the influence of intercalators on the persistence length of DNA, stretching experiments on DNA molecules were performed using only low forces (≤ 2 pN) (Rocha et al. 2007). With increasing concentration c of 6 or 10a, the persistence length A of the ligand-DNA complexes increased up to a critical value ($A_{10a} = 280$ nm for $c_{10a} = 18.3$ µM; $A_6 = 150$ nm for $c_6 = 3.1$ µM). Further addition of the intercalating agents led to a sharp decrease of the persistence length until a mostly constant value was reached. A helix-coil transition, which takes place due to unwinding, may cause DNA denaturing under the stretching conditions and thus lead to the decay of the persistence length after a critical ligand concentration is reached. The contour length of the DNA complex with ethidium bromide (6) increases up to 127% and with the bis-intercalator YOYO-1 (21) even up to 136% as compared with the length of free λ-DNA (Sischka et al. 2005).

Near-infrared Raman spectra of DNA–ethidium complexes with an excitation wavelength of 1064 nm indicated a structural transition of B-DNA to A-DNA with increasing drug concentration (Yuzaki and Hamaguchi 2004). More detailed information about conformational changes was provided in a similar study with an excitation wavelength of 752 nm to excite a complex of one ethidium molecule (6) bound per 10 base pairs of ct DNA (Benevides and Thomas 2005). It was shown that intercalation of 6 converts pyrimidine, as well as purine deoxynucleoside sugar puckers, from the C2'-endo to the C3'-endo conformation; Raman markers of helix unwinding were identified for this complex for the first time. Another study based on Raman spectroscopy indicated that the intercalation of actinomycin D (22) induces an asymmetry in the structure of the examined tetradecamer DNA (Toyama et al. 2001), similar to an asymmetric structure that was observed in the solid state (Kamitori and Takusagawa 1994).

Dual-polarization interferometry was employed to study real-time structural changes of ct DNA upon intercalation of ethidium bromide (6) (Wang et al. 2009). The thickness and the refractive index (RI) of immobilized ct DNA were measured before and after the addition of 6. Binding of the ligand induced a decrease of the thickness of nearly 0.6 nm and a slight increase of the density (ΔRI ≈ 0.003) of the

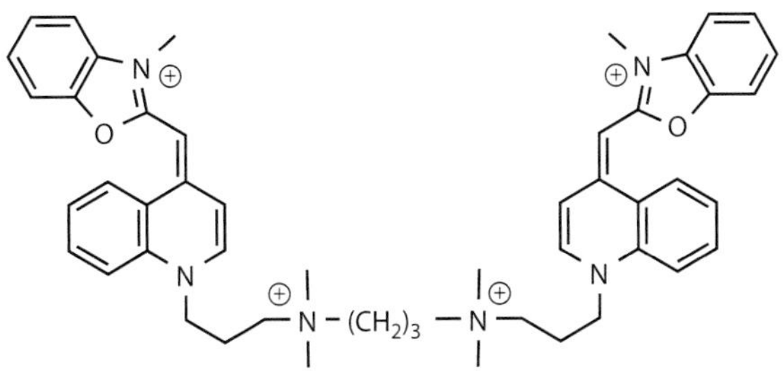

FIGURE 3.13   Structure of DNA bis-intercalators 21 and 22.

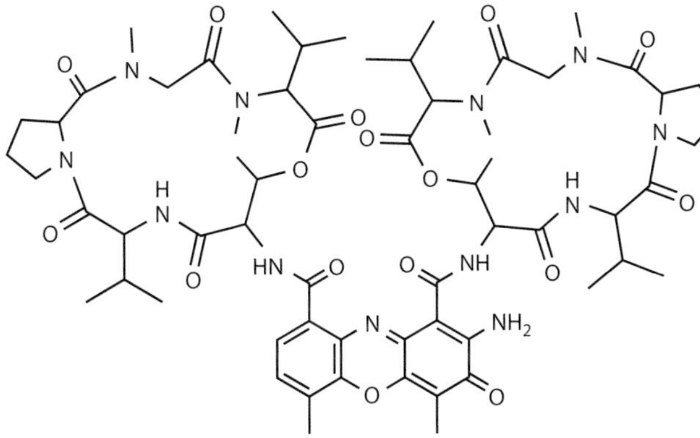

DNA layer. These observations are consistent with an intercalative binding mode since the changes of the DNA structure may be due to the release of water or counter cations from the grooves. A conformational change of the double helix to the A-DNA form may also explain the decrease in thickness because A-DNA has a smaller rise per base pair than B-DNA.

REFERENCES

Adams, A., J. M. Guss, C. A. Collyer, W. A. Denny, and L. P. G. Wakelin. 2000. A novel form of intercalation involving four DNA duplexes in an acridine-4-carboxamide complex of d(CGTACG)₂. *Nucl. Acids Res.* 28: 4244–53.

Andersson, J., S. Li, P. Lincoln, and J. Andréasson. 2008. Photoswitched DNA-binding of a photochromic spiropyran. *J. Am. Chem. Soc.* 130: 11836.

Armitage, B. A. 2005. Cyanine dye-DNA interactions: Intercalation, groove binding, and aggregation. *Top. Curr. Chem.* 253: 55–76.

Aslanoglu, M. 2006. Electrochemical and spectroscopic studies of the interaction of proflavine with DNA. *Anal. Sci.* 22: 439–43.

Bailly, C., P. Colson, J. P. Henichart, and C. Houssier. 1993. The different binding modes of Hoechst-33258 to DNA studied by electric linear dichroism. *Nucl. Acids Res.* 21: 3705–9.

Baraldi, P. G., A. Bovero, F. Fruttarolo, D. Preti, M. A. Tabrizi, M. G. Pavani, and R. Romagnoli. 2004. DNA minor groove binders as potential antitumor and antimicrobial agents. *Med. Res. Rev.* 24: 475–528.

Benevides, J. M., and G. J. Thomas. 2005. Local conformational changes induced in B-DNA by ethidium intercalation. *Biochemistry* 44: 2993–99.

Brana, M. F., M. Cacho, A. Gradillas, B. de Pascual-Teresa, and A. Ramos. 2001. Intercalators as anticancer drugs. *Curr. Pharm. Des.* 7: 1745–80.

Bresloff, J. L., and D. M. Crothers. 1975. DNA-ethidium reaction-kinetics—Demonstration of direct ligand transfer between DNA binding-sites. *J. Mol. Biol.* 95: 103–23.

Cao, R., C. F. Venezia, and B. A. Armitage. 2001. Investigation of DNA binding modes for a symmetrical cyanine dye trication: Effect of DNA sequence and structure. *J. Biomol. Struct. Dyn.* 18: 844–56.

Chaires, J. B. 1995. Molecular recognition of DNA by daunorubicin. In *Anthracycline Antibiotics—New Analogues, Methods of Delivery, and Mechanisms of Action,* ed. W. Priebe, 156–67. Washington: American Chemical Society.

Chaires, J. B., N. Dattagupta, and D. M. Crothers. 1985. Kinetics of the daunomycin-DNA interaction. *Biochemistry* 24: 260–7.

Chu, Y., D. W. Hoffman, and B. L. Iverson. 2009. A pseudocatenane structure formed between DNA and a cyclic bisintercalator. *J. Am. Chem. Soc.* 131: 3499–508.

Cluzel, P., A. Lebrum, C. Heller, R. Lavery, J.-L. Viovy, D. Chatenay, and F. Caron. 1996. DNA: An extensible molecule. *Science* 271: 792–94.

Deligeorgiev, T., and A. Vasilev. 2006. Cyanine dyes as fluorescent non-covalent labels for nucleic acid research. In *Functional Dyes*, ed. S. H. Kim, 137–83. Amsterdam: Elsevier.

Denny, W. A. 2004. Acridine-4-carboxamides and the concept of minimal DNA intercalators. In *Small Molecule DNA and RNA Binders*, ed. M. Demeunynck, C. Bailly and W. D. Wilson, 482–502. Weinheim: Wiley–VCH.

Dervan, P. B. 2001. Molecular recognition of DNA by small molecules. *Bioorg. Med. Chem.* 9: 2215–35.

Du, Y. H., J. Huang, X. C. Weng, and X. Zhou. 2010. Specific recognition of DNA by small molecules. *Curr. Med. Chem.* 17: 173–89.

Eckel, R., R. Ros, A. Ros, S. D. Wilking, N. Sewald, and D. Anselmetti. 2003. Identification of binding mechanisms in single molecule-DNA complexes. *Biophys. J.* 85: 1968–73.

Escudé, C., and J.-S. Sun. 2005. DNA major groove binders: Triple helix-forming oligonucleotides, triple helix-specific DNA ligands and cleaving agents. *Top. Curr. Chem.* 253: 109–48.

Ferguson, L. R., and W. A. Denny. 2007. Genotoxicity of non-covalent interactions: DNA intercalators. *Mutat. Res.* 623: 14–23.

Graves, D. E., and L. M. Velea. 2000. Intercalative binding of small molecules to nucleic acids. *Curr. Org. Chem.* 4: 915–29.

Graves, D.E. 2001. Drug-DNA Interactions. In *Methods in Molecular Biology, Vol. 95: DNA Topoisomerase Protocols: Volume II: Enzymology and Drugs,* ed. N. Osheroff and M. A. Bjornsti, 161–69. Totowa, NJ: Humana Press.

Hannon, M. J. 2007. Supramolecular DNA recognition. *Chem. Soc. Rev.* 36: 280–95.

Haq, I. 2006. Reversible small molecule-nucleic acid interactions. In *Nucleic Acids in Chemistry and Biology*, ed. G. M. Blackburn, M. J. Gait, D. Loakes and D. M. Williams, 341–82. Cambridge: Royal Society of Chemistry.

Hayashi, M., and Y. Harada. 2007. Direct observation of the reversible unwinding of a single DNA molecule caused by the intercalation of ethidium bromide. *Nucl. Acids Res.* 35, e125.

Hendry, L. B., V. B. Mahesh, E. D. Bransome Jr., and D. E. Ewing. 2007. Small molecule intercalation with double stranded DNA: Implications for normal gene regulation and for predicting the biological efficacy and genotoxicity of drugs and other chemicals. *Mutat. Res.* 623: 53–71.

Hogan, M., N. Dattagupta, and D. M. Crothers. 1979. Transient electric dichroism studies of the structure of the DNA complex with intercalated drugs. *Biochemistry* 18: 280–88.

Hossain, M., and G. S. Kumar. 2009. DNA intercalation of methylene blue and quinacrine: New insights into base and sequence specificity from structural and thermodynamic studies with polynucleotides. *Mol. BioSyst.* 5: 1311–22.

Hurley, L. H. 2002. DNA and its associated processes as targets for cancer therapy. *Nat. Rev. Cancer* 2: 188–200.

Ihmels, H., K. Faulhaber, D. Vedaldi, F. Dall'Acqua, and G. Viola. 2005a. Intercalation of organic dye molecules into double-stranded DNA. Part 2: The annelated quinolizinium ion as a structural motif in DNA intercalators. *Photochem. Photobiol.* 81: 1107–15.

Ihmels, H., and D. Otto. 2005b. Intercalation of organic dye molecules into double-stranded DNA—General principles and recent developments. *Top. Curr. Chem.* 258: 161–204.

Kamitori, S., and F. Takusagawa. 1994. Multiple binding modes of anticancer drug actinomycin D: X-ray, molecular modeling, and spectroscopic studies of d(GAAGCTTC)$_2$-actinomycin D complexes and its host DNA. *J. Am. Chem. Soc.* 116: 4154–65.

Kim, S. K., S. Eriksson, M. Kubista, and B. Norden. 1993. Interaction of 4',6-diamidino-2-phenylindole (DAPI) with poly[d(G-C)2] and poly[d(G-m5C)2]: Evidence for major groove binding of a DNA probe. *J. Am. Chem. Soc.* 115: 3441–7.

Kokoschka, M., J.-A. Bangert, R. Stoll, and W. S. Sheldrick. 2010. Sequence-selective organoiridium DNA bis-intercalators with flexible dithiaalkane linker chains. *Eur. J. Inorg. Chem.* 1507–15.

Kubar, T., M. Hanus, F. Ryjacek, and P. Hobza. 2006. Binding of cationic and neutral phenanthridine intercalators to a DNA oligomer is controlled by dispersion energy: Quantum chemical calculations and molecular mechanics simulations. *Chem. Eur. J.* 12: 280–90.

Kudrev, A. G. 2001. Calculation of the cooperativity and anticooperativity parameters of the interaction of a ligand with an infinite homogeneous polymer. *J. Anal. Chem.* 56: 232–7.

Kwon, Y.-W., C. H. Lee, D.-H. Choi, and J.-I. Jin. 2009. Materials science of DNA. *J. Mater. Chem.* 19: 1353–80.

Kypr, J., I. Kejnovska, D. Renciuk, and M. Vorlickova. 2009. Circular dichroism and conformational polymorphism of DNA. *Nucl. Acids Res.* 37: 1713–25.

Lakowicz, J. R. 2006. *Principles of Fluorescence Spectroscopy*, 3rd Ed. New York: Springer.

Lemke, K., M. Wojciechowski, W. Laine, C. Bailly, P. Colson, M. Baginski, A. K. Larsen, and A. Skladanowski. 2005. Induction of unique structural changes in guanine-rich DNA regions by the triazoloacridone C-1305, a topoisomerase II inhibitor with antitumor activities. *Nucl. Acids. Res.* 33: 6034–47.

Lerman, L. S. 1961. Structural considerations in the interaction of DNA and acridines. *J. Mol. Biol.* 3: 18–30.

Lerman, L. S. 1963. The structure of the DNA-acridine complex. *Proc. Natl. Acad. Sci. USA.* 49: 94–102.

Li, S., V. R. Cooper, T. Thonhauser, B. I. Lundqvist, and D. C. Langreth. 2009. Stacking interactions and DNA intercalation. *J. Phys. Chem. B* 113: 11166–72.

Liang, X., H. Nishioka, N. Takenaka, and H. Asanuma. 2008. A DNA nanomachine powered by light irradiation. *ChemBioChem.* 9: 702–705.

Lisgarten, J. N., M. Coil, J. Portugal, C. W. Wright, and J. Aymami. 2002. The antimalarial and cytotoxic drug cryptolepine intercalates into DNA at cytosine-cytosine sites. *Nat. Struct. Biol.* 9: 57–60.

Liu, X. D., Y. Murayama, M. Yamada, M. Nomizu, M. Matsunaga, and N. Nishi. 2003. DNA aqueous solution used for dialytical removal and enrichment of dioxin derivatives. *Int. J. Biol. Macromol.* 32: 121–7.

Luedtke, N. W., Q. Liu, and Y. Tor. 2005. On the electronic structure of ethidium. *Chem. Eur. J.* 11: 495–508.

Luzzati, V., F. Masson, and L. S. Lerman. 1961. Interaction of DNA and proflavine: A small-angle X-ray scattering study. *J. Mol. Biol.* 3: 634–39.

McCauley, M. J., and M. C. Williams. 2006. Mechanisms of DNA binding determined in optical tweezers experiments. *Biopolymers* 85: 154–68.

McClendon, A. K., and N. Osheroff. 2007. DNA topoisomerase II, genotoxicity, and cancer. *Mutat. Res.* 623: 83–97.

McGhee, J. D., and P. H. von Hippel. 1974. Theoretical aspects of DNA-Protein interactions: Cooperative and non-cooperative binding of large ligands to a one-dimensional homogeneous lattice. *J. Mol. Biol.* 86: 469–89.

Mergny, J.-L., and L. Lacroix. 2003. Analysis of thermal melting curves. *Oligonucleotides* 13: 515–37.

Meyer-Almes, F. J., and D. Pörschke. 1993. Mechanism of intercalation into the DNA double helix by ethidium. *Biochemistry* 32: 4246–53.

Mihailovic, A., I. Vladescu, M. McCauley, E. Ly, M. C. Williams, E. M. Spain, and M. E. Nunez. 2006. Exploring the interaction of ruthenium(II) polypyridyl complexes with DNA using single-molecule techniques. *Langmuir* 22: 4699–709.

Modukuru, N. K., K. J. Snow, B. S. Perrin Jr., J. Thota, and C. V. Kumar. 2005. Contributions of a long side chain to the binding affinity of an anthracene derivative to DNA. *J. Phys. Chem. B* 109: 11810–8.

Monaco, R. R. 2007. A novel major groove binding site in B-form DNA for ethidium cation. *J. Biomol. Struct. Dyn.* 25: 119–25.

Mukherjee, A., R. Lavery, B. Bagchi, and J. T. Hynes. 2008. On the molecular mechanism of drug intercalation into DNA: a simulation study of the intercalation pathway, free energy, and DNA structural changes. *J. Am. Chem. Soc.* 130: 9747–55.

Murade, C. U., V. Subramaniam, C. Otto, and M. L. Bennink. 2010. Force spectroscopy and fluorescence microscopy of dsDNA-YOYO-1 complexes: Implications for the structure of dsDNA in the overstreching region. *Nucl. Acids Res.* 38: 3423–31.

Nakamoto, K., M. Tsuboi, and G. D. Strahan. 2008. *Drug-DNA Interactions: Structures and Spectra*, 119–208. Hoboken: Wiley.

Neidle, S. 2001. DNA minor-groove recognition by small molecules. *Nat. Prod. Rep.* 18: 291–309.

Nelson, S. M., L. R. Ferguson, and W. A. Denny. 2007. Non-covalent ligand/DNA interactions: Minor groove binding agents. *Mutat. Res.* 623: 24–40.

Neto, B. A. D., and A. A. M. Lapis. 2009. Recent developments in the chemistry of deoxyribonucleic acid (DNA) intercalators: principles, design, synthesis, applications and trends. *Molecules* 14: 1725–46.

Nordén, B., M. Kubista, and T. Kurucsev. 1992. Linear dichroism spectroscopy of nucleic-acids. *Quart. Rev. Biophys.* 25: 51–170.

Nordén, B., and T. Kurucsev. 1994. Analysing DNA complexes by circular and linear dichroism. *J. Mol. Recognit.* 7: 141–55.

Ogul'chansky, T. Y., M. Y. Losytskyy, V. B. Kovalska, V. M. Yashchuk, and S. M. Yarmoluk. 2001. Interactions of cyanine dyes with nucleic acids. XXIV. Aggregation of monomethine cyanine dyes in presence of DNA and its manifestation in absorption and fluorescence spectra. *Spectrochim. Acta A* 57: 1525–32.

Pace, T. C. S., and C. Bohne. 2008. Dynamics of guest binding to supramolecular systems: Techniques and selected examples. *Adv. Phys. Org. Chem.* 42: 167–223.

Persil, Ö., and N. V. Hud. 2007. Harnessing DNA intercalation. *Trends Biotech.* 25: 433–6.

Pommier, Y. 2009. DNA topoisomerase I inhibitors: Chemistry, biology, and interfacial inhibition. *Chem. Rev.* 109: 2894–902.

Reddy, B. S. P., S. M. Sondhi, and J. W. Lown. 1999. Synthetic DNA minor groove-binding drugs. *Pharmacol. Ther.* 84: 1–111.

Reha, D., M. Kabelac, F. Ryjacek, J. Sponer, J. E. Sponer, M. Elstner, S. Suhai, and P. Hobza. 2002. Intercalators. 1. Nature of stacking interactions between intercalators (ethidium, daunomycin, ellipticine, and 4',6-diaminide-2-phenylindole) and DNA base pairs. Ab initio quantum chemical, density functional theory, and empirical potential study. *J. Am. Chem. Soc.* 124: 3366–76.

Reynisson, J., G. B. Schuster, S. B. Howerton, L. D. Williams, R. N. Barnett, C. L. Cleveland, U. Landman, N. Harrit, and J. B. Chaires. 2003. Intercalation of trioxatriangulenium ion in DNA: Binding, electron transfer, x-ray crystallography, and electronic structure. *J. Am. Chem. Soc.* 125: 2072–83.

Rice, P. A., and C. C. Correll. 2008. *Protein-Nucleic Acid Interactions*. Cambridge: Royal Society of Chemistry.

Richards, A. D., and A. Rodger. 2007. Synthetic metallomolecules as agents for the control of DNA structure. *Chem. Soc. Rev.* 36: 471–83.

Rizzo, V., N. Sacchi, and M. Menozzi. 1989. Kinetic studies of anthracycline-DNA interaction by fluorescence stopped flow confirm a complex association mechanism. *Biochemistry* 28: 274–82.

Rocha, M. S., M. C. Ferreira, and O. N. Mesquita. 2007. Transition on the entropic elasticity of DNA induced by intercalating molecules. *J. Chem. Phys.* 127: 105108.

Rocha, M. S. 2010. Revisiting the neighbor exclusion model and its applications. *Biopolymers* 93: 1–7.

Rouzina, I., and V. A. Bloomfield. 2001a. Force-induced melting of the DNA double helix 1. Thermodynamic analysis. *Biophys. J.* 80: 882–93.

Rouzina, I., and V. A. Bloomfield. 2001b. Force-induced melting of the DNA double helix 2. Effect of solution conditions. *Biophys. J.* 80: 894–900.

Sakore, T. D., S. C. Jain, C.-C. Tsai, and H. M. Sobell. 1977. Mutagen-nucleic acid intercalative binding: Structure of a 9-aminoacridine:5-iodocytidylyl(3'-5')guanosine crystalline complex. *Proc. Natl. Acad. Sci. USA* 74: 188–92.

Schipper, P. E., B. Norden, and F. Tjerneld. 1980. Determination of binding geometry of DNA-adduct systems through induced circular dichroism. *Chem. Phys. Lett.* 70: 17–21.

Schipper, P. E., and A. Rodger. 1983. Symmetry rules for the determination of the intercalation geometry of host/guest systems using circular dichroism: A symmetry-adapted coupled-oscillator model. *J. Am. Chem. Soc.* 105: 4541–50.

Sischka, A., R. Eckel, K. Toensing, R. Ros, and D. Anselmetti. 2003. Compact microscope-based optical tweezers system for molecular manipulation. *Rev. Sci. Instrum.* 74: 4827–31.

Sischka, A., K. Toensing, R. Eckel, S. D. Wilking, N. Sewald, R. Ros, and D. Anselmetti. 2005. Molecular mechanisms and kinetics between DNA and DNA binding ligands. *Biophys. J.* 88: 404–11.

Smith, S. B., Y. Cui, and C. Bustamante. 1996. Overstreching B-DNA: The elastic response of individual double-stranded and single-stranded DNA molecules. *Science* 271: 795–99.

Stootman, F. H., D. M. Fisher, A. Rodger, and J. R. Aldrich-Wright. 2006. Improved curve fitting procedures to determine equilibrium binding constants. *Analyst* 131: 1145–51.

Strekowski, L., and B. Wilson. 2007. Noncovalent interactions with DNA: An overview. *Mutat. Res.* 623: 3–13.

Suh, D., and J. B. Chaires. 1995. Criteria for the mode of binding of DNA-binding agents. *Bioorg. Med. Chem.* 3: 723–8.

Toyama, A., Y. Miyagawa, A. Yoshimura, N. Fujimoto, and H. Takeuchi. 2001. Characterization of individual adenine residues in DNA by a combination of site-selective C8-deuteration and UV resonance Raman difference spectroscopy. *J. Mol. Struct.* 598: 85–91.

Trotta, E., E. D'Ambrosio, G. Ravagnan, and M. Paci. 1996. Simultaneous and different binding mechanisms of 4',6-diamidino-2-phenylindole to DNA hexamer (d(CGATCG))$_2$. *J. Biol. Chem.* 271: 27608–14.

Tsai, C.-C., S. C. Jain, and H. M. Sobell. 1975. X-ray crystallographic visualization of drug-nucleic acid intercalative binding: Structure of an ethidium-dinucleoside monophosphate crystalline complex, ethidium: 5-iodouridylyl(3'-5')adenosine. *Proc. Natl. Acad. Sci. USA* 72: 628–32.

Tsuboi, M., J. M. Benevides, and G. J. Thomas Jr. 2007. The complex of ethidium bromide with genomic DNA: Structure analysis by polarized Raman spectroscopy. *Biophys. J.* 92: 928–34.

Tumir, L. M., and I. Piantanida. 2010. Recognition of single stranded and double stranded DNA/RNA sequences in aqueous medium by small bis-aromatic derivatives. *Mini-Rev. Med. Chem.* 10: 299–308.

Vladescu, I. D., M. J. McCauley, I. Rouzina, and M. C. Williams. 2005. Mapping the phase diagram of single DNA molecule force-induced melting in the presence of ethidium. *Phys. Rev. Lett.* 95: 158102.

Wakelin, L. P. G., and M. J. Waring. 1980. Kinetics of drug-DNA interaction—dependence of the binding mechanism on structure of the ligand. *J. Mol. Biol.* 144: 183–214.

Wang, J. C. 1974. The degree of unwinding of the DNA helix by ethidium I. Titration of twisted PM2 DNA molecules in alkaline cesium chloride density gradients. *J. Mol. Biol.* 89: 783–801.

Wang, J., X. Xu, Z. Zhang, F. Yang, and X. Yang. 2009. Real-time study of genomic DNA structural changes upon intercalation with small molecules using dual-polarization interferometry. *Anal. Chem.* 81: 4914–21.

Waring, M. J. 1970. Variation of the supercoils in closed circular DNA by binding of antibiotics and drugs: Evidence for molecular models involving intercalation. *J. Mol. Biol.* 54: 247–79.

Waring, M. J., and S. M. Henley. 1975. Stereochemical aspects of the interaction between steroidal diamines and DNA. *Nucl. Acids Res.* 2: 567–86.

Waring, M. J., A. Gonzalez, A. Jimenez, and D. Vazquez. 1979. Intercalative binding to DNA of antitumour drugs derived from 3-nitro-1,8-naphthalic acid. *Nucl. Acids Res.* 7: 217–30.

Waring, M. J. 2006. *Sequence-Specific DNA Binding Agents.* Cambridge: Royal Society of Chemistry.

Waywell, P., V. Gonzalez, M. R. Gill, H. Adams, A. J. H. M. Meijer, M. P. Williamson, and J. A. Thomas. 2010a. Structure of the complex of [Ru(tpm)(dppz)py]$^{2+}$ with a B-DNA oligonucleotide—a single-substituent binding switch for a metallo-intercalator. *Chem. Eur. J.* 16: 2407–17.

Waywell, P., J. A. Thomas, and M. P. Williamson. 2010b. Structural analysis of the binding of the diquaternary pyridophenazine derivative dqdppn to B-DNA oligonucleotides. *Org. Biomol. Chem.* 8: 648–54.

Wheate, N. J., C. R. Brodie, J. G. Collins, S. Kemp, and J. R. Aldrich-Wright. 2007. DNA intercalators in cancer therapy: Organic and inorganic drugs and their spectroscopic tools of analysis. *Mini-Rev. Med. Chem.* 7: 627–48.

Wemmer, D. E. 2001. Ligands recognizing the minor groove of DNA: Development and applications. *Biopolymers* 52: 197–211.

Wenner, J. R., M. C. Williams, I. Rouzina, and V. A. Bloomfield. 2002. Salt dependence of the elasticity and overstretching transition of single DNA molecules. *Biophys. J.* 82: 3160–69.

Williams, M. C., J. R. Wenner, I. Rouzina, and V. A. Bloomfield. 2001a. Effect of pH on the overstretching transition of double-stranded DNA: Evidence of force-induced DNA melting. *Biophys. J.* 80: 874–81.

Williams, M. C., J. R. Wenner, I. Rouzina, and V. A. Bloomfield. 2001b. Entropy and heat capacity of DNA melting from temperature depence of single molecule stretching. *Biophys. J.* 80: 1932–39.

Wilson, W. D., B. Nguyen, F. A. Tanious, A. Mathis, J. E. Hall, C. E. Stephens, and D. W. Boykin. 2005. Dications that target the DNA minor groove: Compound design and preparation, DNA interactions, cellular distribution and biological activity. *Curr. Med. Chem. Anti-Cancer Agents.* 5: 389–408.

Xie, Y., V. K. Tam, and Y. Tor. 2010. The interaction of small molecules with DNA and RNA. In *The Chemical Biology of Nucleic Acids*, ed. G. Mayer, 115–40. Chichester: Wiley.

Yamada, M., K. Kato, K. Shindo, M. Nomizu, M. Haruki, N. Sakairi, K. Ohkawa, H. Yamamoto, and N. Nishi. 2001. UV-irradiation-induced DNA immobilization and functional utilization of DNA on nonwoven cellulose fabric. *Biomaterials* 22: 3121–6.

Yamada, M., K. Kato, M. Nomizu, K. Ohkawa, H. Yamamoto, and N. Nishi. 2002. UV-irradiated DNA matrixes selectively bind endocrine disruptors with a planar structure. *Environ. Sci. Technol.* 36: 949–54.

Yarmoluk, S. M., S. S. Lukashov, T. Y. Ogul'chansky, M. Y. Losytskyy, and O. S. Kornyushyna. 2001. Interaction of cyanine dyes with nucleic acids. XXI. Arguments for half-intercalation model of interaction. *Biopolymers* 62: 219–27.

Yuzaki, K., and H. Hamaguchi. 2004. Intercalation-induced structural change of DNA as studied by 1064 nm near-infrared multichannel Raman spectroscopy. *J. Raman Spectrosc.* 35: 1013–15.

Zeglis, B. M., V. C. Pierre, and J. K. Barton. 2007. Metallo-intercalators and metallo-insertors. *Chem. Commun.* 4565–79.

4 DNA and Carbon-Based Nanomaterials

Preparation and Properties of Their Composites

Thathan Premkumar and Kurt E. Geckeler

CONTENTS

4.1 Introduction 77
4.2 Deoxyribonucleic Acid (DNA) 79
4.3 Carbon Nanotubes (CNTs) 80
4.4 DNA–CNT Hybrids 82
    4.4.1 Covalent Linkage 83
    4.4.2 Encapsulation of DNA in CNTs 91
    4.4.3 Noncovalent Interactions Between DNA and CNTs 93
4.5 Graphene (GRP) 102
4.6 DNA–Graphene Hybrids 104
4.7 Advantages and Disadvantages of Synthetic Approaches 109
4.8 Conclusions 110
Acknowledgments 112
References 112

4.1 INTRODUCTION

One of the most inspiring confronts in modern science are fundamental studies and applications in association with those of molecular, inorganic, and biological material components in interdisciplinary research involving chemistry, physics, biology, materials science, and medicine. Biomolecules are of interest to a wide spectrum of scientists ranging from biologists, chemists, physicists, and material scientists to nanotechnologists. In the current scenario, the manipulations and assembly of molecules in general and biomolecules in particular, at the nanoscale, are of prime interest to the scientists, as their properties are completely different from their bulk counterparts. Among the various biomolecules, mainly biomacromolecules (large polymeric biomolecules) such as proteins, nucleic acids, and polysaccharides, the nucleic acids gained special attention, owing to their important functions in living organisms, such as carrying genetic information from cell to cell and their unique structural arrangements, as well as their attractive properties. Among the most common nucleic acids, the ribonucleic acid (RNA) and deoxyribonucleic acid (DNA), DNA has materialized as a tempting biomacromolecule for fundamental functional materials due to its biocompatibility and renewability. In addition, the well-known and very interesting double helix structure guarantees several exceptional properties that is difficult to find in other molecules and polymers. Further, DNA has been demonstrated to be a promising material in the electronic, optical, and biomedical fields and for use as a catalyst, as well as a material for cleaning the environment. The majority of these applications are achieved based on combining DNA with other chemicals, especially nanomaterials by different interactions such as covalent linking and/or supramolecular interactions, namely electrostatic binding, intercalation or groove binding or wrapping of nanomaterials through van der Waals interactions. As far as nanomaterials are concerned, carbon-based materials such as carbon nanotubes (CNTs), fullerenes (C_{60} and others), and graphene are attractive owing to their exceptional properties and extraordinary potential utilizations. Especially, CNTs comprise a novel class of nanomaterials with significant applications in various domains. Hence, the attachment of DNA to new carbon allotropes, namely CNTs and graphene by covalent linking or by supramolecular interactions, is a challenging area for achieving novel nanobiomaterials with exceptional characteristic properties for a variety of potential applications such as biosensors and catalysts, as well as electronic and optical devices (Geckeler and Mueller 1993).

Owing to the stimulating properties of CNTs such as high mechanical strength, electrical conductivity, flexibility, and accessibility of chemical functionalization, extra attention has been focused to exploit them for biological applications, both at the molecular as well as cellular levels. In this respect so far, studies on the interface between CNTs and biology have been largely concentrated on the interactions of CNTs and biological molecules, especially biological macromolecules such as DNA, peptides, and proteins, to form novel nanobio-composite materials. Further, the electrical properties of CNTs in general, and single-walled carbon nanotubes (SWNTs) in particular, are highly susceptible to surface charge transfer and are altered in the surrounding electrostatic environments, undergoing radical changes by easy adsorption of definite biomacromolecules including DNA or synthetic polymers. Therefore, SWNTs are potential candidates for the development of chemical sensors for identifying molecules in the gas phase, as well as biosensors for investigating biological processes in solution. DNA is regarded as an ideal macromolecule for creating new functional materials. Hence, the exciting properties of carbon-based materials in general, and CNTs in particular, such as the high surface area to volume ratio, the excellent mechanical and electronic properties, thermal and electrical conductivity, chemical stability and inertness, and the lack of porosity make them suitable as support materials for DNA in many potential applications. They range from advanced biomedical systems through very sensitive electrochemical sensors and biosensors to highly efficient electronic and optical-based biochips. In this chapter, we focus on the recent developments in utilizing DNA as a functional material in combination with a different new class of carbon-based nanomaterials by exploring the various synthetic approaches and types of assemblies and interactions, in which DNA can be connected or linked to carbon nanomaterials, especially CNTs and graphene. Also, the various applications of the resulting new materials are highlighted.

4.2 DEOXYRIBONUCLEIC ACID (DNA)

DNA, deoxyribonucleic acid, is the genetic material of cells, the fundamental, hereditary material of almost all known living organisms. DNA is made up of four types of nucleotides. A nucleotide is composed of a nucleobase (nitrogenous base) such as adenine, guanine, cytosine, and thymine. The bases are generally referred to by their first letters, that, is A, G, C, and T, respectively. Chemically, DNA is composed of two long-chain polymers of simple nucleobases, with backbones made of sugars and phosphate groups connected through ester bonds. In 1953, Watson and Crick proposed a double helix structure for DNA in which two strands run in opposite directions to each other (Watson and Crick 1953). This model states that the only two pairs that are possible are AT and CG, which indirectly indicates that in a DNA molecule, always A = T and G = C. For their revolutionary innovation Watson and Crick with Maurice Wilkins shared the Nobel Prize in physiology and medicine in 1962 (Watson and Crick 1953; Wilkins et al. 1953).

In the past two decades, DNA molecules have been exploited in the field of materials science and nanotechnology as influential synthetic building blocks to generate nanoscale architectures, as well as versatile programmable templates for the assembly of nanomaterials (Mirkin et al. 1996; Alivisatos et al. 1996; Storhoff and Mirkin 1999; Seeman 2003a; Katz and Willner 2004; Feldkamp and Niemeyer 2006; Lu and Liu 2006; Kwon et al. 2009). DNA molecules are highly stable owing to their reduced vulnerability to hydrolysis when compared with the other nucleic acid, RNA, as well as with proteins. These characteristic features make DNA a unique biomacromolecule with extremely predictable sequence-dependent properties, which have been utilized to build DNA-based geometric and topologic structures (Seeman 2003a; Simmel and Dittmer 2005).

Being a biocompatible and renewable material, DNA has been playing a vital role in the field of interdisciplinary research areas including materials, nanoscience, and nanotechnology (Mirkin et al. 1996; Alivisatos et al. 1996; Storhoff and Mirkin 1999; Seeman 2003a; Katz and Willner 2004; Feldkamp and Niemeyer 2006; Lu and Liu 2006; Kwon et al. 2009; Simmel and Dittmer 2005; Scharer and Jiricny 2001; Seeman 2003b; Schrer 2003; Condon 2006; Merino et al. 2008; Liu, Diao et al. 2008; Endo and Sugiyama 2009; Sanchez-Pomales et al. 2009). This is obviously evident from the literature survey of published articles on DNA with respect to years (see www.isiwebofknowledge.com). An analysis of the survey indicates that reports on DNA increased almost every year (Figure 4.1a), which signifies extraordinary interest from both fundamental and applied technology aspects has been paid to the studies on DNA including the synthesis and design, characterization as well as potential application in various domains. The authority of DNA as a structural molecule, a scaffold, and as a template in developing nanotechnology has been established by conjugating the DNA with inorganic nanoparticles, which can then be arranged in a sequenced way to build structures containing either an inadequate number of particles or cross-linked nanoparticle agglomerates (Geckeler and Nishide 2009;

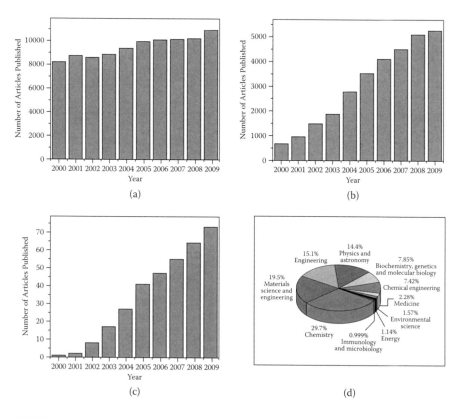

FIGURE 4.1 Statistical survey analysis, for the year 2000–2009, of articles published on (a) DNA, (b) CNTs, and (c) DNA-CNT hybrids. (d) Studies on DNA-CNT hybrids, covered in the top ten major subject areas (in %).

Geckeler and Rosenberg 2006; Thomas and Kamat 2003; Daniel and Astruc 2004; Sonnichsen et al. 2005; Zhao et al. 2006; Lin et al. 2009; Hung et al. 2010). Further, DNA molecules, also called DNA aptamers (usually consisting of short strands of oligonucleotides), have been isolated that are capable of interacting with a wide range of functional molecules, ranging from small organic molecules to proteins, cells, and even intact viral particles with high affinity and specificity (Lu and Liu 2006; Endo and Sugiyama 2009). Further, DNA was also proved to act as a catalyst for the first time in 1994 (Endo and Sugiyama 2009). The catalytic DNA molecules are generally known as DNA enzymes or DNAzymes, and DNAzymes and DNA aptamers are collectively called *functional DNAs*. These developments show that the chemical functions of DNA have been extended well beyond the DNA double helix.

4.3 CARBON NANOTUBES (CNTs)

Among the various types of tubes made of organic, inorganic, or polymeric materials displaying attractive properties, CNTs, a family of nanomaterials with remarkable properties, are very promising as applications in materials science and nanotechnology. The history of CNTs may be traced back to the production of very tiny diameter (<10 nm) carbon filaments in 1976 (Oberlin et al. 1976a, 1976b). However, the significance of the as-prepared carbon materials was not well recognized until the breakthrough innovation of fullerene chemistry (1985) (Kroto et al. 1985). The exciting discoveries and continuous progress in fullerene chemistry encouraged research in the field of carbon materials, especially the development of carbon filaments; this activity paved the way for the invention of CNTs.

CNTs are a new class of carbon allotropes having tube-like structures, which may be imagined as rolled-up sheets of graphene. They were introduced in 1991 by S. Iijima (Iijima 1991), who first reported the CNT structure, using high-resolution transmission electron microscopy (HR-TEM). CNTs can be mainly classified into two types, according to their high structural perfection. They are single-walled carbon nanotubes (SWNTs) and multiwalled carbon nanotubes (MWNTs). However, currently, the double-walled carbon nanotubes are also slowly attracting the researchers' attention. In fact, the first determined CNTs were MWNTs, and subsequently, SWNTs with a smaller diameter were independently discovered in 1993 by Iijima (Iijima and Ichihashi 1993) and Bethune (Bethune et al. 1993). CNTs consist of benzene-type hexagonal rings of carbon atoms, with the ends capped by half-fullerene-shaped molecules (Geckeler 2003); each carbon atom is covalently linked to its three adjacent carbon atoms, and the fourth electron is in the delocalized state. SWNTs are made of a single graphene sheet rolled up into a hollow cylindrical form (Ajayan 1999) with a typical diameter ranging approximately from 0.4 to 3 nm. In contrast, MWNTs are formed by an array of multiple concentric cylinders with a diameter varying from ~1.4 nm to 100 nm, and both of these two types are up to several micrometers to millimeters in length (Tang et al. 2001). Between the two types of CNTs, SWNTs are an extremely significant type of CNTs as they display noteworthy electric properties that are not communal by the MWNTs variants. According to their conductivity property, SWNTs may be further divided into two types such as metallic and semiconducting tubes, which are highly dependent on the arrangement of the hexagon rings along the tubular surface.

Although, until now no systematic methods have been established to synthesize pure CNTs (without impurities such as amorphous carbon, carbon nanoparticles, metal catalysts, carbon-coated metal, graphite, and structural defects), both SWNTs and MWNTs are commonly produced mainly by arc-discharge (Endo and Sugiyama 2009; Journet et al. 1997), laser ablation (Sanchez-Pomales et al. 2009; Rinzler et al. 1998), chemical vapor deposition (CVD) (Endo et al. 1995), or the gas-phase catalytic process (HiPco) method (Nikolaev et al. 1999). Irrespective of the synthetic procedure and approach, the CNTs form bundles, which are entangled together in the solid state, giving rise to a highly complex network. Hence, all of the presently known forms of CNT material are almost insoluble or poorly dispersible in most of the known common solvents including organic (Thess et al. 1996; Kim, Nepal et al. 2005) as well as water, making it tricky to explore and understand the chemistry of CNTs at the molecular level studies and device applications. One of the important ways to solve this problem is the use of polymers (Tasis et al. 2003, 2006; Kim et al. 2007) or bio-polymers (Nepal and Geckeler 2006; Yang, Wang et al. 2006; Nepal and Geckeler 2007) or surfactants (Tasis et al. 2003, 2006; Priya and Byrne 2008;

Shin et al. 2008) as dispersing agents, and, so far, a range of polymers and surfactants were used to disperse CNTs.

Polymers have, nevertheless, a strong tendency to arbitrarily wrap around CNT bundles, unless they have an unambiguous binding interaction and an encapsulation process with CNTs. There are three main categories for the modification of the quasi one-dimensional structures such as CNTs to improve their dispersibility in the specific medium: (1) the covalent connection of functional materials via reactions onto the π-conjugated skeleton of CNT; (2) the endohedral filling of their inner hollow cavity; and (3) supramolecular interaction or noncovalent wrapping of different functional molecules. However, the fact is that the dispersion of CNTs in a specific solvent mostly depends strongly on the approach followed for their growth, purity, the method used for purifying these materials as well as the materials and procedure applied for the dispersion techniques. Techniques such as oxidation of contaminants (Chiang et al. 2001), flocculation and selective sedimentation (Bonard et al. 1997), filtration (Bandow et al. 1997), size-exclusion chromatography (Duesberg et al. 1998), selective interaction with organic polymers (Coleman et al. 2000), microwave irradiation (Vazquez et al. 2002) have been generally applied to remove impurities; however, each method has its own merits and demerits and, hence, the further investigation for the improvement of new purification methodologies or developments in the existing techniques needs to be addressed so as to facilitate the use of CNTs as standard laboratory reagents.

Recently, after CNTs could be synthesized with extreme aspect ratios (length to diameter), it was found they were particularly appropriate for application in electronics and semiconductors devices. Further, because of the efficient methods available for the synthesis of large quantities (Ebbesen and Ajayan 1992) of high-quality CNTs (Thess et al. 1996) and also because of their stimulating electronic, mechanical, and structural properties, CNTs were seen as the obvious choice for diverse nanotechnological applications (Baughman et al. 2002) such as fillers in polymer matrixes, molecular tanks, (bio)sensors, nanoelectronic circuits, field emitters, nanoelectronic devices, nanotube aquators, batteries, probe tips for scanning probe microscopy, nanotube-reinforced materials, nanoelectro-mechanical systems (NEMS), nanorobotic systems (Dong et al. 2007), and several others (Geckeler and Nishide 2009; Geckeler and Rosenberg 2006). As a result, CNT development is now being rigorously pursued at all levels pertinent to the application and scope of these materials. As clearly seen from the high number of citations (Figure 4.1b), studies on and scientific attention given to CNTs have radically increased and continue to expand since their emergence. Their inherent size and hollow geometry with extraordinary electronic properties make CNTs promising building blocks for molecular and nanoscale devices.

4.4 DNA–CNT HYBRIDS

Generally, hybrid materials can be achieved by combining two or more different materials via some kind of attractive forces. Such hybrid materials have recently received substantial attention in both academic and industrial domains owing to their exceptional mechanical, thermal, electronic, and catalytic properties, which may not be found in their individual counterparts. Hybrid structures obtained by the combination of inorganic nanomaterials with biological molecules are finding applications in electronics, medicine, and security. In the last two decades, the study of functionalizing nanostructures with biological molecules for the biomedical applications of nanosystems has developed drastically due to their compatibility in forming interesting hybrid structures. These hybrid systems show the distinctive properties of nanostructures and the extraordinary recognition ability of biological moieties. One particularly promising hybrid structure arises when a strand of DNA is combined with CNTs by means of covalent or noncovalent interactions. The unique binding specificity of DNA, owing to sequence-specific pairing interactions between complementary strands, can be attached to the exceptional properties of CNTs to make novel nanosystems, particularly electronic nanodevices, biosensors, and biochips. Furthermore, the formation of a hybrid structure using two special kinds of materials such as DNA and CNTs is certainly different from the hybrid systems of CNTs with other molecules, because DNA provides the advantage of defined length and sequence, biocompatibility, and well-developed chemistry for further functionalization, which cannot be provided by other polymers or molecules. Particularly, the integration of the conducting properties of CNT and the recognition properties of DNA can open a new door to bioelectronic systems, especially sensitive biosensors and biochips.

The combination of CNT with biological systems in general and DNA in particular to form functional assemblies is an innovative and little explored domain of research. However, there are different ingenious techniques to construct DNA–CNT hybrid structures in the literature (Figure 4.1c,d), each contributing varying degrees of dispersion of CNTs and stability of the final hybrid materials. There are different ways to build DNA–CNT hybrid materials. One approach is either naked DNA or DNA modified with appropriate functional groups and connected to functionalized CNTs through covalent bonding. The other way is the adsorption (or wrapping) of DNA on the surface of bare CNTs through supramolecular interactions. The alternative, least probable, approach is to encapsulate or insert the DNA molecules into CNTs' empty cavity. Hence, the new kind of hybrid materials derived from the aforesaid methods are significant not only for fundamental and academic studies of the interactions between biomacromolecules and nanomaterials, but also for diverse potential applications such as catalysts, as well as electronic and sensor devices.

4.4.1 Covalent Linkage

As CNTs are chemically inert, and because they have a hydrophobic exterior, activating their surface is a crucial requirement for connecting DNA to them. In general, chemical functionalization is the most frequent and extensively used way to introduce the linkers, as well as to improve the dispersibility of CNTs, which is an important requirement to make CNTs suitable for practical potential application purposes. Further, functionalization of CNTs by using DNA is still paramount to improve their dispersibility, biocompatibility, and bioselectivity. The as-synthesized DNA–CNT hybrid materials can be readily applied as sensors for medicinal testing, environmental monitoring, and food safety. The functionalization of SWNTs by the chemical method (covalent linkage) to enhance their dispersibility was introduced by Chen and coworkers (Chen et al. 1998). The formation of DNA–CNT hybrids using covalent interactions has the advantage of being able to place the DNA strand to avoid desorption, and to increase accessibility, stability, selectivity, and signal amplification in sensing, when compared to hybrid structures obtained by other approaches such as supramolecular interactions and encapsulations.

There are different kinds of chemistry that have been followed to attach DNA to CNTs via covalent interaction. Among them, the chemical oxidation method was commonly used to introduce functional groups on either ends or the sidewall of the CNTs. Even though various functional groups such as carboxylic acids, quinone, phenol, esters, and alcohols were introduced to CNTs, CNTs with carboxylic acids functional groups have been extensively used as templates to attach biomolecules including DNA.

A multistep approach has been introduced to link DNA with SWNTs via covalent bonding. The multistep procedure includes the introduction of carboxylic acid groups at the ends and sidewalls of purified SWNTs by acid oxidation. They were subsequently treated with thionyl chloride and then ethylenediamine to form amine-terminated sites. The heterobifunctional cross-linker succinimidyl 4-(N-maleimidomethyl)cyclohexane-1-carboxylate, (SMCC) was then reacted with the amine-terminated CNTs to modify the CNT surface terminated with the maleimide groups, which were finally interacted with thiol-modified DNA to produce DNA–SWNTs composites (Figure 4.2). The method followed a multistep procedure; it was observed that the as-prepared covalently linked DNA–SWNT composites were highly dispersed, as well as accessible to hybridization. Thus, selective hybridization occurred with molecules having complementary sequences with only minimal interaction with noncomplementary sequences (Baker et al. 2002). Hence, it is expected that these composites may find a variety of potential applications such as building blocks for more complex supramolecular structures and in highly selective, reversible biosensors.

An interesting two-step approach for covalently anchoring amino-modified DNA to the SWNT surface-bound carboxylic groups via the carbodiimide-assisted amidation was reported (Hazani et al. 2003). In the first step, the oxidized SWNTs were sonicated in the presence of 1-ethyl-3-(3-dimethyl aminopropyl) carbodiimide and N-hydroxy succinimide in order to form a labile intermediate. In the second step, the amino-modified DNA was added to form an amide bond between the primary amine located at the 3′ of the DNA and the SWNT. It was observed that the resulting DNA–SWNT composites were highly dispersible in water, and hence, the specific and nonspecific interactions between DNA and SWNTs were demonstrated by UV-vis spectroscopy and confocal fluorescence microscopy. It was demonstrated that the DNA–SWNT composites hybridize selectively with complementary strands with minimal nonspecific interactions with noncomplementary sequences, which was examined clearly using confocal fluorescence microscopy images. The important advantage of this method is to obtain highly dispersible DNA–SWNT composites and to visualize them using confocal microscopy. This is in contrast to other approaches, where atomic force or scanning or transmission electron microscopes were used for the visualization of SWNTs. These methods of higher resolution

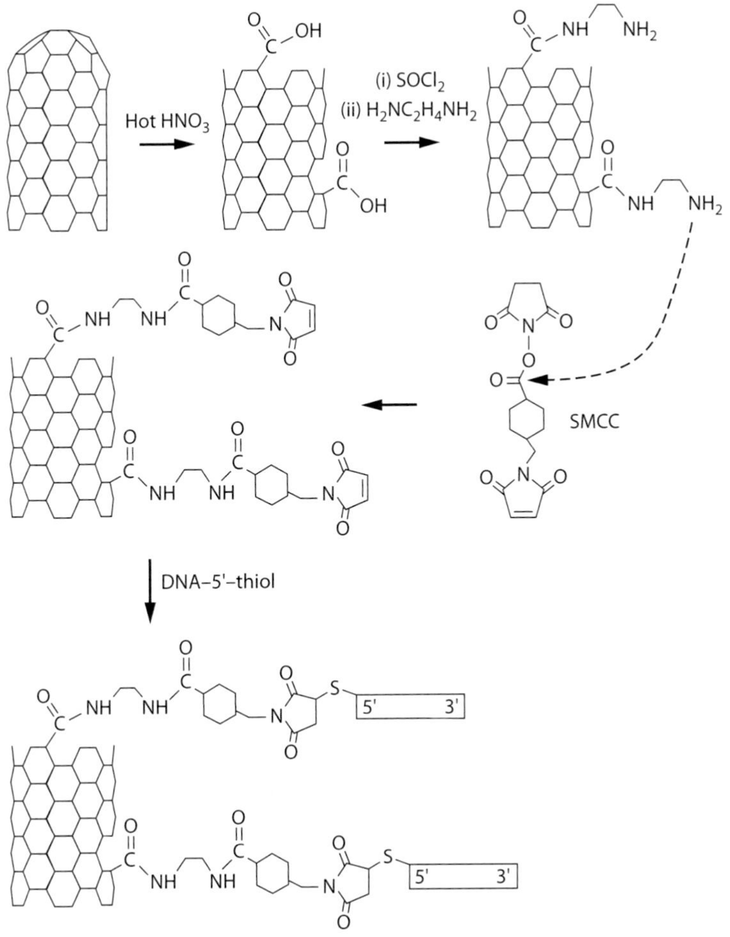

FIGURE 4.2 Overall scheme for the fabrication of covalently linked DNA–SWNT hybrids. (From Baker, S.E., *Nano Lett.*, 2, 1413, 2002. With permission.)

imaging may cause damage for the samples and also need special attention to prepare the samples for the analysis. Hence, these merits open new promises for using SWNT for different potential practical applications.

In contrast, a simple and facile method was established to covalently attach DNA and SWNT by an amino-terminated DNA strand in functionalizing the open ends and defect sites of oxidatively prepared SWNTs using coupling agents such as 1-ethyl-3-(3-dimethylaminopropyl) carbodiimide hydrochloride (EDC) (Figure 4.3). This kind of chemical modification is an important first step toward applying a DNA-guided, self-assembling process capable of directing the placement of SWNTs (Dwyer et al. 2002).

Chen et al. developed a multistep method to covalently link functionalized MWNTs to DNA oligonucleotides (Chen et al. 2005). Appreciably, the chemical analysis of the MWNT functionalization as well as direct visualization of the as-prepared DNA–MWNT hybrid materials were achieved by the combined use of x-ray

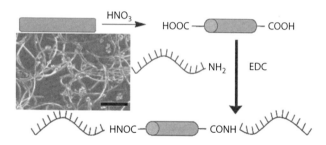

FIGURE 4.3 Schematic representation of the formation of DNA–SWNT composites. The inset shows the SEM image demonstrating the lambda-DNA clusters on the SWNT bundles. (From Dwyer, C., *Nanotechnology,* 13, 601, 2002. With permission.)

photoelectron spectroscopy (XPS) and atomic force microscopy (AFM). While the initial chemical modification to form the amine-terminated MWNTs, which were then covalently combined with DNA, was noticed from the XPS characterization, the morphology of the resulting DNA–MWNT composites, showing DNA strands that can be assembled to both ends and sidewalls of the MWNTs, was achieved by AFM analysis. This combined use of XPS and AFM for the characterization of DNA–MWNT composites may offer a new direction for the definitive characterization of the nanotube functionalization and nanotube-biomolecule-based systems.

An excellent approach was introduced to attach DNA on SWNTs by coupling SWNTs to peptide nucleic acid (PNA, an uncharged DNA analogue) (Nielsen et al. 1991) and hybridizing these macromolecular wires with complementary DNA. Different steps with different chemicals including the coupling agent (Figure 4.4a–c) were introduced following a special experimental procedure to obtain the final DNA/PNA–SWNT adducts (Williams et al. 2002). The DNA/PNA–SWNT hybrids were characterized by using AFM (Figure 4.4d,e) under ambient conditions. From the examinations of several samples it was concluded that the DNA attachment occurs predominantly at or near the SWNT ends. The results also further indicated that there is a sequence-specific PNA–DNA base-pairing rather than nonspecific interaction. Therefore, this method offers a new, flexible means of incorporating SWNTs into larger electronic devices by recognition-based assembly, and of using SWNTs as probes in biological systems by sequence-specific attachment. In addition, there several advantages to couple PNA rather than DNA directly to SWNT: (1) PNA is compatible with most of the suitable solvents such as DMF; (2) PNA is not vulnerable to enzymatic degradation; and (3) PNA–DNA duplexes are more thermally stable than their DNA–DNA counterparts, because there is no electrostatic repulsion as the PNA backbones are uncharged.

In general, manipulating the material properties and engineering of nanotube-based devices are mostly governed by the functionalization chemistry of the open ends and defect sites on the sidewalls of SWNTs. As an example of the application of SWNT to biotechnology, it was noticed that the covalent bonding of DNA-functionalized SWNT multilayer films exhibited an exceptional chemical stability and specificity under the conditions of DNA hybridization (Jung, Kim et al. 2004). The fabrication of patterned SWNT multilayer films were created through successive condensation

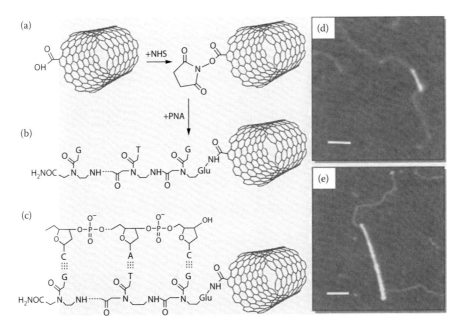

FIGURE 4.4 Attachment of DNA to carbon nanotubes. (a), (b) *N*-hydroxysuccinimide (NHS) esters formed on carboxylated SWNTs are displaced by peptide nucleic acid (PNA), forming an amide linkage. (c) A DNA fragment with a single-stranded, "sticky" end hybridizes by Watson–Crick base-pairing to the PNA–SWNT. (d), (e) Atomic-force microscopic (tapping mode) images of the PNA–SWNTs. SWNTs appear as bright lines; the paler strands represent the bound DNA. Scale bars: 100 nm; nanotube diameters: (d) 0.9 nm; (e) 1.6 nm. (From Williams, K.A., *Nature*, 420, 761, 2002. With permission.)

reactions, making stacks of modified SWNT layers covalently linked together via the use of an appropriate linker molecule and condensation agent (Jung, D.-H., Jung, M.S. et al. 2004; Pellois et al. 2000; Komolpis et al. 2002). Then, the DNA oligonucleotides were covalently immobilized to prepatterned SWNT multilayer films by linking aminated or carboxylated DNA oligonucleotides with the respective carboxylated or aminated SWNT through amide bond formation using 1-ethyl-3-(3-dimethylaminopropyl)carbodiimide hydrochloride (Figure 4.5).

The conjugation of single-stranded DNA (ssDNA) oligonucleotides and SWNT multilayer films was confirmed by XPS and fluorescence-based measurements, and their specificity was established through hybridization experiments with their corresponding DNA complements (Jung, Kim et al. 2004). The fluorescence optical micrograph image of the hybridized oligonucleotide regions (Figure 4.6b) demonstrates a significant similarity to that of the SWNT multilayer film pattern (Figure 4.6a), indicating that a high quantity of DNA assemblies are specifically attached to the SWNT sites. Consequently, the ssDNA–SWNT multilayer films on solid substrates symbolize a novel and new biosubstrate with outstanding potential for both immobilization and hybridization applications and for use in bioelectronic devices.

Recently, Roy and coworkers fabricated a unique nanoelectronic device (Figure 4.7a) based on SWNT for measuring the dc conductivity in single molecule DNA (Roy et

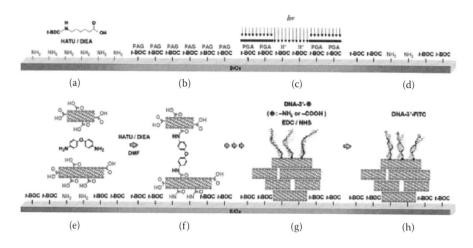

FIGURE 4.5 **(See color insert.)** Overall scheme for the fabrication of patterned SWNT multilayer films and the selective immobilization of DNA: (a) surface treatment with *t*-BOC (*tert*-butyloxycarbonyl) protecting group, (b) deposition of a layer of a PAG (photoacid generator), (c) UV exposure (405 nm, 25 mW) using a photomask for patterning, (d) development process, (e) selective immobilization of SWNTs onto the aminated regions of the substrate, (f) chemical attachment of additional SWNT layers using an ODA (4,4′-oxydianiline) linker molecule and a condensation agent (HATU, *O*-(7-azabenzotriazol-1-yl)-*N,N,N′,N′*-tetramethyluronium hexafluorophosphate), (g) covalent immobilization of oligonucleotide probes onto the patterned SWNT multilayer regions [●, functional groups (NH₂ or COOH) terminated to the oligonucleotide], and (h) hybridization of the FITC (fluorescein isothiocyanate)-labeled complementary oligonucleotide. (From Jung, D.–H., *Langmuir* 20, 8886, 2004. With permission.)

al. 2008). For this purpose, the amine-terminated, single- or double-stranded DNA (ssDNA and dsDNA, respectively) were covalently linked to the carboxy-functionalized SWNTs (Figure 4.7b), where the functional group is acting as a charge transfer bridge. A dielectrophoresis (DEP) field was used for stretching the coiled-shaped DNA molecules. Several devices were used to measure the current values for both the ssDNA and the dsDNA devices under both ambient and vacuum conditions at various temperatures. The results demonstrated that the dsDNA showed a current value in the range of 25–40 pA (at 1 V), whereas the ssDNA displayed a much lower current of ~1 pA or less at the same voltage. The lower current value observed for the ssDNA may be attributed due to the absence of regular stacking of the nucleotide bases. Further, the back-gate voltage characteristics verified that the bridging dsDNA molecule forms a p-type semiconducting channel between the SWNT source and drain electrodes. Also, it was confirmed that the measured electrical signals indeed originated from the covalently linked DNA molecules between the SWNTs nanoelectrodes (Nguyen et al. 2002; Cai et al. 2003; Star et al. 2006), which was proved by performing several blank experiments (Roy et al. 2008). This approach can be used for the identification of specific genes based on the hybridization-induced change in the electrical signal.

Further, different types of chemistry such as photochemistry were applied to attach the ssDNA to the sidewalls and tips of carbon nanotubes via a gentle and easy approach (Moghaddam et al. 2004). DNA oligonucleotides were synthesized in situ

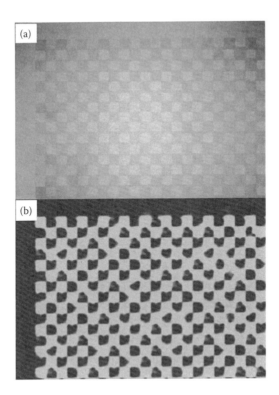

FIGURE 4.6 (a) Optical micrograph image of a chessboard-like patterned array of a SWNT multilayer film, and (b) fluorescence image revealing the successful hybridization of the DNA oligonucleotide conjugated SWNT multilayer film. The fluorescence regions correspond to the areas covered by SWNTs. (From Jung, D.-H., *Langmuir* 20, 8886, 2004. With permission.)

from the reactive group on each photoadduct to form water-dispersible DNA–MWNT hybrid materials. In another case, a combination of chemical and electrochemical reactions was followed to achieve electrically addressable DNA modification of SWNT electrodes and of submicron bundles of vertically aligned carbon nanofibers (Lee et al. 2004). Also, yet another facile approach for the attachment of biomolecules—in general, bovine serum albumin (BSA) and DNA, in particular—to amino-functionalized MWNTs was established (Awasthi et al. 2009). While the initial functionalization of MWNTs and subsequent attachment of biomolecules were confirmed by Fourier transform infrared spectroscopy (FTIR) (Santiago-Rodriguez et al. 2007), the morphology of the resulting DNA–MWNT/BSA–MWNT hybrids was studied by transmission electron microscopy (TEM).

Various facile and highly sensitive electrochemical DNA biosensors and biochips based on CNT-functionalization, generally with a carboxylic acid group, for covalent DNA immobilization and enhanced hybridization detection were studied by different groups independently (Zhu et al. 2005; Koehne, Chen et al. 2004; Li et al. 2005; Taft et al. 2004; Koehne et al. 2003; Koehne, Li et al. 2004; Guo et al. 2004; Dwyer et al. 2004; Zhang, Ma et al. 2009; Zhu et al. 2010). In the majority of the cases, a vertically aligned CNT array or the DNA-directed self-assembling of CNTs

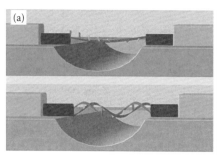
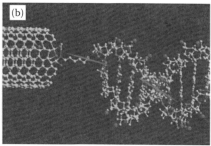

FIGURE 4.7 Schematic illustrations. (a) The upper figure represents an arbitrarily shaped ssDNA molecule, which is stretched and attached to a pair of functionalized SWNT electrodes in the presence of a dielectrophoresis field. The lower figure depicts a covalently attached dsDNA molecule, which has a definite conformation. The nanoelectrodes are separated by a gap of 27 ± 2 nm (the contour length of the 80 bp DNA molecule is ~27 nm). The bridging DNA molecule is suspended over a trench without touching the silicon dioxide surface. (b) Molecular diagram highlighting the covalent bonding between an amine-terminated ssDNA and a carboxyl-functionalized SWNT nanoelectrode via a $(-CH_2-)_6$ spacer. Charge transport takes place through the stacked base pairs in a helical duplex. (From Roy, S., *Nano Lett.*, 8, 26, 2008. With permission.)

was fabricated as a nanoelectrode material for biosensor and biochip development. Generally, cyclic voltammetry (CV) and pulse voltammetry (PV) were employed to characterize the electrochemical properties of the CNT array. This platform can be widely used in analytical applications, as well as fundamental electrochemical studies. Further, this can be employed for increasing highly automated electronic chips with multiplex nanoelectrode arrays for quick DNA analysis.

Guo and coworkers introduced an interesting general method to integrate an amine-functionalized single duplex DNA between oxidatively cut SWNTs as the source and drain electrodes, studying their electrical properties (Guo et al. 2008). It was confirmed that the DNA molecules were bonded covalently to SWNTs via amide linkage that are stable under different experimental conditions. It is noteworthy that the SWNTs were demonstrating semiconductive or metallic behavior before they were cut; however, they were showing insulating behavior after cutting and the fabrication of this type of cut SWNT devices was well established earlier (Guo et al. 2006, 2007; Whalley et al. 2007). Nevertheless, the semiconductive or metallic behavior was observed during the current flow between the SWNT electrodes, when the SWNT electrodes were connected via DNA molecules, indicating the intrinsic conductivity behavior of the DNA. Two different approaches were carried out to link DNA molecules in between the SWNTs' gap. In one case, the SWNTs' termini were connected with each end of the two strands of the DNA duplex, whereas in another method, a single strand was connected between the termini of the SWNT electrodes (Figure 4.8). The ambient condition measurement showed that the latter approach allows for dehybridization/rehybridization with mismatched strands. These types of nanodevices may serve as exceptionally influential reporters to transduce biochemical events into electrical signals at the single-molecule level.

Hence, various approaches and diverse functional groups, as well as coupling agents, different kinds of CNTs and DNA, reaction solvents, and conditions are

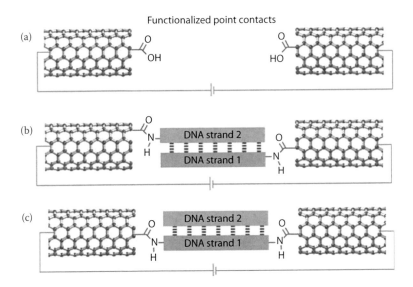

FIGURE 4.8 A method to cut and functionalize individual SWNTs with DNA strands. (a) Functionalized point contacts made through the oxidative cutting of a SWNT wired into a device. (b) Bridging by functionalization of both strands with amine functionality. (c) Bridging by functionalization of one strand with amines on either end. (From Guo, X., *Nat. Nanotech.*, 3, 163, 2008. With permission.)

playing a vital role in the fabrication of covalently attached DNA–CNT hybrids and enhance the dispersity of the resulting final composites to exploit them for the potential practical applications. Generally, carboxylic acid- or amine-functionalized CNTs were mostly used to covalently link the DNA to the CNTs. In many of the cases, covalent linking may be achieved by using different kinds of coupling agents or linkers, which can trigger the carboxylic acid or amine groups of modified CNTs to react with primary amines or the carboxylic acid end of the modified DNA strand, respectively. There are several coupling agents that were used, depending on the reaction conditions and type of functionalization. Among them, 1-ethyl-3-(3-dimethyl-aminopropyl)carbodiimide hydrochloride (EDC) (Dwyer et al. 2002; Chen et al. 2005; Williams et al. 2002), N,N'-dicyclohexyl carbodiimide (DCC) (Li et al. 2003), succinimidyl 4-(N-maleimidomethyl)cyclohexane-1-carboxylate, (SMCC) (Baker et al. 2002), 4,4'-oxydianiline (ODA) (Jung, Kim et al. 2004; Jung, D.H., Jung, M.S. et al. 2004), and O-(7-azabenzotriazol-1-yl)-N,N,N',N'-tetramethyluronium hexafluo-rophosphate (HATU) (Jung, Kim et al. 2004; Jung, D.H., Jung, M.S. et al. 2004) were used frequently. As far as the CNTs are concerned, both SWNTs and MWNTs were used, and in the case of DNA, the ssDNA and dsDNA were used commonly for the formation of covalently linked DNA–CNT hybrids.

4.4.2 ENCAPSULATION OF DNA IN CNTs

In contrast to the traditional approach of covalently linking small molecules or nano-structured materials to the sidewalls or tips of the CNTs, the insertion of suitably

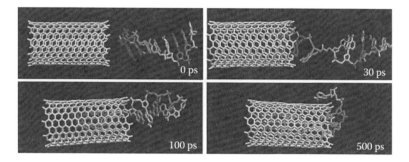

FIGURE 4.9 Simulation snapshots of a DNA oligonucleotide (8 adenine bases) interacting with a (10,10) carbon nanotube at 0, 30, 100, and 500 ps. Water molecules are not displayed for clarity. (From Gao, H., *Nano Lett.*, 3, 471, 2003. With permission.)

sized molecules into the cavity of CNTs is an alternative way to functionalize CNTs with suitable properties for potential applications. It has been reported that small molecules such as C_{60} (Smith et al. 1998) metallofullerences, (Hirahara et al. 2000) water, (Hummer et al. 2001) or gas molecules (Gogotsi et al. 2001) and even metal nanoparticles (Jiang and Gao 2003; Tsang et al. 1994) can be encapsulated inside the CNT cavity.

In this series, an attempt has been taken to address the theoretical expectation for the dynamic processes of encapsulating DNA into the empty cavity of CNTs in a water-solute environment by molecular dynamics simulations (Gao et al. 2003). Thus, it was found (Figure 4.9) that a DNA molecule could be spontaneously encapsulated inside the empty voids of CNTs. Although both the van der Waals and hydrophobic forces were found to be important for the encapsulation process, the van der Waals forces play a vital and dominant role in the DNA–CNT interactions. Hydrophobic forces could be generated by strong hydrogen-bond interactions among the water molecules, which cause hydrophobic solutes (both CNT and DNA) to agglomerate in order to decrease the solvent-solute interface energy. However, it was found that this hydrophobic effect alone was insufficient to insert the DNA into the CNT cavity owing to the resistance from the water molecules present in the cavity of the CNT. Besides, when the distance between the DNA and CNTs is less than ~1 nm, they can also experience an attractive force from each other, owing to van der Waals interactions. Hence, this attractive force is playing a crucial role to insert DNA into the cavity of CNTs. This study has given a new direction in the chemistry of CNTs, in which nanostructured materials such as metal nanoparticles and important biomolecules such as protein and DNA can be encapsulated into the cavity of CNTs. Therefore, it is expected that this kind of CNT-based materials can be further utilized for potential applications in the field of molecular electronics, molecular sensor devices, electronic DNA sequencing, and the nanotechnology of gene delivery systems.

Ito and coworkers observed the DNA transport via a membrane containing a single MWNT using fluorescence microscopy (Ito et al. 2003). Typically, 40 consecutive fluorescence images were documented at regular intervals of ~1 s (exposure time: 300 ms) to monitor the transport of DNA through a MWNT channel. The inner radius and length of the MWNT channel were determined by using TEM and

CV, and determined to be 77 nm and 0.84 μm, respectively. After careful analysis of the results, it was suggested that the DNA molecules will be trapped at the MWNT channel entrance when the radius of gyration (R_g) of a DNA molecule is larger than the channel radius (~77 nm); however, by the application of an adequately high membrane potential, eventually the DNA molecule may be transported through the channel. On the other hand, this behavior was not noticed when the R_g of the DNA molecule was smaller than the MWNT channel radius. This type of hybrid material may become significant for investigating and creating a rich variety of electrical and sensor devices (Daniel et al. 2007).

4.4.3 NONCOVALENT INTERACTIONS BETWEEN DNA AND CNTs

Another excellent strategy to prepare DNA–CNT hybrid materials is supramolecular or noncovalent functionalization of CNTs using DNA. Noncovalent functionalization may have more advantages over covalent functionalization, in which there is a possibility to perturb the extended π-networks on the CNTs surfaces, and hence, change their intrinsic mechanical and electronic properties. In contrast, supramolecular strategies can preserve the CNT structure and, in turn, maintain the peculiar properties of the tubes. Further, noncovalent interaction or wrapping of CNTs using biopolymers, polymers, surfactants, polyaromatic compounds, and small organic macrocyclic molecules can increase the dispersibility of the CNTs in water as well as in organic solvents, so that the final composite materials can be applied for the potential application purposes without altering the CNTs' unique properties that can enhance the efficiency of the systems (Geckeler and Nishide 2009; Geckeler and Rosenberg 2006; Tasis et al. 2003, 2006). Usually, the noncovalent or supramolecular complexation of CNTs can be achieved by hydrophobic–hydrophobic interaction, π–π stacking interaction, weak hydrogen bond linkage, and electrostatic attraction between the materials used for the functionalization.

In general, while hydrophobic forces are defined as the attractive forces between molecules due to the close positioning of hydrophobic (nonhydrophilic) part of the two or more molecules, hydrogen bonds are a type of weak attractive (dipole–dipole) interaction between an electronegative atom and a hydrogen atom bonded to another electronegative atom such as nitrogen, oxygen, or fluorine. Hydrogen bonding can take place between molecules or within parts of a single molecule. In the DNA, the double helical structure is mainly stabilized owing to hydrogen bonding between the base pairs, which connect one complementary strand to the other and enable replication. Hydrogen bonds tend to be stronger than van der Waals forces, but weaker than covalent bonds or ionic bonds. The π–π stacking is a kind of supramolecular interaction, which is generally happening by intermolecular overlapping of p-orbitals in π–conjugated systems. In contrast to a variety of aromatic compounds (Tasis et al. 2003, 2006), pyrene derivatives are more susceptible to stack on the surface of the CNTs via π–π stacking interactions (Liu et al. 2003; Yang et al. 2006; Guldi et al. 2005). However, electrostatic interactions are noncovalent dipole–dipole or induced dipole–dipole interactions that can be stabilizing or destabilizing.

Especially, supramolecular functionalization of CNT using biomolecules, in general, and DNA specifically, holds attractive prospects in a variety of domains including dispersion in aqueous media, nucleic acid sensing, gene-therapy, and controlled deposition on conducting or semiconducting substrates. Further, DNA molecules can increase the CNT dispersion in most of the solvents including water, and also be used to differentiate or separate metallic CNTs from semiconducting CNTs, which are highly useful to utilize the final materials to fabricate sensitive biosensor, biochips, and electronic devices.

An extremely effective and facile approach was introduced to disperse bundled SWNTs in water by sonication in the presence of ssDNA. The resulting DNA-wrapped SWNTs were purified using centrifugation and separated into fractions with different electronic structures by ion-exchange chromatography (Zheng, Jagota, Semke et al. 2003). The conversion of SWNT bundles into individual tubes by using DNA was well described by electronic absorption and fluorescence spectroscopy and AFM measurements. It was noted that the dispersion ability of SWNTs were improved greatly and much more efficient than the other polymers used for the same purpose. Further it was proved by molecular modeling (Figure 4.10a,b) that the ssDNA can adhere to the SWNTs surface via π-stacking in such a way that the aromatic nucleotide DNA bases can associate with the surface of the CNTs, whereas

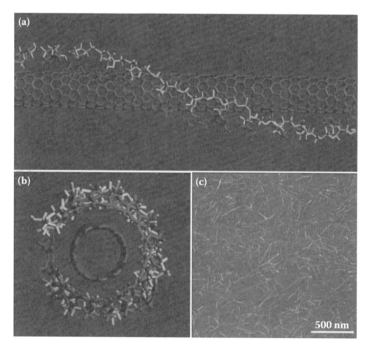

FIGURE 4.10 (a, b) Binding model of a (10,0) CNT wrapped by a DNA. (a)The right-handed helical structure shown here is one of several binding structures found, including left-handed helices and linearly adsorbed structures. In all cases, the bases (red) orient to stack with the surface of the nanotube and extend away from the sugar-phosphate backbone (yellow). (b) The DNA wraps to provide a tube within which the CNT can reside, hence converting it into a water-dispersible object. (c) AFM image of DNA-CNT hybrids. (From Zheng, M., *Nat. Mater.* 2, 338, 2003. With permission.)

the sugar–phosphate backbone (the hydrophilic fraction) of the DNA strand is left exposed to interact with the solvent. It was evidenced from AFM measurements (Figure 4.10c) that the DNA–SWNTs have a length distribution from 50 to 1,000 nm (the average length of SWNTs is 117 ± 68 nm), with tube diameters ranging from 1 to 2 nm. The highlight of this approach is to obtain a high degree of SWNT dispersibility in water, which could be possible due to combined advantages of DNA chain length and sequence with flexibility, its backbone charge as well as well-developed chemistries for the functionalization. In addition, the DNA–SWNT hybrids could be purified and separated on the basis of length, diameter, and conductivity-type (semiconducting or metallic tubes). Hence, this approach provides an entry to CNT-based applications in biotechnology. Further, the specific advantages and unique properties of the as-prepared DNA–SWNT hybrid material offer numerous ways to influence CNTs in aqueous media, including separation and surface modification.

Our group introduced an unprecedented, facile, one-pot synthetic strategy for preparing short-length, highly water-dispersible, DNA-wrapped CNTs (both MWNTs and SWNTs) by a solid-state mechanochemical reaction (Nepal et al. 2005). During this process, the mechanical energy creates highly reactive centers in the solid phase, enabling the nanotubes to interact with the DNA (Figure 4.11). It was perceived that both the MWNTs and SWNTs were cut into shorter lengths and were fully covered with DNA after reacting the specified amount of respective CNTs and DNA. The naked CNT surface (without any surface modification or functionalization) was fully wrapped with DNA via noncovalent interaction, and the resulting hybrid material was highly dispersible in aqueous medium. Interestingly, the careful analysis showed that the nanotubes were cut also with uniform distribution, where >90% of the MWNT products were 500 nm to 3 μm and 80% of the SWNT products were 250 nm to 1 μm in length, respectively (Figure 4.12). More significantly, the resulting supramolecularly masked, water-dispersible DNA–CNTs hybrids show a good stability of >6 months in aqueous medium. This high stability and good dispersibility are owing to the π–π interactions between the nanotube surface and the backbone of the DNA, which afford a strong binding, and the phosphate side-groups to interact with the solvent to make the tubes more dispersible in the medium. Here, the solid-

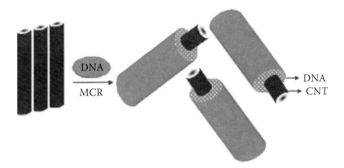

FIGURE 4.11 Formation of supramolecular conjugates of CNT and DNA by a solid-state mechanochemical reaction. (From Nepal, D., *Biomacromolecules*, 6, 2919, 2005. With permission.)

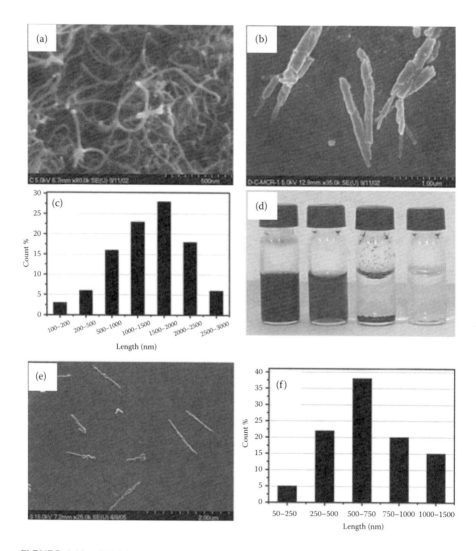

FIGURE 4.12 SEM images and length distribution of MWNT-DNA (a) Pristine MWNT, (b) MWNT-DNA conjugates, (c) length distribution, and (d) photographs of solutions of MWNT-DNA conjugates: 5 mg/mL, 0.25 mg/mL, pristine MWNT in water, and DNA (from left to right). (e) SEM image of SWNT-DNA conjugates, and (f) length distribution. (From Nepal, D., *Biomacromolecules,* 6, 2919, 2005. With permission.)

state technique plays a dual role in shortening the length of the CNTs and the formation of supramolecular adducts of DNA–CNT conjugates. The nanotube dispersion, length distribution, and adduct formation were confirmed by UV-vis absorption spectroscopy, scanning electron microscopy (SEM), and fluorescence microscopy. As short-length, highly water-dispersible with a high stability CNT-based products are very high in demand to improve their compatibility with biological systems; this one-pot approach to prepare supramolecular DNA–CNT conjugates paves the way for both biological and nonbiological applications.

Many methods were proposed for the dispersion of CNTs using DNA by noncovalent functionalization (Zheng, Jagota, Semke et al. 2003; Nepal et al. 2005; Nakashima et al. 2003; Dovbeshko et al. 2003a, 2003b), which were mostly performed by the sonication of CNTs in the presence of DNA at room temperature followed by the purification of the DNA–CNT hybrids by centrifugation (Zheng, Jagota, Semke et al. 2003; Sanchez-Pomales et al. 2007; Zheng, Jagota, Semke et al. 2003; Cheung et al. 2009). In most of the cases, the detailed interaction between DNA and CNT was not identified clearly, however, π–π interactions between the nanotube surface and bases of DNA was proposed for the formation of DNA–CNT hybrids via noncovalent interactions, which is mainly responsible for the dispersion of CNT in a medium containing DNA (Nakashima et al. 2003; Guo et al. 1998).

An excellent method was demonstrated to disperse entangled SWNTs in water in the presence of ssDNA, resulting in ssDNA-helically wrapped SWNTs with both negatively charged DNA phosphate backbone and bare SWNT graphite regions exposed to the solution (Figure 4.13; Ma et al. 2006). It was determined from the

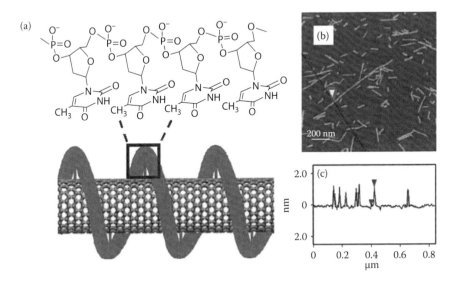

FIGURE 4.13 (a) Cartoon showing ssDNA-wrapped SWNT (the schematic is only a graphical representation and does not represent the precise way ss-DNA binds on SWNTs), (b) Tapping mode AFM image of ssDNA-SWNTs, and (c) its corresponding section analysis showing the thickness of the nanowires. (From Ma, Y., *J. Phys. Chem. B*, 110, 16359, 2006. With permission.)

tapping-mode AFM images of the DNA–SWNTs hybrid on mica surfaces that the diameter of the DNA–SWNTs was 1.3 nm ± 0.4 nm, measured from their cross-sections in the AFM image and averaged from measurements of 30 nanotubes (Figure 4.13). However, the detailed investigation of the DNA-wrapped MWNTs was also studied using the tapping mode AFM analysis (Takahashi et al. 2006). Further, for the first time it was shown that the as-prepared DNA–SWNTs hybrid can perform as a template for in situ electrochemical polymerization of self-doped polyaniline. Hence, it is expected that the exceptional properties of the resulting DNA–SWNTs hybrids observed in this study may also apply during their integration into other polymers, in addition to substituted polyanilines, to fabricate new class of functional nanomaterials.

Further, from the structure of DNA, it was speculated that there may also be some kind of weak interaction between the CNTs and the major and minor grooves of the DNA. This speculation was addressed by Lu and coworkers (Lu et al. 2005), who investigated in a theoretical simulation study that a system consisting of B-DNA and an array of (10,0) CNTs periodically arranged to fit into the major groove of the DNA (Figure 4.14). Three electron transport mechanisms were identified from the electronic structure of the combined system. This interesting theoretical prediction may result in a very sensitive nanoscale electronic device and a means for ultra-fast DNA sequencing. Also, in the control experiment, it was demonstrated from different groups that the wrapping of CNTs by ssDNA was dependent on the specific DNA sequence (Zheng, Jagota, Strano et al. 2003; Humphrey et al. 1996; Tu

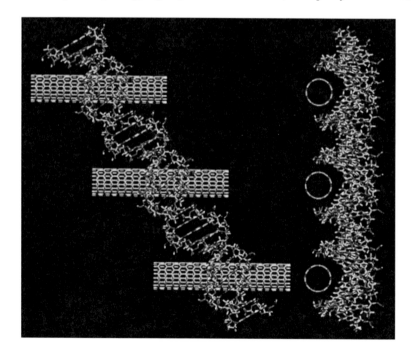

FIGURE 4.14 Proposed DNA-CNT system in top and side views, the latter along the CNT axis (created with VMD). (From Lu, G., *Nano Lett.* 5, 897, 2005. With permission.)

and Zheng 2008; Enyashin et al. 2007; Vogel et al. 2007; Li et al. 2007; Chen, Liu et al. 2007). In contrast to a previous work (Zheng, Jagota, Strano et al. 2003), in which short oligonucleotides were used, a systematic study demonstrated with long genomic ssDNA (>>100 bases) of a completely random sequence of bases to disperse CNTs proficiently because of the ssDNA's capability to form tight helices around the CNTs with distinct periodic pitches (Gigliotti et al. 2006).

In another approach, ssDNA was shown as an efficient noncovalent dispersant for individual SWNTs in water, forming a DNA–SWNT hybrid material, which has different advantages for CNT separation and practical applications. Further, the location of DNA on the DNA–SWNT nanomaterial was identified using core-shell (CdSe/ZnS) quantum dots with the help of AFM analysis and revealed that the wrapped DNA strands were closely arranged end-to-end in a single layer along the SWNTs, and the obtained periodic pattern was independent of the length and sequence of the wrapping DNA (Campbell et al. 2008). The insight gained about the DNA location and wrapping structure are anticipated to benefit the separation of CNT mixtures into homogeneous samples, which is a fundamental step in the majority of the envisioned applications for CNTs (He et al. 2006; Yogeswaran et al. 2008).

Cathcart and coworkers reported an extensive study on the time-dependence of DNA (natural salmon testes) wrapping of SWNT and monitored over a three-month period (Cathcart et al. 2008). It was found that the fraction of DNA bound to the SWNTs increases with time, and the progressive formation of a coating of DNA on the sidewalls of the SWNTs over a three-month period. Albeit, periodic wrapping of CNTs by ssDNA was observed using AFM analysis (Zheng, Jagota, Strano et al. 2003; He et al. 2006), no evidence for helical wrapping was reported for SWNTs dispersed with dsDNA, which was clearly observed in this study. The HR-TEM images clearly demonstrated the DNA wrapping helically around the SWNTs in a surprisingly ordered fashion (Figure 4.15). Interestingly, based on the analysis of a number of images (Figure 4.15), it was noted that on average, the one monolayer of DNA coating on the SWNTs was almost complete by day 35, coinciding with the time at which the absorbance spectra and photoluminescence spectra started to improve. The rate of DNA wrapping was investigated with respect to the sample temperature. It was established that the time needed for a complete DNA monolayer to form on the SWNTs was controlled by a rate-limiting process with an activation enthalpy of 41 kJ mol^{-1} (0.43 eV). This low energy barrier is attributed to the final important step in the wrapping mechanism, which engages the transformation of the disordered population of DNA at the surface into a strongly bound array approximating a monolayer coating. Although AFM and SEM analysis were used to demonstrate the DNA wrapping of CNTs, the effort has not been attempted to show the CNTs wrapping of either ss- or ds-DNA using HR-TEM analysis.

Though DNA-CNT hybrids showed a remarkable set of technologically useful properties such as facilitation of SWNT separation, chemical sensing, and detection of DNA hybridization, its microscopic structure and physical properties have not been well understood. However, Johnson and coworkers (Johnson et al. 2008) took an effort to address this issue by performing the classical all-atom molecular dynamics (MD) simulation studies exploring the self-assembly mechanisms (Maune et al. 2010), the existence structure, and energetic properties of this novel hybrid material.

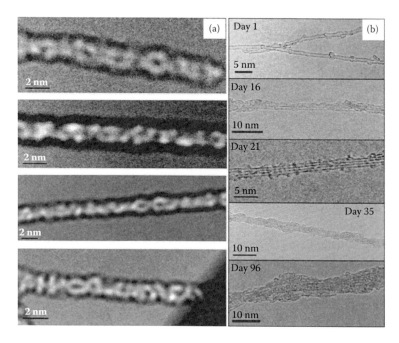

FIGURE 4.15 (a) Various HR-TEM images showing helical DNA wrapping of nanotubes in a 32-day-old sample. (b) Representative HR-TEM images showing both single nanotubes and small bundles at various times. The time-dependent formation of the DNA coating on the nanotubes can be seen clearly. (From Cathcart, H., *J. Am. Chem. Soc.*, 130, 12734, 2008. With permission.)

It was found that in aqueous medium, the existence of SWNT induces ssDNA to undergo a spontaneous conformational change that enables oligonucleotides of general sequence to adsorb via the π–π stacking interaction. The flexibility of ssDNA facilitates the spontaneous wrapping about SWNT including right- and left-handed helices within a few nanoseconds. It was observed that the helix formation for adsorbed oligonucleotides is driven by electrostatic and torsional interactions within the sugar–phosphate backbone, which results in ssDNA wrapping around the SWNT from the 3′ end to the 5′ end. The theoretical studies on wrapping of DNA to CNTs were reported independently by different groups, and the binding energies of DNA to SWNTs have also been examined by different levels of calculation (Frischknecht and Martin 2007; Gowtham et al. 2008; Meng et al. 2007; Johnson et al. 2010). The binding energies are mainly derived from the π–π stacking interaction, although the sugar–phosphate backbone also has some assistance.

In addition to the molecular dynamics simulation studies, the first topographic images of the DNA–CNT hybrids with significant morphological details by using scanning tunneling microscopy (STM) were reported by Yarotski and coworkers (Yarotski et al. 2009). The STM images demonstrate the direct observation of strands of DNA wrapping around a single CNT at ~63° angle with a coiling period of 3.3 nm, which is in good agreement with predictions based on theoretical simulations (3.2 nm period). The STM image in Figure 4.16a clearly shows the DNA strand

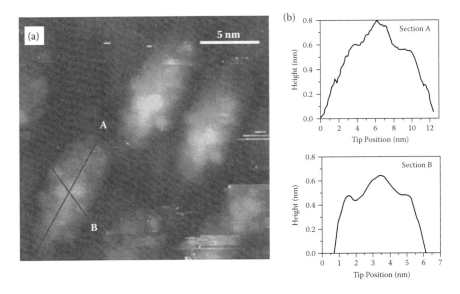

FIGURE 4.16 (a) STM topographic image (21 × 21 nm) of DNA–CNT hybrid on a Si(110) substrate acquired at $It = 10$ pA, $Ub = 3$ V, and (b) corresponding height profiles along the line A and B. (From Yarotski, D.A., *Nano Lett.*, 9, 12, 2009. With permission.)

wrapped around the CNT; regular height modulations of the DNA-covered segments of the CNTs are also clearly seen in the image. Two sections, such as section A and B, of the hybrid profile emphasize the periodic nature of these modulations both along the nanotube and across it, respectively (Figure 4.16). Further, the periodic distortions of 1.9 and 2.6 nm along the DNA strand were also noticed on diverse types of CNTs, which remain unexplained. The application of STM to structural and electronic characterization of DNA–CNT hybrids could pave the way toward more details of their formation and offer further direction for optimization of DNA-assisted CNTs separation strategies via the assembly of the most stable DNA–CNT hybrids with the optimal π-staking geometries (Yu et al. 2010; Muller et al. 2010).

Though the main two types of CNTs such as the SWNTs and MWNTs have been studied extensively due to their interesting properties and potential applications in various domains, recently, another type of CNTs known as double-walled carbon nanotubes (DWNTs) have attracted considerable attention owing to their intrinsic coaxial structure, which makes them mechanically, thermally, and structurally more stable than SWNTs. As the dispersion of CNTs in aqueous medium is the prime requirement for their processability, Kim and coworkers (Kim et al. 2010) established a method to disperse DWNTs using ssDNA as the dispersing agent and performed detailed resonant raman/fluorescence spectroscopic studies on ssDNA-dispersed DWNT solutions at different dispersion states in order to understand the interactions between the DNA and the outer tubes, and the effect of these different DWNT environments on vibrational and luminescence behavior. From several controlled experiments, it was observed that by increasing the dispersion state of the DWNTs in an aqueous DNA solution, the intensity of luminescence peaks was increased and the peak positions shifted toward longer wavelengths, indicating that the DWNTs were individually dispersed in an aqueous medium. This is strongly supported by the sharp absorption spectra as well as HR-TEM observations. After analyzing the results obtained by several techniques, the authors concluded that the circumferentially wrapped DNA on the outer tubes of individually isolated DWNTs in an aqueous medium gives rise to a strong charge transfer to the semiconducting and metallic outer tubes, as well as generating physical strain in the outer tubes. Therefore, it is anticipated that the DNA–DWNTs hybrids are highly promising for creating strong and stable luminescence signals, as well as for high-yield optoelectronic applications (Li, Kaneko et al. 2010; Khamis et al. 2010). Also, freestanding, thin, and bendable electrodes for supercapacitors were fabricated by using a DNA–DWNTs thin film (Cooper et al. 2009). In this system, the DNA plays an important role in dispersing the strongly bundled DWNTs; its conversion to phosphorus-enriched carbons gave rise to strong redox peaks at around 0.4 V. The combined effects of the large capacitance from the DNA-derived carbons and the high electrical conductivity properties of CNTs make this system (DNA–DWNT hybrid films) useable as a potential electrode material for supercapacitors.

In general, the ssDNA acts as an effective and compatible noncovalent dispersant for the dispersion of different types of CNTs. In most cases, the ultrasonication technique facilitates the separation of entangled CNTs, producing individually dispersed, DNA-wrapped CNTs (DNA–CNT hybrid). The dispersion is accomplished by the aromatic nucleotides of DNA interact via π-stacking with the hydrophobic CNT surface, while the polyanionic DNA backbone interacts with the solvent (water) to make the tubes more dispersed in the water. The strength of the binding interactions between CNTs and ssDNA is shown by the capability of the DNA to disrupt the strong intertube interactions responsible for CNT agglomeration into bundles.

4.5 GRAPHENE (GRP)

Graphene, an important class of carbon-based materials, is a flat monolayer of sp^2-bonded carbon atoms that are tightly packed in a honeycomb lattice (Geim and Novoselov 2007). The bond length of adjacent carbon–carbon atoms in GRP is around 142 pm. As GRP is the basic structural element or building blocks of the new classes of graphitic materials of all other dimensionalities, it can be called as the "mother of all graphitic forms". GRP can be wrapped up into 0D fullerenes; a single sheet GRP can be rolled up into 1D nanotubes or several GRP layers stacked into 3D graphite (Figure 4.17). Compared to the one-dimensional CNTs, the two-dimensional GRP is expected to exceed the CNTs' versatility, as it provides a large detection area, biocompatibility, and unique electronic properties such as ultra-high mobility and an ambipolar field-effect. Further, GRP can be considered as a good counterpart to CNTs and fullerenes but with much lower synthetic costs.

Recently, new research on GRP became a hot topic in the materials science and nanotechnology field after it became known that a free-standing GRP monolayer can be prepared on a SiO_2/Si substrate by using a micromechanical cleavage method (Novoselov et al. 2004). Owing to the unusual electronic properties of GRP such as high charge transport mobility (Geim and MacDonald 2007), it has been identified

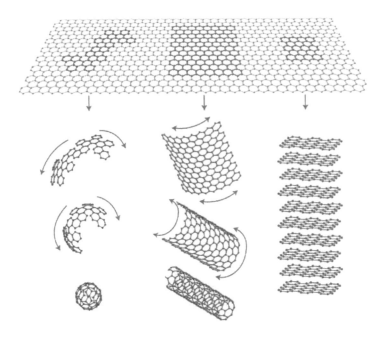

FIGURE 4.17 The "mother" of all graphitic forms. Graphene is a 2D building material for carbon materials of all other dimensionalities. It can be wrapped up into 0D buckyballs, rolled into 1D nanotubes, or stacked into 3D graphite. (From Geim, A.K., *Nat. Mater.* 6, 183, 2007. With permission.)

as a potential material for a variety of applications, especially in the fabrication of sensitive nanodevices, nanoelectronics, and nanosensors (Li et al. 2008). Recently, chemical (Gilje et al. 2007; Stankovich et al. 2006; Hod et al. 2007; Gomez-Navarro et al. 2007; Park et al. 2008) and geometric (Yan et al. 2007; Obradovic et al. 2006; Barone et al. 2006) manipulation of GRP has shown huge potential in being able to control its band gap between semimetallic and semiconducting. Further, GRP is a promising building block for high performance electronics (Sordan et al. 2009; Yoo et al. 2008; Lin et al. 2009), sensors (Huang et al. 2008; Schedin et al. 2007; Sakhaee-Pour et al. 2008; Lu et al. 2009) and energy storage devices (Wu et al. 2008; Stoller et al. 2008; Zhang, Li et al. 2009; Wang et al. 2008). Furthermore, GRP nanostructures have been exceptionally integrated into a variety of electronic and optoelectronic applications (Geim and Novoselov 2007) such as gas sensors (Schedin et al. 2007), transistors (Wu et al. 2008; Chen, Ishigami et al. 2007), solar cells (Wang et al. 2008; Liu, Q., Liu, Z. et al. 2008; Wang 2008; Becerril et al. 2008), and liquid-crystal elements (Blake et al. 2008), due to its low electrical noise (and low charge-scattering) (Novoselov et al. 2004, 2005; Berger et al. 2006) and ballistic transport (Novoselov et al. 2005; Zhou et al. 2006), properties. However, it is surprising that there are very limited studies on the application of GRP in biological devices. One of the reasons may be the lacking of experimental protocols to disperse large-scale GRP in an aqueous medium without disruption of the sp^2 GRP structure, which is a significant feature for biological applications of GRP (Hao et al. 2008).

There are several fabrication methods that have been proposed to produce high-quality GRP crystallites, namely, micromechanical cleavage (Geim and MacDonald 2007), epitaxial growth by chemical vapor deposition (Kim et al. 2009), microwave irradiation (Murugan et al. 2009; Li, Yao et al. 2010), chemical exfoliation (Xu et al. 2009), and so on. Especially, solution-based techniques represent easy and cost-effective procedures for the large-scale production of GRP sheets by using either polar organic solvents (whose interaction energy with GRP is comparable to the van der Waals force of GRP layers) (Hernandez et al. 2008; Qian et al. 2009), or water-dispersible graphene oxide (GO) as an intermediate, which is then converted to GRP via a reduction process (Tung et al. 2009). However, in most cases, the typically used reducing agents such as hydrazine and dimethylhydrazine are highly toxic and potentially carcinogenic, which is also an inhibiting issue for applying GRP in biological applications. Therefore, finding an easy, economical, as well as effective procedure to prepare GRP, and modifying its surface using biologically important materials under benign conditions is important to the potential application in the biological and biomedical fields. The DNA is a unique biopolymer showing exceptional binding specificity due to sequence-specific pairing interactions between complementary strands. These can serve as effective dispersants to increase the dispersibility of GRP in water and in turn, can form novel DNA–GRP hybrid materials. Hence, there is no doubt that the exceptional and unique electronic properties (ultra-high mobility and ambipolar field-effect) of GRP coupled with the specific recognition properties of the immobilized DNA (exceptional binding specificity) would therefore construct sensitive miniaturized biosensors and nonodevices.

4.6 DNA–GRAPHENE HYBRIDS

The integration of GRP with DNA to form functional assemblies is new, and not much work has been done in this area of research. The novel DNA–GRP hybrids can be obtained by linking the DNA with GRP or chemically modified GRP, with their two-dimensional nanostructure and adjustable surface chemistry via covalent or noncovalent interaction.

Inspired by earlier approaches in which the hydrophobic CNTs were successfully functionalized using ssDNA in an aqueous medium to fabricate novel nanocomposites, Patil and coworkers (Patil et al. 2009) demonstrated for the first time the use of ssDNA in the preparation of high concentration (2.5 mg mL^{-1}) and stable aqueous suspensions of graphene single sheets (Figure 4.18a). Further, they have extended their experiments by exploiting the DNA–GRP dispersed sheets as highly negative charged materials for the fabrication of the self-assembly of novel graphene-based layered bionanocomposite materials containing intercalated DNA molecules, or cointercalated mixtures of DNA and the positively charged redox protein cytochrome C (Figure 4.18b). The composites have ordered lamellar nanostructures, in which the entrapped biomolecules are located specifically within the gallery regions of a stacked array of graphene sheets. The TEM image of the DNA–GRP hybrid material shows ultra-thin flat sheets ranging from 500 nm to more than 1 mm in size, and with occasional folds, wrinkles, and rolled edges (Figure 4.18c). However, the

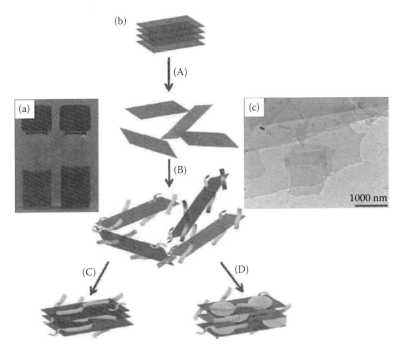

FIGURE 4.18 (a) Photographic image showing stable dispersions of ssDNA-GRP sheets prepared at graphene and DNA concentrations of (i) 1 and 2 mg mL^{-1}, or (ii) 5 and 20 mg mL^{-1}, respectively, followed by a 20-fold dilution. (b) Schematic showing the synthesis of aqueous, DNA-stabilized graphene suspensions and the fabrication of lamellar, multifunctional nanocomposites. (A) Oxidative treatment of graphite (grey-black) yields delaminated nanometer-thick sheets of graphite oxide (GO) (brown). (B) Chemical reduction of GO sols with hydrazine in the presence of freshly prepared ssDNA produces a stable aqueous suspension of ssDNA-functionalized graphene sheets (ssDNA-GRP). (C, D) Processing of ssDNA-GRP dispersions to produce ordered layered nanocomposites; (C) evaporation-induced deposition and self-assembly on flat substrates results in lamellar nanocomposite films with intercalated ssDNA molecules and (D) co-assembly of negatively charged ssDNA-GRP sheets and positively charged cytochrome C produces co-intercalated multifunctional layered nanocomposites. (c) TEM image of ssDNA-GRP sheets. (From Patil, A.J., *Adv. Mater.*, 21, 3159, 2009. With permission.)

height of the DNA–GRP sheets was found to be 2–2.5 nm using AFM height profile analysis, which was considerably greater than the theoretical thickness of a single graphene layer (0.34 nm), may indicate that the DNA was wrapped over the graphene sheet. This approach has the potential to be considered as a general method to fabricate GRP-based lamellar nanocomposite films with coassembly of a wide range of functional molecules, and the final nanocomposites may find potential applications in biosensing, nanoelectronics, and biotechnology.

In an approach, the versatility of chemically modified graphene was demonstrated by utilizing it as sensitive building blocks for bioelectronics, especially for the fabrication of a novel chemically modified graphene-based biodevice, label-free DNA sensor, and bacterial DNA-protein and polyelectrolyte chemical transistor (Mohanty and Berry 2008). In the fabrication process of the DNA sensor, the GO sheets immobilized on silica were acting as templates to selectively and covalently tether the ssDNA to build the DNA–GRP hybrid. It was observed that ssDNA tethered on graphene hybridizes reversibly with its complementary DNA strand. This study opens the door to the area of graphene-related nanoelectronics and motivates the development of the next-generation electronic systems and devices. It was also demonstrated that large-sized graphene films consisting of monolayered and few-layered graphene domains were used to fabricate liquid-gated transistors for DNA sensing applications (Dong et al. 2010).

As the research on graphene-related functional materials in general and functional biomaterials in particular is an emerging area, different types of investigations were carried out to understand and explore the chemistry of graphene. In this aspect, an effort was taken to understand the interaction of graphene with DNA nucleobases and nucleosides by using isothermal titration calorimetry (ITC). After several controlled experiments, it was observed that the DNA nucleobases and nucleosides bind to graphene, and the order of relative binding energy of the nucleobases was found to be G>A>C>T in an aqueous medium, although the positions of C and T seem to be interchangeable (Varghese et al. 2009). The same tendency was found with the nucleosides; however, the interaction energies of the base pairs were generally between those of the component bases. The experimentally observed trend was also verified by theoretical calculations, which show the significance of solvation effects.

Recently, the noncovalent DNA decorations on the basal planes of GO and reduced graphene oxide (RGO) nanosheets were prepared, and the as-prepared DNA–GO or DNA–RGO hybrids bearing multiple thiol groups labeled on DNA strands were then used to scaffold the two-dimensional self-assembly of gold nanoparticles (AuNPs) into metal–carbon hybrid nanostructures (namely, DNA–GO-AuNP or DNA–RGO-AuNP), which are expected to be excellent materials for many different potential applications in diverse areas (Liu, Li et al. 2010). The final DNA-carbon-metal hetero-nanostructures were found to be highly stable and well dispersible in water and could be separated with high purity using gel electrophoresis. First, the thiolated DNA oligos [d(GT)$_{29}$SH, containing 29 GT repeats] were allowed to adsorb on the graphene oxide nanosheets. The as-prepared DNA–GO was further subjected to chemical reduction to prepare the DNA–RGO (Figure 4.19). The resulting highly water-dispersible DNA–carbon bioconjugates were isolated from the free DNA by multiple centrifugation–redispersion cycles. The final DNA–GO-AuNP or DNA–RGO-AuNP nanocoposites were prepared by adding an excess amount of AuNPs to the dispersion mixture of the corresponding DNA–GO or DNA–RGO products. It was suggested that the interactions between the DNA and RGO could be the strong π–π stacking forces, which is similar to the case of the previously noticed DNA wrapping on CNTs. In the case of the DNA–GO hybrid, in addition to the π–π stacking, hydrogen bonding between the DNA and some oxygen-containing groups on GO might be an extra driving force for the DNA coating on the GO nanosheets. This approach represents a general, easy, highly specific, nondestructive, and environmentally friendly process for nanoconstruction and composite materials (Kim, Lee et al. 2005), that can be used in fields such as catalysis, magnetism, battery materials, optoelectronics, field-effect devices, and biodetection platforms. The good

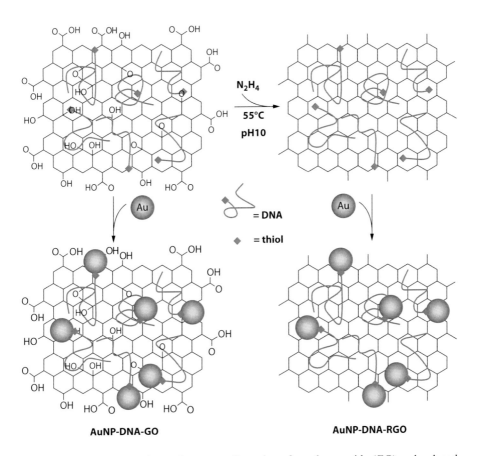

FIGURE 4.19 DNA coating and aqueous dispersion of graphene oxide (GO) and reduced graphene oxide (RGO), which were then used as two-dimensional nanobio-interfaces for a homogeneous assembly of metal–carbon heteronanostructures. (From Liu, J., *J. Mater. Chem.*, 20, 900, 2010. With permission.)

water-dispersibility of the as-prepared nanostructures also seems to be compatible with ink-jet or micro-contact printing-based device fabrications.

Different approaches have been carried out to design novel graphene-based molecular beacon that could sensitively and selectively detect specific DNA sequences (Li, Huang et al. 2010; Lu et al. 2010). The DNA–GO-based hybrids may be used as biosensors and can be applied in a variety of bio-analytical fields such as nanomedicine, nanobiotechnology, and immunoassay (Lu et al. 2010; Liu et al. 2010b). Also, a technique was suggested to use GRP nanogaps for DNA sequencing, using GRP as the electrode as well as the membrane material (Postma 2010). GRP is a perfect object for building nanogaps for sequencing owing to three different advantages: (1) its single-atom thickness, which facilitates transverse conductance measurements with single-base resolution, (2) its capability to survive large transmembrane pressures, and (3) its intrinsic conducting properties. Especially, the third property is more advantageous than the other two, because the membrane *is* the electrode, automatically solving the problem of having to fabricate nanoelectrodes that are exactly aligned with a nanogap.

Liu and coworkers (Liu et al. 2010a) introduced another facile yet versatile and effective method to fabricate mono- and bilayered graphene sheets by sonication of graphite flakes in the presence of pyrene-tagged ssDNAs (Py–ssDNA). In fact, the Py–ssDNAs was prepared by the reaction of the *N*-hydroxysuccinimide ester of 1-pyrenebutyric acid with an amino-modified DNA and the as-prepared Py–ssDNAs was confirmed by spectroscopic analysis. Originally, the energy of tightly bound graphene was disrupted due to π–π interactions between the surface of graphene and aromatic moieties of the DNA and pyrene, leading to a water-dispersible single-layered graphene (Figure 4.20a). The colloidal nature and high stability of the Py–ssDNA–graphene hybrid was corroborated by the presence of a Tyndall effect in the form of a characteristic visible red line that arises from light scattering,

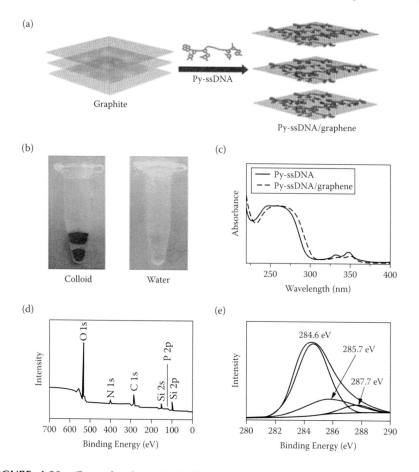

FIGURE 4.20 (See color insert.) (a) Schematics of the Py-ssDNA-GRP hybrid. (b) Tyndall effect of a Py-ssDNA-GRP colloid. (c) UV-vis absorption spectra of Py-ssDNA and Py-ssDNA-GRP. XPS data for (d) Py-ssDNA-GRP hybrids and (e) core C 1s level. (From Liu, F., *Chem. Commun.*, 46, 2844, 2010. With permission.)

when a laser beam is passed through the Py–ssDNA–GRP mixture (Figure 4.20b). Further, the formation of Py–ssDNA–GRP hybrid was confirmed by different analysis such as UV-vis absorption (Figure 4.20c) and x-ray photoelectron spectroscopy (Figure 4.20d,e), TEM, AFM, and SEM measurements. It was observed that the GRP hybrids with different sizes, ranging from 100 nm to 4 mm, and both mono- and bilayered graphene were obtained as the major forms. The merit of this approach was further to show by hybridizing an AuNP-labeled complementary DNA with the Py–ssDNA–GRP surface to produce Py–ssDNA–GRP–AuNP nanocomposites. Hence, the Py–ssDNA, acts in a dual role, not only improving the dispersion of GRP, but it also plays a vital role for specific DNA–DNA hybridization (Tang et al. 2010). This new and facile approach offers stable, pristine GRP in an aqueous medium, which is an indispensable prerequisite for biological applications of GRP.

4.7 ADVANTAGES AND DISADVANTAGES OF SYNTHETIC APPROACHES

Different synthetic strategies have been followed to fabricate DNA–nanocarbon hybrids, namely by chemical linking via covalent interaction, physical adsorption through supramolecular interaction, and encapsulation related to cavity and materials size. It is a fact that each and every system has its own advantages and disadvantages; similarly, different strategies have their advantages and disadvantages. Various ingenious techniques have been described in the literature to construct novel DNA–carbon-based hybrid materials, and the preferred approaches can be followed, depending on the requirements and specific application purpose.

The fabrication of hybrid materials by using covalent attachment offers the benefit of sticking the DNA strand to avoid desorption, and to increase accessibility, stability, selectivity, and signal amplification in sensing, when compared to hybrid structures obtained by other approaches. As the linking can be achieved by using already functionalized pristine materials (DNA and carbon-based materials), in some cases, the connection between DNA and carbon-based materials is under certain conditions reversible. However, usually the approach can be successful only by using coupling agents or suitable linkers. Hence, this approach is rather expensive and followed by multistep and complicated procedures. Furthermore, due to the surface modification of carbon materials and the functionalization of DNA, as well as the connection of two components using linkers under different experimental conditions, this approach is in reality a time-consuming process. The experiment needs more equipment, as well as characterization analysis at various stages. In addition, as the number of modified DNA that are linked to the surface of the carbon materials through covalent linking is generally associated with the number of carboxyl groups or other chemical functionalities of the carbon surface, the formation of hybrid materials by using covalent linking is further complex and difficult.

Commonly, noncovalent functionalization may have more advantages over covalent functionalization, in which there is a possibility to perturb the extended π-networks on the carbon-based material surface, and, hence, their intrinsic mechanical and electronic properties may be altered. In contrast, supramolecular approaches can conserve the carbon material surface structure, and, in turn, maintain the peculiar properties of the materials. Also, it is an easy and inexpensive technique, when compared to the covalent approach, and it needs no special manipulation skills.

Further, the supramolecular functionalization of carbon materials using DNA holds attractive prospects in a variety of domains, including the dispersion in aqueous media, nucleic acid sensing, gene therapy, and the controlled deposition on conducting or semiconducting substrates. Also, DNA molecules can enhance the dispersion of carbon materials in most of the solvents including water. In addition, they can be used to separate metallic CNTs from semiconducting CNTs, which is highly useful to utilize the final materials to fabricate sensitive biosensors, biochips, and electronic devices. The isolation and purification of the final hybrid materials are rather easy. On the other hand, controlling the amount of DNA coated on the surface of the carbon materials is tricky, according to this pathway. As the interaction between the DNA and carbon-based materials are very weak in the noncovalent approaches, extra care has to be taken while handing the materials for sensitive characterization and application processes.

4.8   CONCLUSIONS

The combination of novel carbon-based nanomaterials with biological molecules such as DNA to form functional materials is an innovative and challenging area of research. Due to its inimitable double helix geometry, exceptional properties, biocompatibility, and unique binding specificity, DNA is an excellent candidate as a base material for a new class of carbon-based novel nanomaterials such as CNTs and graphene. In addition to different exceptional properties of carbon-based materials, specifically the integration of the electronic properties of carbon-based materials and the recognition properties of the DNA can open a new door to create novel nanobio-systems such as electronic nanodevices, biosensors, and biochips. Hence, the blend of DNA and carbon nanomaterials demonstrates a new type of hybrid materials, which, in turn, leads to a successful incorporation of the properties of the two different components in the new hybrid materials that present important features for potential applications. These range from advanced biomedical systems through very sensitive electrochemical sensors and biosensors to highly efficient electronic and optical-based biochips.

The fabrication of these novel nanobio-hybrid materials may be achieved via different ways, normally by either covalent or noncovalent interactions between the DNA and carbon nanomaterials. In addition, in the case of CNTs, the encapsulation of DNA in nanotubes or entrapping the CNTs in the major grooves of DNA is also another ingenious technique to form hybrid materials. Similarly, the insertion of DNA into the nanogaps of graphene was also suggested for the formation of hybrid structures. In fact, there are different types of chemistry, various approaches, diverse functional groups as well as coupling agents, different kinds of carbon nanomaterials (e.g., SWNTs, MWNTs, DWNTs, mono- and bilayer graphene, and graphene oxide sheets), and DNA (ssDNA and dsDNA), reaction solvents and conditions to be followed in order to obtain the hybrid materials. Considering the CNTs, generally, carboxylic acid or amine-functionalized CNTs were extensively used as templates to covalently link the DNA. In many of the cases, covalent linking may be achieved by using different kinds of coupling agents or linkers, which can trigger the carboxylic acid or amine groups of modified CNTs to react with primary amines or carboxylic acid termini of the modified DNA strand, respectively. The description of the synthetic strategies and types of assemblies and interactions related to DNA–carbon nanomaterials hybrids and their potential applications has been given and discussed. Obviously, depending on the requirements and applications, the preparation methods for the formation of hybrid materials have been varied. The final hybrid materials show numerous potential applications such as in advanced bioelectronic systems, biosensors, electronic nanodevices, field effect transistors, electrochemical sensors, and gas sensors. The high surface-area-to-volume ratio, mechanical stiffness, fracture strength, optical and electronic transport properties, unusual thermal and electrical conductivity, the chemical stability and inertness, and the lack of porosity are the common significant advantageous rewards of carbon-based nanomaterials as a supporting material for DNA. Further, wrapping CNTs with DNA also permits them to be chromatographically sorted by nanotube length, diameter, and type (metallic and semiconducting tubes), as well as increasing the dispersibility of CNTs in aqueous media, which is an essential requirement for biological applications of the hybrid materials. Considerable achievements have been gained in the area of sensors using simple yet efficient DNA with nanocarbon-based hybrid materials that demonstrate their own performance of good stability, fine selectivity, and high sensitivity. In addition, their electrical and thermal properties are also excellent. Therefore, definitely there will be an attention for proficient technological use of these hybrid materials as different sensors. No doubt, these hybrid materials provide biorecognition electrodes sensitive enough to detect selectively even a few target molecules. Nevertheless, the most important and vital issue for the future improvement of these novel hybrid materials for industrial and technological applications is to establish powerful synthetic procedures and methodologies to produce hybrid materials with superb reproducible properties and performances. Further, many other resourceful techniques for the preparation of sensitive hybrid materials for electrochemical and biosensor applications will be explored soon, and efforts will be focused more on the fabrication of devices for more reliable diagnostics of cancer and other diseases. Also, in most of the cases, the interaction between DNA and nanocarbon materials has not been identified clearly. Hence, a deep attention has to be paid to understanding the reaction mechanism or chemistry of these hybrid materials. In addition, the potential of differential binding by DNA to carbon-based materials of different diameter and chirality remains to be explored. Also, the heterogeneity of CNTs samples is one of the major barriers to nanotube research and applications. Furthermore, to employ DNA to disperse CNTs makes it feasible to apply molecular biology tools to influence CNTs and to exploit the sequence-specific nature of DNA base pairing to direct the assembly of CNTs into useful architectures. In these DNA–CNT hybrid separation and assembly applications, the structure of DNA on the CNTs is an important feature that is not entirely understood and has to be addressed unambiguously. Since the purification and isolation of DNA–carbon-based hybrids are critical to most of their proposed applications, further investigation of the hybrid structure is expected to prove valuable. Further, the common impurities such as catalyst particles, amorphous graphite, carbon nanoparticles, and carbon-coated metals are the major problem in carbon-based nanomaterials, which affects the sensitivity of the final hybrid materials. Hence, special care needs to be taken to remove the impurities mentioned under mild conditions that should not disrupt the electronic properties of the main carbon materials. Further, it will be paramount that, for additional applications, especially in nanotechnology, DNA structures should also be controllable via the use of sequence-designed DNA strands and building blocks. As both pristine materials (DNA and carbon-based materials) have their own interesting and unique properties, obviously it is envisaged that the corresponding hybrid materials will also have improved properties. Hence, there is no doubt that these interesting hybrid materials will persuade researchers and scientists to further explore their properties and practical applications. Furthermore, these hybrid materials will have the potential to be identified as future generation materials in the nano- and biotechnology research fields.

ACKNOWLEDGMENTS

This work was supported by the World-Class University (WCU) program at the Gwangju Institute of Science and Technology (GIST) through a grant provided by the Ministry of Education, Science and Technology (MEST) of Korea (Project No. R31-10026).

REFERENCES

Ajayan, P.M. 1999. Nanotubes from carbon. *Chem. Rev.* 99: 1787–1800.
Alivisatos, A.P., Johnsson, K.P., and Peng, X. et al. 1996. Organization of "nanocrystal molecules" using DNA. *Nature* 382: 609–11.
Awasthi, K., Singh, D.P., Singh, S.K., Dash, D., and Srivastava, O.N. 2009. Attachment of biomolecules (protein and DNA) to amino-functionalized carbon nanotubes. *New Carbon Mater.* 24: 301–06.
Baker, S.E., Cai, W., Lasseter, T.L., Weidkamp, K.P., and Hamers, R.J. 2002. Covalently bonded adducts of deoxyribonucleic acid (DNA) oligonucleotides with single-wall carbon nanotubes: Synthesis and hybridization. *Nano Lett.* 2: 1413–17.
Bandow, S., Rao, A.M., Williams, K.A., Thess, A., Smalley, R.E., and Eklund, P.C. 1997. Purification of single-wall carbon nanotubes by microfiltration. *J. Phys. Chem. B* 101: 8839–42.
Barone, V., Hod, O., and Scuseria, G.E. 2006. Electronic structure and stability of semiconducting graphene nanoribbons. *Nano Lett.* 6: 2748–54.
Baughman, R.H., Zakhidov, A.A., and de Heer, W.A. 2002. Carbon nanotubes—the route toward applications. *Science* 297: 787–92.
Becerril, H.A., Mao, J., Liu, Z., Stoltenberg, R.M., Bao, Z., and Chen, Y. 2008. Evaluation of solution-processed reduced graphene oxide films as transparent conductors. *ACS Nano* 2: 463–70.
Berger, C., Song, Z., Li, X. et al. 2006. Electronic confinement and coherence in patterned epitaxial graphene. *Science* 312: 1191–96.
Bethune, D.S., Kiang, C.H., and de Vries, M.S. et al. 1993. Cobalt-catalysed growth of carbon nanotubes with single-atomic-layer walls. *Nature* 363: 605–07.
Blake, P., Brimicombe, P.D., Nair, R.R. et al. 2008. Graphene-based liquid crystal device. *Nano Lett.* 8: 1704–08.
Bonard, J.M., Stora, T., Salvetat, J.-P. et al. 1997. Purification and size-selection of carbon nanotubes. *Adv. Mater.* 9: 827–31.

Cai, H., Cao, X., Jiang, Y., He, P., and Fang, Y. 2003. Carbon nanotube-enhanced electro-chemical DNA biosensor for DNA hybridization detection. *Anal. Bioanal. Chem.* 375: 287–93.

Campbell, J.F., Tessmer, I., Thorp, H.H., and Erie, D.A. 2008. Atomic force microscopy stud-ies of DNA-wrapped carbon nanotube structure and binding to quantum dots. *J. Am. Chem. Soc.* 130: 10648–55.

Cathcart, H., Nicolosi, V., Hughes, J.M. et al. 2008. Ordered DNA wrapping switches on lumi-nescence in single-walled nanotube dispersions. *J. Am. Chem. Soc.* 130: 12734–44.

Chen, J., Hamon, M.A., Hu, H. et al. 1998. Solution properties of single-walled carbon nano-tubes. *Science* 282: 95–98.

Chen, J.-H., Ishigami, M., Jang, C., Hines, D.R., Fuhrer, M.S., and Williams, E.D. 2007. Printed graphene circuits. *Adv. Mater.* 19: 3623–27.

Chen, W., Tzang, C.H., Tang, J., Yang, M., and Lee, S.T. 2005. Covalently linked deoxyri-bonucleic acid with multiwall carbon nanotubes: Synthesis and characterization. *Appl. Phys. Lett.* 86: 103114-1–3.

Chen, Y., Liu, H., Ye, T., Kim, J., and Mao, C. 2007. DNA-directed assembly of single-wall carbon nanotubes. *J. Am. Chem. Soc.* 129: 8696–97.

Cheung, W., Chiu, P.L., Parajuli, R.R., Ma, Y., Ali, S.R., and He, H. 2009. Fabrication of high performance conducting polymer nanocomposites for biosensors and flexible electron-ics: Summary of the multiple roles of DNA dispersed and functionalized single walled carbon nanotubes. *J. Mater. Chem.* 19: 6465–80.

Chiang, I.W., Brinson, B.E., Huang, A.Y. et al. 2001. Purification and characterization of sin-gle-wall carbon nanotubes (SWNTs) obtained from the gas-phase decomposition of CO (HiPco process). *J. Phys. Chem. B* 105: 8297–8301.

Coleman, J.N., Dalton, A.B., Curran, S. et al. 2000. Phase separation of carbon nanotubes and turbostratic graphite using a functional organic polymer. *Adv. Mater.* 12: 213–16.

Condon, A. 2006. Designed DNA molecules: principles and applications of molecular nano-technology. *Nature Rev., Genetics* 7: 565–75.

Cooper, L., Amano, H., Hiraide, M. et al. 2009. Freestanding, bendable thin film for super-capacitors using DNA-dispersed double walled carbon nanotubes. *Appl. Phys. Lett.* 95: 233104-1–3.

Daniel, M.-C. and Astruc, D. 2004. Gold nanoparticles: Assembly, supramolecular chemistry, quantum-size-related properties, and applications toward biology, catalysis, and nano-technology. *Chem. Rev.* 104: 293–346.

Daniel, S., Rao, T.P., Rao, K.S. et al. 2007. A review of DNA functionalized/grafted carbon nanotubes, and their characterization. *Sens. Actuat. B* 122: 672–82.

Dong, L., Subramanian, A., and Nelson, B.J. 2007. Carbon nanotubes for nanorobot-ics. *Nanotoday* 2: 12–21.

Dong, X., Shi, Y., Huang, W., Chen, P., and Li, L.-J. 2010. Electrical detection of DNA hybrid-ization with single-base specificity using transistors based on CVD-grown graphene sheets. *Adv. Mater.* 22: 1649–53.

Dovbeshko, G.I., Repnytska, O.P., Obraztsova, E.D., and Shtogun, Y.V. 2003a. DNA inter-action with single-walled carbon nanotubes: A SEIRA study. *Chem. Phys. Lett.* 372: 432–37.

Dovbeshko, G.I., Repnytska, O.P., Obraztsova, E.D., Shtogun, Y.V., and Andreev, E.O. 2003b. Study of DNA interaction with carbon nanotubes. *Semicond. Phys. Quantum. Electron. & Optoelectron.* 6: 105–08.

Duesberg, G.S., Burghard, M., Muster, J., Philipp, G., and Roth, S. 1998. Separation of carbon nanotubes by size exclusion chromatography. *Chem. Commun.* 435–36.

Dwyer, C., Guthold, M., Falvo, M., Washburn, S., Superfine, R., and Erie, D. 2002. DNA-functionalized single-walled carbon nanotubes. *Nanotechnology* 13: 601–04.

Dwyer, C., Johri, V., Cheung, M., Patwardhan, J., Lebeck, A. and Sorin, D. 2004. Design tools for a DNA-guided self-assembling carbon nanotube technology. *Nanotechnology* 15: 1240–45.

Ebbesen, T.W. and Ajayan, P.M. 1992. Large-scale synthesis of carbon nanotubes. *Nature* 358: 220–22.

Endo, M. and Sugiyama, H. 2009. Chemical approaches to DNA nanotechnology. *ChemBioChem* 10: 2420–43.

Endo, M., Takeuchi, K., Kobori, K., Takahashi, K., Kroto, H.W., and Sarkar, A. 1995. Pyrolytic carbon nanotubes from vapor-grown carbon fibers. *Carbon* 33: 873–81.

Enyashin, A.N., Gemming, S., and Seifert, G. 2007. DNA-wrapped carbon nanotubes. *Nanotechnology* 18: 245702-1–10.

Feldkamp, U. and Niemeyer, C.M. 2006. Rational design of DNA nanoarchitectures. *Angew. Chem. Int. Ed. Engl.* 45: 1856–76.

Frischknecht, A.L. and Martin, M.G. 2007. Simulation of the adsorption of nucleotide monophosphates on carbon nanotubes in aqueous solution. *J. Phys. Chem. C.* 112: 6271–78.

Gao, H., Kong, Y., Cui, D., and Ozkan, C.S. 2003. Spontaneous insertion of DNA oligonucleotides into carbon nanotubes. *Nano Lett.* 3: 471–73.

Geckeler, K.E. (Ed.). 2003. *Advanced Macromolecular and Supramolecular Materials and Processes*, Kluwer Academic/Plenum, New York.

Geckeler, K.E. and Mueller, B. 1993. Polymer materials in biosensors. *Naturwissenschaften* 80: 18–24.

Geckeler, K.E. and Nishide, H. (Eds.). 2009. *Advanced Nanomaterials*, Wiley-VCH, Weinheim.

Geckeler, K.E. and Rosenberg, E. (Eds.). 2006. *Functional Nanomaterials*, American Scientific Publ, Valencia, USA.

Geim, A.K. and MacDonald, A.H. 2007. Graphene: Exploring carbon flatland. *Phys. Today* 60: 35–41.

Geim, A.K. and Novoselov, K.S. 2007. The rise of graphene. *Nat. Mater.* 6: 183–191.

Gigliotti, B., Sakizzie, B., Bethune, D.S., Shelby, R.M., and Cha, J.N. 2006. Sequence-independent helical wrapping of single-walled carbon nanotubes by long genomic DNA. *Nano Lett.* 6: 159–164.

Gilje, S., Han, S., Wang, M., Wang, K.L., and Kaner, R.B.A. 2007. Chemical route to graphene for device applications. *Nano Lett.* 7: 3394–98.

Gogotsi, Y., Libera, J.A., Guvenc-Yazicioglu, A., and Megaridis, C.M. 2001. In situ multiphase fluid experiments in hydrothermal carbon nanotubes. *Appl. Phys. Lett.* 79: 1021–23.

Gomez-Navarro, C., Weitz, R.T., Bittner, A.M. et al. 2007. Electronic transport properties of individual chemically reduced graphene oxide sheets. *Nano Lett.* 7: 3499–3503.

Gowtham, S., Scheicher, R.H., Pandey, R.K., Karnas, D., and Ahuja, R. 2008. First-principles study of physisorption of nucleic acid bases on small-diameter carbon nanotubes. *Nanotechnology* 19: 125701-1–6.

Guldi, D.M., Rahman, G.M.A., Jux, N., Prato, M., Qin, S.H., and Ford, W. 2005. Single-wall carbon nanotubes as integrative building blocks for solar-energy conversion. *Angew. Chem. Int. Ed.* 44: 2015–18.

Guo, M., Chen, J., Liu, D., Nie, L., and Yao, S. 2004. Electrochemical characteristics of the immobilization of calf thymus DNA molecules on multi-walled carbon nanotubes. *Bioelectrochem.* 62: 29–35.

Guo, X., Small, J.P., Klare, J.E. et al. 2006. Covalently bridging gaps in single-walled carbon nanotubes with conducting molecules. *Science* 311: 356–59.

Guo, X., Gorodetsky, A.A., Hone, J., Barton J.K., and Nuckolls, C. 2008. Conductivity of a single DNA duplex bridging a carbon nanotube gap. *Nat. Nanotech.* 3: 163–67.

Guo, X., Whalley, A., Klare J.E. et al. 2007. Single-molecule devices as scaffolding for multi-component nanostructure assembly. *Nano Lett.* 7: 1119–22.

Guo, Z., Sadler, P.J., and Tsang, S.C. 1998. Immobilization and visualization of DNA and proteins on carbon nanotubes. *Adv. Mater.* 10: 701–03.

Hao, R., Qian, W., Zhang, L., and Hou, Y. 2008. Aqueous dispersions of TCNQ-anion-stabilized graphene sheets. *Chem. Commun.* 6576–78.

Hazani, M., Naaman, R., Hennrich, F., and Kappes, M.M. 2003. Confocal fluorescence imaging of DNA-functionalized carbon nanotubes. *Nano Lett.* 3: 153–55.

He, P., Xu, Y., and Fang, Y. 2006. Applications of carbon nanotubes in electrochemical DNA biosensors. *Microchim Acta* 152: 175–186.

Hernandez, Y., Nicolosi, V., Lotya, M. et al. 2008. High-yield production of graphene by liquid-phase exfoliation of graphite. *Nat. Nanotechnol.* 3: 563–68.

Hirahara, K., Suenaga, K., Bandow, S. et al. 2000. One-dimensional metallofullerene crystal generated inside single-walled carbon nanotubes. *Phys. Rev. Lett.* 85: 5384–87.

Hod, O., Peralta, J.E., and Scuseria, G.E. 2007. Edge effects in finite elongated graphene nanoribbons. *Phys. Rev. B* 76: 233401-1–4.

Huang, B., Li, Z.Y., Liu, Z.R. et al. 2008. Adsorption of gas molecules on graphene nanoribbons and its implication for nanoscale molecule sensor. *J. Phys. Chem. C* 112: 13442–46.

Hummer, G., Rasalah, J.C., and Noworyta, J.P. 2001. Water conduction through the hydrophobic channel of a carbon nanotube. *Nature* 414: 188–190.

Humphrey, W., Dalke, A., and Schulten, K. 1996. VMD: Visual molecular dynamics. *J. Mol. Graph.* 14: 33–38.

Hung, A.M., Micheel, C.M., Bozano, L.D., Osterbur, L.W., Wallraff, G.M., and Cha, J.N. 2010. Large-area spatially ordered arrays of gold nanoparticles directed by lithographically confined DNA origami. *Nat. Nanotechnol.* 5: 121–26.

Iijima, S. 1991. Helical microtubules of graphitic carbon. *Nature* 354: 56–58.

Iijima, S. and Ichihashi, T. 1993. Single-shell carbon nanotubes of 1-nm diameter. *Nature* 363: 603–05.

Ito, T., Sun, L., and Crooks, R.M. 2003. Observation of DNA transport through a single carbon nanotube channel using fluorescence microscopy. *Chem. Commun.* 1482–83.

Jiang, L. and Gao, L. 2003. Modified carbon nanotubes: An effective way to selective attachment of gold nanoparticles. *Carbon,* 41: 2923–29.

Johnson, R.R., Johnson, A.T.C., and Klein, M.L. 2008. Probing the structure of DNA-carbon nanotube hybrids with molecular dynamics. *Nano Lett.* 8: 69–75.

Johnson, R.R., Johnson, A.T.C., and Klein, M.L. 2010. The nature of DNA-base–carbon-nanotube interactions. *Small* 6: 31–34.

Journet, C., Maser, W.K., Bernier, P. et al. 1997. Large-scale production of single-walled carbon nanotubes by the electric-arc technique. *Nature* 388: 756–58.

Jung, D.-H., Jung, M.S., Ko, Y.K., Seo, S.J., and Jung, H.-T. 2004a. Carbon nanotube conducting arrays by consecutive amidation reactions. *Chem.Commun.* 526–27.

Jung, D.-H., Kim, B.H., Ko, Y.K. et al. 2004b. Covalent attachment and hybridization of DNA oligonucleotides on patterned single-walled carbon nanotube films. *Langmuir* 20: 8886–91.

Katz, E. and Willner, I. 2004. Nanobiotechnology: Integrated nanoparticle–biomolecule hybrid systems: Synthesis, properties, and applications. *Angew. Chem. Int. Ed. Engl.* 43: 6042–6108.

Khamis, S.M., Johnson, R., Luo, Z., and Johnson, A.T.C. 2010. Homo-DNA functionalized carbon nanotube chemical sensors. *J. Phys. Chem. Solids* 71: 476–79.

Kim, D., Lee, T., and Geckeler, K.E. 2005. Hole-doped single-walled carbon nanotubes: Ornamenting with gold nanoparticles in water. *Angew. Chem Int. Ed.* 45: 104–07.

Kim, D.S., Nepal, D., and Geckeler, K.E. 2005. Individualization of single-walled carbon nanotubes: Is the solvent important? *Small* 1: 1117–24.

Kim, J.-B., Premkumar, T., Giani, O., Robin, J.-J., Schue, F., and Geckeler, K. E. 2007. *Macromol. Rapid Commun.* 28: 767–71.

Kim, J.H., Kataoka, M., Shimamoto, D. et al. 2010. Raman and fluorescence spectroscopic studies of a DNA-dispersed double-walled carbon nanotube solution. *ACS Nano* 4: 1060–66.

Kim, K.S., Zhao, Y., Jang, H. et al. 2009. Large-scale pattern growth of graphene films for stretchable transparent electrodes. *Nature* 457: 706–10.

Koehne, J., Chen, H., Li, J. et al. 2003. Ultrasensitive label-free DNA analysis using an electronic chip based on carbon nanotube nanoelectrode arrays. *Nanotechnol.* 14: 1239–45.

Koehne, J., Li, J., Cassell, A.M. et al. 2004. The fabrication and electrochemical characterization of carbon nanotube nanoelectrode arrays. *J. Mater. Chem.* 14: 676–84.

Koehne, J.E., Chen, H., Cassell, A.M. et al. 2004. Miniaturized multiplex label-free electronic chip for rapid nucleic acid analysis based on carbon nanotube nanoelectrode arrays. *Clinical Chem.* 50: 1886–93.

Komolpis, K., Srinannavit, O., and Gulari, E. 2002. Light-irected simultaneous synthesis of oligopeptides on microarray substrate using a photogenerated acid. *Biotechnol. Prog.* 18: 641–46.

Kroto, H.W., Heath, J.R., O'Brien, S.C., Curl, R.F., and Smalley, R.E. 1985. C60: Buckminsterfullerene. *Nature* 318: 162–63.

Kwon, Y. -W., Lee, C. H., Choi, D.-H., and Jin, J.-I. 2009. Materials science of DNA. *J. Mater. Chem.* 19: 1353–80.

Lee, C.-S., Baker, S.E., Marcus, M.S., Yang, W., Eriksson, M.A., and Hamers, R.J. 2004. Electrically addressable biomolecular functionalization of carbon nanotube and carbon nanofiber electrodes. *Nano Lett.* 4: 1713–16.

Li, D., Muller, M.B., Gilje, S., Kaner, R.B., and Wallace, G.G. 2008. Processable aqueous dispersions of graphene nanosheets. *Nat. Nanotechnol.* 3: 101–05.

Li, F., Huang, Y., Yang, Q. et al. 2010. A graphene-enhanced molecular beacon for homogeneous DNA detection. *Nanoscale* 2: 1021–26.

Li, J., Ng, H.T., Cassell, A. et al. 2003. Carbon nanotube nanoelectrode array for ultrasensitive DNA detection. *Nano Lett.* 3: 597–602.

Li, S., He, P., Dong, J., Guo, Z., and Dai, L. 2005. DNA-directed self-assembling of carbon nanotubes. *J. Am. Chem. Soc.* 127: 14–15.

Li, Y., Han, X., and Deng, Z. 2007. Grafting single-walled carbon nanotubes with highly hybridizable DNA sequences: Potential building blocks for DNA-programmed material assembly. *Angew. Chem. Int. Ed.* 46: 7481–84.

Li, Y., Kaneko, T., Hirotsu, Y., and Hatakeyama, R. 2010. Light-induced electron transfer through DNA-decorated single-walled carbon nanotubes. *Small* 6: 27–30.

Li, Z., Yao, Y., Lin, Z., Moon, K.–S., Lin, W., and Wong, C. 2010. Ultrafast, dry microwave synthesis of graphene sheets. *J. Mater. Chem.* 20: 4781–83.

Lin, Y.-M., Jenkins, K.A., Valdes-Garcia, A., Small, J.P., Farmer, D.B., and Avouris, P. 2009. Operation of graphene transistors at gigahertz frequencies. *Nano Lett.* 9: 422–26.

Lin, Y.-W., Liu, C.-W. and Chang, H.-T. 2009. DNA functionalized gold nanoparticles for bioanalysis. *Anal. Methods* 1: 14–24.

Liu, F., Choi, J.Y., and Seo, T.S. 2010a. DNA mediated water-dispersible graphene fabrication and gold nanoparticle-graphene hybrid. *Chem. Commun.* 46: 2844–46.

Liu, F., Choi, J.Y., and Seo, T.S. 2010b. Graphene oxide arrays for detecting specific DNA hybridization by fluorescence resonance energy transfer. *Biosens. Bioelect.* 25: 2361–65.

Liu, J., Li, Y., Li, Y., Li, J., and Deng, Z. 2010. Noncovalent DNA decorations of graphene oxide and reduced graphene oxide toward water-soluble metal–carbon hybrid nanostructures via self-assembly. *J. Mater. Chem.* 20: 900–06.

Liu, L., Wang, T., Li, J. et al. 2003. Self-assembly of gold nanoparticles to carbon nanotubes using a thiol-terminated pyrene as interlinker. *Chem. Phys. Lett.* 367: 747–52.

Liu, Q., Liu, Z., Zhang, X. et al. 2008a. Organic photovoltaic cells based on an acceptor of soluble graphene. *Appl. Phys. Lett.* 92: 223303-1–3.

Liu, X.D., Diao, H. Y., and Nishi, N. 2008b. Applied chemistry of natural DNA. *Chem. Soc. Rev.* 37: 2745–57.

Lu, C.-H., Zhu, C.-L., Li, J., Liu, J.-J., Chen, X., and Yang, H.-H. 2010. Using graphene to protect DNA from cleavage during cellular delivery. *Chem. Commun.* 46: 3116–18.

Lu, G., Maragakis, P., and Kaxiras, E. 2005. Carbon nanotube interaction with DNA. *Nano Lett.* 5: 897–900.

Lu, G., Ocola, L.E., and Chen, J. 2009. Reduced graphene oxide for room-temperature gas sensors. *Nanotechnol.* 20: 445502-1–9.

Lu, Y. and Liu, J. 2006. Functional DNA nanotechnology: emerging applications of DNAzymes and aptamers. *Curr. Opin. Biotechnol.* 17: 580–88.

Ma, Y., Ali, S. R., Dodoo, A.S., and He, H. 2006. Enhanced sensitivity for biosensors: Multiple functions of DNA-wrapped single-walled carbon nanotubes in self-doped polyaniline nanocomposites. *J. Phys. Chem. B* 110: 16359–65.

Maune, H.T., Han, S.-P., Barish, R.D. et al. 2010. Self-assembly of carbon nanotubes into two-dimensional geometries using DNA origami templates. *Nat. Nanotech.* 5: 61–66.

Meng, S., Maragakis, P., Papaloukas, C., and Kaxiras, E. 2007. DNA nucleoside interaction and identification with carbon nanotubes. *Nano Lett.* 7: 45–50.

Merino, E.J., Boal, A.K., and Barton, J.K. 2008. Biological contexts for DNA charge transport chemistry. *Curr. Opin. Chem. Biol.* 12: 229–37.

Mirkin, C.A., Letsinger, R.L., Mucic, R.C., and Storhoff, J.J. 1996. A DNA-based method for rationally assembling nanoparticles into macroscopic materials. *Nature* 382: 607–09.

Moghaddam, M.J., Taylor, S., Gao, M., Huang, S., Dai, L., and McCall, M.J. 2004. Highly efficient binding of DNA on the sidewalls and tips of carbon nanotubes using photochemistry. *Nano Lett.* 4: 89–93.

Mohanty, N. and Berry, V. 2008. Graphene-based single-bacterium resolution biodevice and DNA transistor: Interfacing graphene derivatives with nanoscale and microscale biocomponents. *Nano Lett.* 8: 4469–76.

Muller, K., Malik, S., and Richert, C. 2010. Sequence-specifically addressable hairpin DNA-single-walled carbon nanotube complexes for nanoconstruction. *ACS Nano* 4: 649–56.

Murugan, A.V., Muraliganth, T., and Manthiram, A. 2009. Rapid, facile microwave-solvothermal synthesis of graphene nanosheets and their polyaniline nanocomposites for energy storage. *Chem. Mater.* 21: 5004–06.

Nakashima, N., Okuzono, S., Murakami, H., Nakai, T., and Yoshikawa, K. 2003. DNA dissolves single-walled carbon nanotubes in water. *Chem. Lett.* 5: 456–57.

Nepal, D. and Geckeler, K.E. 2006. pH-sensitive dispersion and debundling of single-walled carbon nanotubes: Lysozyme as a tool. *Small* 2: 406–12.

Nepal, D. and Geckeler, K. E. 2007. Proteins and carbon nanotubes: close encounter in water. *Small* 3: 1259–65.

Nepal, D., Sohn, J.-I., Aicher, W.K., Lee, S., and Geckeler, K.E. 2005. Supramolecular conjugates of carbon nanotubes and DNA by a solid-state reaction. *Biomacromolecules* 6: 2919–22.

Nguyen, C.V., Delzeit, L., Cassell, A.M., Li, J, Han, J., and Meyyappan, M. 2002. Preparation of nucleic acid functionalized carbon nanotube arrays. *Nano Lett.* 2: 1079–81.

Nielsen, P. E., Engholm, M., Berg, R.H., and Buchardt, O. 1991. Sequence-selective recognition of DNA by strand displacement with a thymine-substituted polyamide. *Science* 254: 1497–1500.

Nikolaev, P., Bronikowski, M., Bradley, R. et al. 1999. Gas-phase catalytic growth of single-walled carbon nanotubes from carbon monoxide. *Chem. Phys. Lett.* 313: 91–97.

Novoselov, K.S., Geim, A.K., Morozov, S.V. et al. 2004. Electric field effect in atomically thin carbon films. *Science* 306: 666–69.

Novoselov, K.S., Geim, A.K., Morozov, S.V. et al. 2005. Two-dimensional gas of massless Dirac fermions in graphene. *Nature* 438: 197–200.

Oberlin, A., Endo, M., and Koyama, T. 1976a. Filamentous growth of carbon through benzene decomposition. *J. Cryst. Growth* 32: 335–49.

Oberlin, A., Endo, M., and Koyama, T. 1976b. High resolution electron microscope observations of graphitized carbon fibers. *Carbon* 14: 133–35.

Obradovic, B., Kotlyar, R., Heinz, F. et al. 2006. Analysis of graphene nanoribbons as a channel material for field-effect transistors. *Appl. Phys. Lett.* 88: 142102-1–3.

Park, S., Lee, K.-S., Bozoklu, G., Cai, W., Nguyen, S.T., and Ruoff, R.S. 2008. Graphene oxide papers modified by divalent ions—Enhancing mechanical properties via chemical cross-linking. *ACS Nano* 2: 572–78.

Patil, A.J., Vickery, J.L., Scott, T.B., and Mann, S. 2009. Aqueous stabilization and self-assembly of graphene sheets into layered bio-nanocomposites using DNA. *Adv. Mater.* 21: 3159–64.

Pellois, J.P., Wang, W., and Gao, X. 2000. Peptide synthesis based on *t*-Boc chemistry and solution photogenerated acids. *J. Comb. Chem.* 2: 355–60.

Postma, H.W.Ch. 2010. Rapid sequencing of individual DNA molecules in graphene nanogaps. *Nano Lett.* 10: 420–25.

Priya, B.R. and Byrne, H.J. 2008. Investigation of sodium dodecyl benzene sulfonate assisted dispersion and debundling of single-wall carbon nanotubes. *J. Phys. Chem. C* 112: 332–37.

Qian, W., Hao, R., Hou, Y. et al. 2009. Solvothermal-assisted exfoliation process to produce graphene with high yield and high quality. *Nano Res.* 2: 706–12.

Rinzler, A.G., Liu, J., Dai, H. et al. 1998. Large-scale purification of single-wall carbon nanotubes: Process, product, and characterization. *Appl. Phys. A* 67: 29–37.

Roy, S., Vedala, H., Roy, A.D. et al. 2008. Direct electrical measurements on single-molecule genomic DNA using single-walled carbon nanotubes. *Nano Lett.* 8: 26–30.

Sakhaee-Pour, A., Ahmadian, M.T., and Vafai, A. 2008. Potential application of single-layered graphene sheet as strain sensor. *Solid State Commun.* 147: 336–40.

Sanchez-Pomales, G., Santiago-Rodriguez, L., and Cabrera, C.R. 2009. DNA-functionalized carbon nanotubes for biosensing applications. *J. Nanosci. Nanotechnol.* 9: 2175–88.

Sanchez-Pomales, G., Santiago-Rodriguez, L., Rivera-Velez, N.E., and Cabrera, C.R. 2007. Characterization of the DNA-assisted purification of single-walled carbon nanotubes. *Phys. Stat. Sol. (A)* 204: 1791–96.

Santiago-Rodriguez, L., Sanchez-Pomales, G., Rios-Pagan, A., and Cabrera, C.R. 2007. Electrochemical study and preparation of gold substrates functionalized with single-walled carbon nanotubes for DNA biosensor application. *ECS Trans.* 3: 15–26.

Scharer, O.D. and Jiricny, J. 2001. Recent progress in the biology, chemistry and structural biology, of DNA glycosylases. *BioEssays* 23: 270–81.

Schedin, F., Geim, A.K., Morozov, S.V. et al. 2007. Detection of individual gas molecules adsorbed on graphene. *Nat. Mater.* 6: 652–55.

Schrer, O.D. 2003. Chemistry and biology of DNA repair. *Angew. Chem. Int. Ed.* 42: 2946–74.

Seeman, N.C. 2003a. DNA in a material world. *Nature* 421: 427–31.

Seeman, N.C. 2003b. At the crossroads of chemistry, biology, and materials: Structural DNA nanotechnology. *Chem. Biol.* 10: 1151–59.

Shin, J.-Y., Premkumar, T., and Geckeler, K.E. 2008. Dispersion of single-walled carbon nanotubes by using surfactants: Are the type and concentration important? *Chem. Eur. J.* 14: 6044–48.

Simmel, F.C. and Dittmer, W.U. 2005. DNA nanodevices. *Small* 1: 284–99.

Smith, B. W., Monthioux, M., and Luzzi, D. E. 1998. Encapsulated C_{60} in carbon nanotubes. *Nature* 396: 323–24.

Sonnichsen, C., Reinhard, B.M., Liphardt, J. and Alivisatos, A.P. 2005. A molecular ruler based on plasmon coupling of single gold and silver nanoparticles. *Nat. Biotechnol.* 23: 741–45.

Sordan, R., Traversi, F., and Russo, V. 2009. Logic gates with a single graphene transistor. *Appl. Phys. Lett.* 94: 073305-1-3.

Source—ISI web of knowledge, http://apps.isiknowledge.com/.

Stankovich, S., Dikin, D.A., Dommett, G.H.B. et al. 2006. Graphene-based composite materials. *Nature* 442: 282–86.

Star, A., Tu, E., Niemann, J., Gabriel, J.C.P., Joiner, C.S., and Valcke, C. 2006. Label-free detection of DNA hybridization using carbon nanotube network field-effect transistors. *Proc. Natl. Acad. Sci. USA* 103: 921–26.

Stoller, M.D., Park, S., Zhu, Y., An, J. and Ruoff, R.S. 2008. Graphene-based ultracapacitors. *Nano Lett.* 8: 3498–3502.

Storhoff, J.J. and Mirkin, C.A. 1999. Programmed materials synthesis with DNA. *Chem. Rev.* 99: 1849–62.

Taft, B.J., Lazareck, A.D., Withey, G.D., Yin, A., Xu, J.M., and Kelley, S.O. 2004. Site-specific assembly of DNA and appended cargo on arrayed carbon nanotubes. *J. Am. Chem. Soc.* 126: 12750–51.

Takahashi, H., Namao, S., Bandow, S., and Iijima, S. 2006. AFM imaging of wrapped multiwall carbon nanotube in DNA. *Chem. Phys. Lett.* 418: 535–39.

Tang, Z., Wu, H., Cort, J.R. et al. 2010. Constraint of DNA on functionalized graphene improves its biostability and specificity. *Small* 6: 1205–09.

Tang, Z.K., Zhang, L., Wang, N. et al. 2001. Superconductivity in 4 angstrom single-walled carbon nanotubes. *Science* 292: 2462–65.

Tasis, D., Tagmatarchis, N., Bianco, A., and Prato, M. 2006. Chemistry of carbon nanotubes. *Chem. Rev.* 106: 1105–36.

Tasis, D., Tagmatarchis, N., Georgakilas, V., and Prato, M. 2003. Soluble carbon nanotubes. *Chem. Eur. J.* 9: 4000–08.

Thess, A., Lee, R., Nikolaev, P. et al. 1996. Crystalline ropes of metallic carbon nanotubes. *Science* 273: 483–87.

Thomas, K.G. and Kamat, P.V. 2003. Chromophore-functionalized gold nanoparticles. *Acc. Chem. Res.* 36: 888–89.

Tsang, S.C., Chen, Y.K., Harris, P.J.F., and Green, M.L.H. 1994. A simple chemical method of opening and filling carbon nanotubes. *Nature* 372: 159–62.

Tu, X. and Zheng, M. 2008. A DNA-based approach to the carbon nanotube sorting problem. *Nano Res.* 1: 185–194.

Tung, V.C., Allen, M.J., Yang, Y., and Kaner, R.B. 2009. High-throughput solution processing of large-scale graphene. *Nat. Nanotechnol.* 4: 25–29.

Varghese, N., Mogera, U., Govindaraj, A. et al. 2009. Binding of DNA nucleobases and nucleosides with graphene. *ChemPhysChem.* 10: 206–10.

Vazquez, E., Georgakilas, V., and Prato, M. 2002. Microwave-assisted purification of HIPCO carbon nanotubes. *Chem. Commun.* 2308–09.

Vogel, S.R., Kappes, M.M., Hennrich, F., and Richert, C. 2007. An unexpected new optimum in the structure space of DNA solubilizing, single-walled carbon nanotubes. *Chem. Eur. J.* 13: 1815–20.

Wang, X. 2008. Transparent carbon films as electrodes in organic solar cells. *Angew. Chem. Int. Ed.* 47: 2990–92.

Wang, X., Zhi, L.J., and Mullen, K. 2008. Transparent, conductive graphene electrodes for dye-sensitized solar cells. *Nano Lett.* 8: 323–27.

Watson, J.D. and Crick, F.H.C. 1953. Molecular structure of nucleic acids: A structure for deoxyribose nucleic acid. *Nature* 171: 737–38.

Whalley, A.C., Steigerwald, M.L., Guo, X., and Nuckolls, C. 2007. Reversible switching in molecular electronic devices. *J. Am. Chem. Soc.* 129: 12590–91.

Wilkins, M.H., Stokes, A.R., and Wilson, H.R. 1953. Molecular structure of nucleic acids: Molecular structure of deoxypentose nucleic acids. *Nature* 171: 738–40.

Williams, K.A., Veenhuizen, P.T.M., de la Torre, B.G., Eritja, R., and Dekker, C. 2002. Carbon nanotubes with DNA recognition. *Nature* 420: 761–61.

Wu, J.B., Becerril, H.A., Bao, Z.N., Liu, Z.F., Chen, Y.S. and Peumans, P. 2008. Top-gated graphene field-effect-transistors formed by decomposition of SiC. *Appl. Phys. Lett.* 92: 092102-1–3.

Xu, Y., Liu, Z., Zhang, X. et al. 2009. A graphene hybrid material covalently functionalized with porphyrin: Synthesis and optical limiting property. *Adv. Mater.* 21: 1275–79.

Yan, Q.M., Huang, B., Yu, J. et al. 2007. Intrinsic current-voltage characteristics of graphene nanoribbon transistors and effect of edge doping. *Nano Lett.* 7: 1469–73.

Yang, D.Q., Hennequin, B., and Sacher, E. 2006. XPS Demonstration of π-π interaction between benzyl mercaptan and multiwalled carbon nanotubes and their use in the adhesion of Pt nanoparticles. *Chem. Mater.* 18: 5033–38.

Yang, H., Wang, S.C., Mercier, P., and Akins, D.L. 2006. Diameter-selective dispersion of single-walled carbon nanotubes using a water-soluble, biocompatible polymer. *Chem. Commun.* 1425–27.

Yarotski, D.A., Kilina, S.V., Talin, A.A. et al. 2009. Scanning tunneling microscopy of DNA-wrapped carbon nanotubes. *Nano Lett.* 9: 12–17.

Yogeswaran, U., Thiagarajan, S., and Chen, S.-M. 2008. Recent updates of DNA incorporated in carbon nanotubes and nanoparticles for electrochemical sensors and biosensors. *Sensors* 8: 7191–7212.

Yoo, E.J., Kim, J., Hosono, E., Zhou, H., Kudo, T., and Honma, I. 2008. Large reversible Li storage of graphene nanosheet families for use in rechargeable lithium ion batteries. *Nano Lett.* 8: 2277–82.

Yu, M., Zu, S.-Z., Chen, Y., Liu, Y.-P., Han, B.-H. and Liu, Y. 2010. Spatially controllable DNA condensation by a water-soluble supramolecular hybrid of single-walled carbon nanotubes and β-cyclodextrin-tethered ruthenium complexes. *Chem. Eur. J.* 16: 1168–74.

Zhang, Y., Li, H., Pan, L., Lu, T., and Sun, Z. 2009a. Capacitive behavior of graphene–ZnO composite film for supercapacitors. *J. Electroanal. Chem.* 634: 68–71.

Zhang, Y., Ma, H., Zhang, K., Zhang, S., and Wang, J. 2009b. An improved DNA biosensor built by layer-by-layer covalent attachment of multi-walled carbon nanotubes and gold nanoparticles. *Electrochim. Acta* 54: 2385–91.

Zhao, W., Gao, Y., Kandadai, S.A., Brook, M.A, and Li, Y. 2006. DNA polymerization on gold nanoparticles through rolling circle amplification: Towards novel scaffolds for three-dimensional periodic nanoassemblies. *Angew. Chem. Int. Ed. Engl.* 45: 2409–13.

Zheng, M., Jagota, A., Semke, E.D. et al. 2003a. DNA-assisted dispersion and separation of carbon nanotubes. *Nat. Mater.* 2: 338–42.

Zheng, M., Jagota, A., Strano, M.S. et al. 2003b. Structure-based carbon nanotube sorting by sequence-dependent DNA assembly. *Science* 302: 1545–48.

Zhou, S.Y., Gweon, G.-H., Graf, J. et al. 2006. First direct observation of Dirac fermions in graphite. *Nat. Phys.* 2: 595–99.

Zhu, N., Chang, Z., He, P., and Fang, Y. 2005. Electrochemical DNA biosensors based on platinum nanoparticles combined carbon nanotubes. *Anal. Chim. Acta,* 545: 21–26.

Zhu, N., Gao, H., Xu, Q., Lin, Y., Su, L., and Mao, L. 2010. Sensitive impedimetric DNA biosensor with poly(amidoamine) dendrimer covalently attached onto carbon nanotube electronic transducers as the tether for surface confinement of probe DNA. *Biosens. Bioelect.* 25: 1498–1503.

5 Electrical and Magnetic Properties of DNA

Chang Hoon Lee, Young-Wan Kwon,
and Jung-Il Jin

CONTENTS

5.1 Electrical Properties of DNA 121
    5.1.1 Charge Transport in Dry DNA 121
    5.1.2 Electrical Conductivity of DNA—A Summary 130
5.2 Magnetic Properties of DNA 134
    5.2.1 Historical Recount 134
    5.2.2 DNA Magnetism 135
    5.2.3 Discotic Liquid Crystals as DNA-Mimicking Compounds 152
5.3 Concluding Remarks 153
References 154

5.1 ELECTRICAL PROPERTIES OF DNA

5.1.1 CHARGE TRANSPORT IN DRY DNA

The question as to whether or not DNA can transport electric current is of great importance because, if DNA can indeed transport a charge carrier, such a mechanism may prove useful in understanding a signaling pathway for biological functions (Boal et al. 2005; Boon et al. 2003; Rajski et al. 2000; Kneuer et al. 2000; Shukla et al. 2004; Drummond et al. 2003; Yavin et al. 2005) as well as relevant biocompatible molecular or nanoelectronic applications (Ito and Fukusaki 2004; Wettig et al. 2003; Bhalla et al. 2003; Odenthal and Gooding 2007; Boon et al. 2002; Zwolak and Di Ventra 2002; Tabata et al. 2003). Indeed, the question was raised immediately after Watson and Crick's discovery of the helical structure of DNA (Watson and Crick 1953). Eley and Spivey (Eley and Spivey 1962) theoretically predicted that the previously determined longitudinal distance of 3.4 Å (Watson and Crick 1953) between neighboring basal planes is close enough to make it possible for the P_z orbitals of ring carbon and nitrogen atoms to overlap along the long axis of the DNA helix; this can also guide charge transport, as in the case of graphite, which has a layer separation of 3.35 Å (Delhaes 2000). In the experimental setup established to demonstrate their idea, they sandwiched *calf thymus* DNA between two metal electrodes and measured direct current (DC) resistivity while pressing the two electrodes toward opposite directions. They successfully measured a DC resistivity of 5×10^{11} Ω·cm

121

(Eley and Spivey 1962). This initial report provided sufficient evidence to stimulate subsequent studies by many researchers. In other following studies, however, contradictory experimental results began to appear. As a consequence, the question as to whether or not DNA can transport charge carriers has gradually become increasingly murky (Endres et al. 2004; Taniguchi and Kawai 2006; Ventra and Zwolak 2004; Kwon et al. 2009; Shinwari et al. 2010; Kanvah et al. 2010; Genereux and Barton 2010; Fink and Schönenberger 1999; Priyadarshy et al. 1996; de Pablo et al. 2000; Zhang et al. 2002; Cai et al. 2000; Hwang et al. 2002; Maiya and Ramasarma 2001; Porath et al. 2000). Nonetheless, the early results of DNA researchers have provided an abundance of physical insights into possible charge transport mechanisms. The first candidate is an electron or a hole flowing through the π-stacks of the DNA helical structure (Voityuk et al. 2000; Takada et al. 2004; Wan et al. 1999; Henderson et al. 1999; Hennig et al. 2004; Wang and Chakraborty 2006; Barnett et al. 2001; Feng et al. 2006; Klotsa et al. 2005; Fink 2001; O'Neill and Barton 2004; Kats and Lebedev 2002; Apalkov and Chakraborty 2005; Gutierrez et al. 2005; Wei et al. 2005; Albuquerque et al. 2005; Matsuo et al. 2005; Otsuka et al. 2002). The second one is an ionic conduction of counter ions including Na^+, K^+, etc., which coordinate in a one-to-one manner with the PO_4^- anion of the DNA helix backbone but can also move around from here to there substantially under physiological conditions (Neubert et al. 1985; Okahata et al. 1998; Cai et al. 2000; Otsuka et al. 2002; Taniguchi et al. 2003; Ha et al. 2002; Terawaki et al. 2005). The next factor to be considered is the participation of water molecules in conduction (Kleine-Ostmann et al. 2006; Otsuka et al. 2002; Taniguchi et al. 2003; Ha et al. 2002; Brovchenko et al. 2007; Lee and Jin 2010). The final issue is the possibility that the experimentally determined DC voltage could have been generated by piezoelectricity under strong mechanical compression conditions (Fukada and Ando 1972; Zimmerman 1976; Fukada 1995; Hayakawa and Wada 1973).

It is also important to know which part of the DNA molecule makes contact with the metal electrodes, as electric resistance depends on contact resistance in the two-electrodes method. The overall resistance measured can be governed by the contact resistance, particularly when the sample's resistance is much less than the contact resistance. If the DNA helix is viewed from the top, water molecules occupy the outermost shell, followed by the PO_4^- backbone, sugar moiety, and the central base stack (Fuller, Forsyth, and Mahendrasingam 2004; Bansal 2003). Accordingly, when one lays DNA molecules upon metal electrodes, it is most probable that an electrical contact is created between the metal electrodes and the water layer. However, contact between electrodes and other parts of the DNA helix cannot be ruled out. Under this circumstance, reproducible experimental results are problematic at best. Therefore, a precise experimental design is essential to determining the charge transport mechanism responsible for the conduction of DNA. First of all, one must define the water content and the nature of the counterions of the phosphate group. Additionally, electrical contact between the electrodes and the structural components of DNA such as the base pair, sugar moiety, and phosphate backbone should be precisely controlled. At the same time, a single DNA molecular sample should be used, as either fibrous or amorphous DNAs would produce only an average conductivity, which depends on their morphological state. This is why we would like to limit our discussion to the charge transport mechanism of the single, double-helical DNA molecule in the dry state.

As is now well known, in the early days of DNA research, almost all conductivity values were acquired via the two-probe method (Schroder 1998). In this case, the obtained conductivity values are likely to diverge profoundly from the reliable values, as the overall resistance measured may be due largely to the contact resistance. This is especially true when the sample's intrinsic resistance is much lower than the contact resistance. To overcome this disadvantage, one may utilize the standard 4-probe method (Schroder 1998), which helps to avoid electrical contact errors. Although present nanotechnology can pattern electrodes with a separation of a few tens nanometers, the persistence length of a single- and double-strand (ds) DNA molecule is not sufficient to configure four-different electrodes with three finite separations. As a consequence, many researchers have used a third method—such as a field effect transistor (FET) and diverse electron microscopes (Cohen et al. 2005; Xu, Tsukamoto et al. 2005; Xu, Endres et al. 2005; Shapir et al. 2008)—to measure the DC conductivity of DNA molecules.

Figure 5.1(a and b) is one of the representative I–V curves for a DNA sample measured at room temperature using the FET system (Yoo et al. 2001), which consists of three components: the source (S), drain (D), and gate (G) electrodes, as shown in the inset of Figure 5.1B. The authors prepared two different types of DNA sequences of poly(dG)–poly(dC) and poly(dA)–poly(dT), and situated them between the source and drain electrodes, which were separated by 20 nm. They obtained I–V characteristics for each DNA as a function of temperature and gate voltage. From the gate voltage dependence, they determined that the poly(dA)–poly(dT) sample functioned as an n-type semiconductor showing that the current was suppressed when the gated voltage, V_{gate}, was increased to negative values, but enhanced for positive V_{gate} values (Figure 5.1a). By way of contrast, the poly(dG)–poly(dC) behaved as a p-type one (Figure 5.1b).

The temperature dependence of the two samples could be explained satisfactorily by the small-polaron hopping model (Henderson et al. 1999; Conwell and Rakhmanova 2000; Conwell 2005; Rakhmanova and Conwell 2001; Triberis, Simserides, and Karavolas 2005; Hatcher et al. 2008; Brisker-Klaiman and Peskin 2010), which can be described by $\sigma(T) = \sigma_o \mathrm{Exp}(-E_a/k_B T)$, in which E_a is the activation energy, k_B is the Boltzmann constant, and T is the temperature, respectively (Figure 5.1C). The conductivity depended only a little on the gate and temperature regardless of whether the sample was in vacuum or in ambient conditions. However, such characteristics disappeared completely when the DNAse enzyme decomposed DNA, implying that the semiconductive behavior originated from the DNA itself. From a close examination of the reported experimental results, one can learn the following:

- One usually does not know which part of the DNA molecule was in contact with the metal electrodes (Cuniberti et al. 2001).
- The concentration of water molecules that can potentially act as a scattering center for the charge transport is unknown (Jurchescu et al. 2005; Jo et al. 2003; Berashevich and Chakraborty 2008; Conwell and Basko 2006).
- DNA coil morphology makes it difficult to exclude possible interstrand conduction mechanism established in the coiling process (Bloomfield 1999).

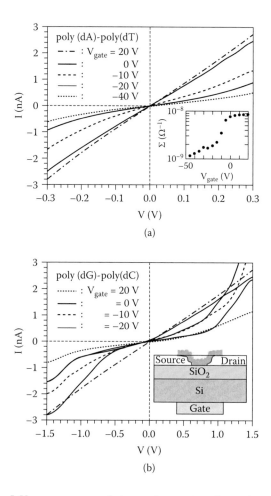

(a)

(b)

FIGURE 5.1 The I–V curves measured at room temperature for various values of the gate voltage (V_{gate}) for poly(dA)–poly(dT) (a) and poly(dG)–poly(dC) (b). In the inset of (a), the conductance at $V = 0$ is plotted as a function of V_{gate} for poly(dA)–poly(dT). The inset of (b) shows the schematic diagram of electrode arrangement for gate-dependent transport experiments. (c) Conductance versus inverse temperature for poly(dA)–poly(dT)(●) and poly(dG)–poly(dC)(○), where the conductance at $V = 0$ was numerically calculated from the I–V curve. The solid curves are the ones calculated using $\sigma(T) = \sigma_0 Exp(-E_a/k_B T)$. (Reprinted with permission from the American Physical Society: Yoo, K. H., D. H. Ha, J. O. Lee, J. W. Park, J. Kim, J. J. Kim, H. Y. Lee, T. Kawai, and H. Y. Choi, *Phys. Rev. Lett.*, 87 (19), 198102, 2001, Copyright (2001).)   *Continued*

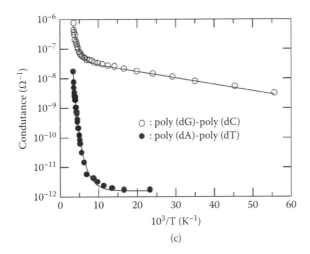

(c)

FIGURE 5.1 (*Continued*) The I–V curves measured at room temperature for various values of the gate voltage (V_{gate}) for poly(dA)–poly(dT) (a) and poly(dG)–poly(dC) (b). In the inset of (a), the conductance at $V = 0$ is plotted as a function of V_{gate} for poly(dA)–poly(dT). The inset of (b) shows the schematic diagram of electrode arrangement for gate-dependent transport experiments. (c) Conductance versus inverse temperature for poly(dA)–poly(dT) (●) and poly(dG)–poly(dC) (○), where the conductance at $V = 0$ was numerically calculated from the *I–V* curve. The solid curves are the ones calculated using $\sigma(T) = \sigma_0 \mathrm{Exp}(-E_a/k_B T)$. (Reprinted with permission from the American Physical Society: Yoo, K. H., D. H. Ha, J. O. Lee, J. W. Park, J. Kim, J. J. Kim, H. Y. Lee, T. Kawai, and H. Y. Choi, *Phys. Rev. Lett.*, 87 (19), 198102, 2001, Copyright (2001).)

The first of these questions was answered by Barton's group (Guo et al. 2008), who used functionalized single-wall carbon nanotube (SWNT) electrodes, as illustrated in Figure 5.2, to achieve the electrical point contact covalently with the amine DNA-modified. Such an amine functionalization of DNA is possible via the base part of the duplex.

In an FET configuration under ambient atmosphere conditions (see I–V curves in Figure 5.3), the SWNT itself is originally semiconductive but becomes insulating after being cut. However, the I–V curve is again semiconductive when the amine-modified DNA is reconnected to the SWNT electrodes, showing that DNA is a semiconductor. Although this approach clearly resolved the electric contact issue, the problem caused by water molecules as scattering centers and disturbed conformational order by interactions between substrate and DNA duplex (Kasumov et al. 2004) remain to be resolved. In an effort to circumvent such issues, Roy et al. (2008) designed a trench on the surface of an SiO_2/Si wafer and, after connecting the amine DNA-modified duplex to the carboxyl functionalized SWNT electrodes, they measured the DC conductivity of the DNA FET devices in the ambient and vacuum conditions (Figure 5.4).

Furthermore, a possible coiling of the DNA molecule was prevented via the dielectrophoresis (DEP) method (Zheng et al. 2003; Zheng, Brody, and Burke 2004).

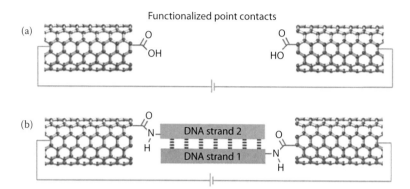

FIGURE 5.2 A method to cut and functionalize individual single-wall carbon nanotube electrodes (SWNTs) with DNA strands. (a) Functionalized point contacts created via the oxidative cutting of a SWNT wired into a device. (b) Bridging by functionalization of both strands with amine functionality. (Reprinted with permission from Macmillan Publishers Ltd: [*Nature*] Guo, X., A. A. Gorodetsky, J. Hone, J. K. Barton, and C. Nuckolls. 2008. *Nat. Nanotech.* 3: 163–167, Copyright (2008).)

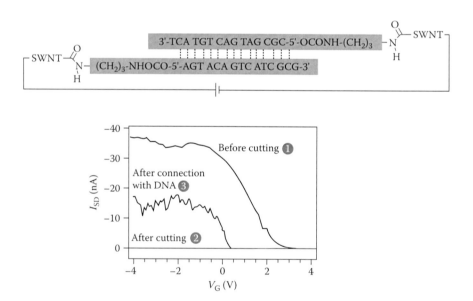

FIGURE 5.3 Device characteristics for individual SWNTs connected with DNA. Source–drain current versus V_G at a constant source–drain voltage (50 mV) before cutting (black curve: 1), after cutting (red curve: 2), and after connection with the DNA sequence shown (green curve: 3), for a semiconducting SWNT device. Guanine, G; cytosine, C; adenine, A; thymine, T. (Reprinted with permission from Macmillan Publishers Ltd: [Nature] Guo, X., A. A. Gorodetsky, J. Hone, J. K. Barton, and C. Nuckolls. 2008. *Nat. Nanotech.* 3: 163–167, Copyright (2008).)

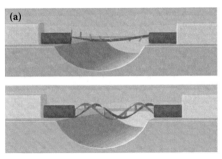
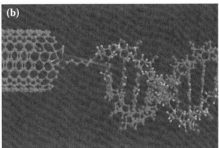

FIGURE 5.4   Schematic illustrations. (a) The upper figure represents an arbitrarily shaped ssDNA molecule, which is stretched and attached to a pair of functionalized SWNT electrodes in the presence of a DEP field. The lower figure depicts a covalently attached dsDNA molecule, which has a definite conformation. The nanoelectrodes are separated by a gap of 27 ± 2 nm (the contour length of the 80 bp DNA molecule is ~27 nm). The bridging DNA molecule is suspended over the trench without touching the silicon dioxide surface. (b) Molecular diagram highlighting the covalent bonding between an amine-terminated ssDNA and a carboxyl-functionalized SWNT nanoelectrode via a -$(CH_2)_6$- spacer. Charge transport occurs via the stacked base pairs in a helical duplex. (Reprinted with permission from the American Chemical Society: Roy, S., H. Vedala, A. D. Roy, D. H. Kim, M. Doud, K. Mathee, H. K. Shin, N. Shimamoto, V. Prasad, and W. B. Choi. 2008. *Nano Lett.* 8 (1):26–30, Copyright (2008).)

The experimental results obtained under both ambient and vacuum conditions are shown in Figure 5.5.

First, all the I–V curves are nonlinear (see Figure 5. 5c–f), which reflects semiconducting behavior. According to the gate voltage (V_g) dependence of the I–V curves, the DNA molecule is a p-type semiconductor that satisfies the back-gate enhancement of conductivity (Figure 5.5f). For the case of temperature dependence, the small-polaron hopping model is supported by the fact that the conductivity increased up to 40°C but decreased at higher temperatures, ultimately disappearing at a melting temperature of 76.5°C. Additionally, the effects of the water molecule could also be confirmed on the I–V curve in Figure 5.5c–d. The DNA conductivity at 10^{-5} torr of a vacuum was lower than that detected at the ambient atmosphere. Unfortunately, it is currently difficult to discriminate whether the origin of the enhanced conductivity at the ambient atmosphere is attributable to the oxygen-doping effect (Lee et al. 2002; Mehrez et al. 2005) or the role of the water molecules functioning as a charge channel. Therefore, further controlled investigations will be necessary to discern between the oxygen-doping and a water-channel effect of the enhanced conductivity under ambient atmosphere.

Scanning tunneling spectroscopy (STS) is a unique technique capable of probing the local density of states of objects deposited on a surface (Xu, Tsukamoto et al. 2005; Xu, Endres et al. 2005). Porath and coworkers (Shapir et al. 2008) employed the STS technique to determine the electronic density of states of the dsDNA duplex, as shown in Figure 5.6. They deposited the poly(G)–poly(C) DNA molecules onto gold and then determined the tunneling current–voltage (I–V) characteristics curves and their derivatives (dI/dV–V) at a cryogenic temperature of 78 K, where the thermal fluctuations of DNA molecules were expected to be substantially diminished.

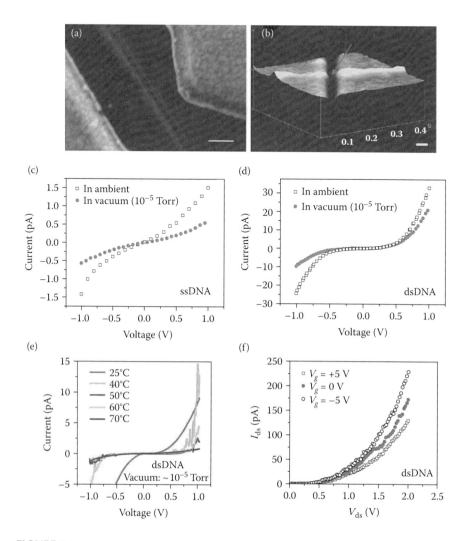

FIGURE 5.5 Device morphology and electrical characterization. (a) SEM image of a pair of FIB-etched SWNT electrodes, connected to the bonding pads by Ti/Au microcontact leads. Scale bar = 100 nm. (b) High-resolution AFM image of the DNA molecule between the SWNT electrodes. Scale bar = 30 nm. The red arrow points to the trapped single DNA molecule. (c) and (d) Current flow at room temperature through single ssDNA and dsDNA molecules, respectively, under ambient and vacuum conditions. (e) Temperature effect on DNA conductivity. The initial increase in the current value (from 25°C to 40°C) is consistent with the small-polaron hopping model. Further increases in the device temperature (50°C–70°C) resulted in a diminishing current signal at a given voltage. Above the melting temperature of the DNA molecule (75.6°C), no signal was detected. (f) The back-gate biasing effect on DNA channel conductivity. Depletion in the positive gate bias region and enhancement in the negative bias region indicates that the DNA molecule with a given sequence functions as a p-type semiconductor. (Reprinted with permission from the American Chemical Society: Roy, S., H. Vedala, A. D. Roy, D. H. Kim, M. Doud, K. Mathee, H. K. Shin, N. Shimamoto, V. Prasad, and W. B. Choi. 2008. *Nano Lett.* 8 (1):26–30, Copyright (2008).)

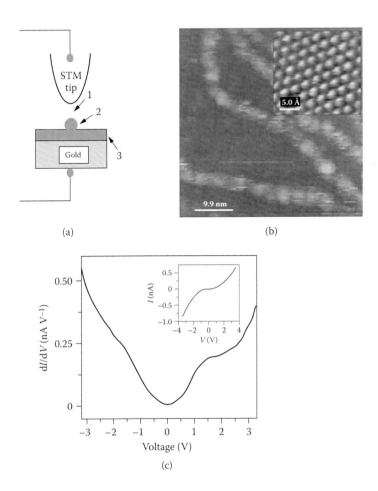

(a)   (b)

(c)

FIGURE 5.6   The measurement scheme, DNA scanning tunneling microscopy (STM) image, and STS on bare gold. (a) Schematic diagram of the double-barrier tunnel junction configuration. The first tunnel junction is formed between the STM tip and the molecule (indicated as 1). The DNA molecule profile is marked as 2, and the second junction (marked as 3) is between the molecule and the gold surface. (b) Room-temperature STM image of poly(G)–poly(C) DNA molecules on which STS measurements were conducted. The full length of the DNA molecules shown here is approximately 1.2 μm. The inset shows an atomic-resolution image of a bare gold part of the surface (scanned at $T \sim 78$ K), confirming the cleanliness of the STM tip and surface. (c) A typical (dI/dV)–V curve measured on bare gold (corresponding I–V curve in the inset), showing the gapless characteristic of the substrate. (Reprinted with permission from Macmillan Publishers Ltd: [*Nature*] Shapir, E., H. Cohen, A. Calzolari, C. Cavazzoni, D. A. Ryndyk, G. Cuniberti, A. Kotlyar, R. Di Felice, and D. Porath. 2008. *Nat. Mater.* 7 (1):68–74, Copyright (2008).)

It is important to note that this was not carried out to determine conductivity along DNA molecules as mentioned previously, but rather to determine the tunneling conductance across the radial direction of poly(G)–poly(C). Since poly(G)–poly(C) had a homogeneous sequence, the STS results were also expected to be homogeneous regardless of the position of the STM tip. However, in practical observations, such expectations are not likely to hold, as the interface junction between the gold surface and the poly(G)–poly(C) DNA as well as the conformational order are not identical to the varying position of the STM tip. This is why the authors collected diverse STS data (Figure 5.7). Based on a statistical analysis of the data (Figure 5.8A), they succeeded in constructing the energy level structures beautifully, with a fundamental energy gap of ~2.5 eV for the poly(G)–poly(C) DNA (Figure 5.8B). In advance, a fully ab initio computer simulation was employed to establish the density of states (DOS) originating from the backbone (PO_4^-), nucleobases, and Na^+ counterions. The simulated total density of states is shown in Figure 5.9.

Surprisingly, Na^+ counterions were found to contribute to new empty electron states in the gap between the G and C peaks. Although it had a very low DOS relative to G and C, a finite number of discrete levels were found scattered throughout the GC gap. This implies that an appropriate substitution of the counterion can modify the total DOS, resulting in a carrier-doping effect as seen in inorganic semiconductors (Furukawa et al. 2007).

As described previously, STS has an advantage in that it is capable of obtaining the DOS of regular DNA molecules such as poly(G)–poly(C) and poly(A)–poly(T), while it simultaneously suffers from the limitation of its inapplicability to random-sequence DNA. Furthermore, from a methodological perspective, STS makes it difficult to detect charge transport along the long axis of a given DNA.

Besides these DC measurements, a variety of alternative current (AC) methods employing either direct contact or noncontact techniques can be employed to identify DNA conductivity as well. Regardless of the selection method used, however, every DNA specimen should be prepared in thin film or in bulk form, thus making it very difficult to determine whether the obtained DNA conductivity value can be attributed to electron/hole carriers, water, or counterions (Bonincontro et al. 1988; Lith et al. 1986; Laudát and Laudát 1992; Ronne et al. 1997; Briman et al. 2004; Tran et al. 2000; Sokolov et al. 1999; Kutnjak et al. 2005; Long et al. 2003; Wang 2008). Therefore, the traditional AC methods are not appropriate for the study of DNA conductivity at the single-molecule level.

5.1.2 ELECTRICAL CONDUCTIVITY OF DNA—A SUMMARY

In the previous section, a variety of factors affecting the carrier transport properties of DNA were discussed. Additionally, several different experimental methods—including their advantages as well as disadvantages—were explained. Table 5.1 summarizes the reported results of the electrical conductivities (Storm et al. 2001; Braun et al. 1998; Zhang et al. 2002; de Pablo et al. 2000; Porath et al. 2000; Rakitin et al. 2001; Yoo et al. 2001; Cai, Tabata, and Kawai 2000; Tran, Alavi, and Graner 2000; Rink and Schönenberger 1999; Kasumov et al. 2001) of various DNA samples (Endres, Cox, and Singh 2004). As one can see from the table, although the diversity

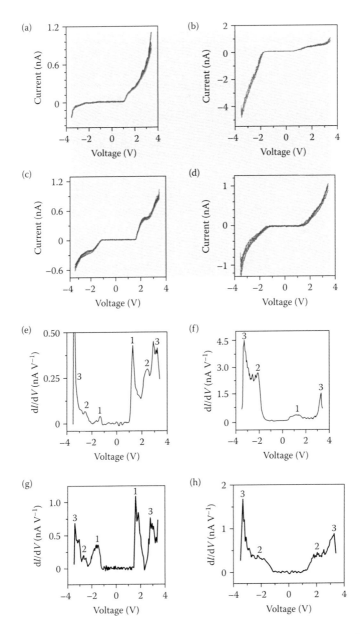

FIGURE 5.7 Current–voltage and conductance curves from STS measurements. STS measurements were presented as I–V and corresponding derivative (dI/dV–V) characteristics. (a–d) I–V sets (containing 10, 12, 16, and 14 curves, respectively) obtained at T~78 K. Each of the I–V sets corresponds to a different poly(G)–poly(C) DNA molecule. (e–h) The corresponding derivatives of the average I–V of each set. The average gap is ~2.5 eV. The measurements were carried out with V_{bias} of 2.8 V and I_{set} of 0.5 nA. (Reprinted with permission from Macmillan Publishers Ltd: [*Nature*] Shapir, E., H. Cohen, A., Calzolari, C. Cavazzoni, D. A. Ryndyk, G. Cuniberti, A. Kotlyar, R. Di Felice, and D. Porath. 2008. *Nat. Mater.* 7(1):68–74, Copyright (2008).)

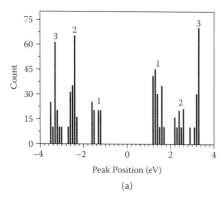

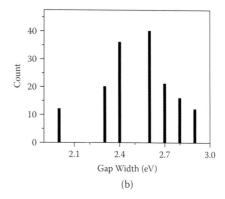

(a)   (b)

FIGURE 5.8 Statistical analysis of the experimental STS results. Statistical analysis of the peak-energy values (a) and the fundamental energy gap (b), obtained from 180 experimental curves (each curve in the statistics is an average of at least 10 consecutive individual similar I–V curves obtained at different molecules). (Reprinted by permission from Macmillan Publishers Ltd: [Nature] Shapir, E., H. Cohen, A. Calzolari, C. Cavazzoni, D. A. Ryndyk, G. Cuniberti, A. Kotlyar, R. Di Felice, and D. Porath. 2008. *Nat. Mater.* 7 (1):68–74, copyright (2008).)

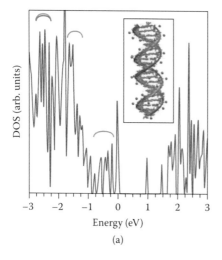

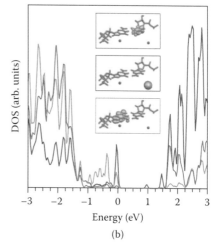

(a)   (b)

FIGURE 5.9 Simulated structures and computed DOS. The energy origin along the horizontal axes of both DOS plots is set at the level of the highest occupied eigenvalue. From a thorough analysis of the character of electron states, similar states can be grouped into gross convoluted features at −0.5 eV, −1.5 eV, and −2.5 eV. (Arches are shown as a visual guide; the origin of the energy scale is set at the top of the occupied levels, and thus the absolute values of the peak energies are shifted with respect to the experimental peaks.) The inset shows a three-dimensional representation of the simulated structure. The periodic unit cell contains 10 GC pairs—2 replicas are shown here. Pink and cyan frames represent guanine and cytosine bases, respectively. The backbone is shown as ribbons and the external green spheres are the Na^+ counterions included in the simulation. (Reprinted with permission from Macmillan Publishers Ltd: [*Nature*] Shapir, E., H. Cohen, A. Calzolari, C. Cavazzoni, D. A. Ryndyk, G. Cuniberti, A. Kotlyar, R. Di Felice, and D. Porath. 2008. *Nat. Mater.* 7 (1):68–74, Copyright (2008).)

TABLE 5.1
Overview of Recent Conductivity Measurements of DNA

Class	Group	DNA sample	Results	Electrodes	Method	Ions
1. Anderson insulator	Storm et al. (2001)	Single λ-DNA/polyG–polyC	Insulating (at RT) (DNA height: 0.5 nm)	Pt/Au	On SiO_2, mica surface	Mg^{2+}
	Braun et al. (1998)	Single λ-DNA	Insulating (at RT)	Au	Free hanging (gluing technique)	Na^+
	Zhang et al. (2002) De Pablo et al. (2000)		(conducting if doped)		SFM, on mica	?
2. Band-gap insulator	Porath et al. (2000)	Single polyG–polyC (only 30 bps)	Wide band-gap Semiconductor (at all RT)	Pt	Free hanging	Na^+
	Rakitin et al. (2001)	Single, short oligomer-λ-DNA	(At RT)	Au	(Gluing technique)	
3. Activated hopping conductor	Rakitin et al (2001)	Bundles of λ-DNA	Narrow "band-gap" semiconductor (at RT)	Au	Free hanging	Na^+
	Yoo et al. (2001)	Supercoiled polyG–polyC polyA–polyT	Linear Ohmic at RT Insulating at low T	Au/Ti	On SiO_2	
	Cai et al. (2000)	Networks of bundles polyG–polyC/polyA–polyT	Linear Ohmic (at RT)	Au	SFM, on mica	
	Tran et al. (2000)	Supercoiled dry and wet λ-DNA	Hopping conductivity	None	Microwave absorption	
	Fink and Schönenberger (1999)	Bundles of λ-DNA	Conducting (doped) (at RT)	Au	Free hanging	
4. Conductor	Kasumov et al. (2001)	Few λ-DNA molecules	Induced superconductivity (T<1K)	Re/C	On mica	Mg^{2+}

Note: Four class, labeled 1–4, of different experimental outcomes can be identified, although this classification is by no means unique. Notation: RT = room temperature.

Source: Reprinted with permission from the American Physical Society: Endres, R. G., D. L. Cox, and R. P. Singh. *Rev. Modern Phys.*, 76 (1), 195–214, 2004, Copyright (2004).).

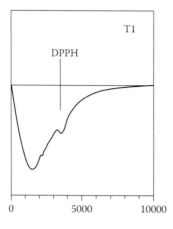
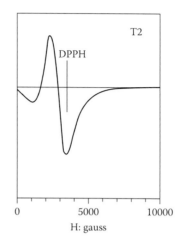

FIGURE 5.10 Typical example of derivatives of EMR absorption at 9.5 GHz in highly purified phage T1 and T2 compared with the absorption signal of the free radical diphenylpicryl-
hydrazyl (DPPH). (Reprinted from *Biochem. Biophys. Res. Commun.* Muller, A., G. Hotz,
and K. G. Zimmer, Electron spin resonances in bacteriophage: alive, dead and irradiated.
1961. 4 (3):214–217, Copyright (1961), with permission from Elsevier.)

of the DNA samples dealt with in each study is rather vast, one can become quite
confused when attempting to draw any firm conclusions as to the electrical conductivity of DNA.

The preponderance of experimental results, however, implies strongly that the
electrical conductivity of dry DNA along the long helical axis is poorly semiconductive. This is consistent with the theoretical expectation that there exists a certain
degree of π-overlap between the basal planes along the axis of DNA (Mac Naughton
et al. 2005; Matta et al. 2006; Hua, Gao et al. 2010; Hua, Yamane et al. 2010; Kummer
et al. 2010; Mehrez and Anantram 2005), as in the case of graphite. It has been well
established that the electrical conductivity of graphite in a direction perpendicular to
the facial plane is substantive, ca. $5 \times 10^2 \ \Omega^{-1}m^{-1}$ (Delhaes 2000); this analogy may
be applicable to DNA. The solenoidal motion of charge carriers is proposed in the
next section to explain, at least in part, the electrical conductivity of double-stranded,
helical DNA.

5.2 MAGNETIC PROPERTIES OF DNA

5.2.1 HISTORICAL RECOUNT

DNA magnetism became a matter of interest among scientists when Russian and other
researchers (Bliumenfeld and Benderskii 1960; Muller et al. 1961) noted an unexpected huge electron magnetic resonance peak over the very low magnetic-field range,
together with the conventional electron magnetic resonance signal (Figure 5.10).

Immediately, this observation was regarded as a sign of ferromagnetic property,
and successive studies were conducted by other researchers. One of the most interesting issues was the complete exclusion of possible magnetic impurities such as ions of

Fe, Mn, and Co, and their oxides. However, at that time, it was somewhat difficult to remove completely the impurities from samples; thus, scientists failed to show that the magnetic properties were intrinsic to DNA. However, six decades later, this question regarding the magnetic properties of DNA was again raised by French workers (Nakamae et al. 2005) who employed λ-DNA in B-form and identified sigmoid S-shaped magnetization (M)-magnetic field (H) curves under a controlled hydration level. Interestingly, they interpreted their results to mean that the S-shaped magnetic curves were attributable to a possible loop current along the circular shape of λ-DNA. This finding represented a great progress because, during the early stages of DNA magnetic research, most researchers were unable to generate a plausible physical model to explain DNA magnetism. However, this model was soon challenged by a Japanese group (Mizoguchi et al. 2006) who could reproduce the same M–H curves by allowing oxygen molecules absorbed on the surface of quartz wool. This will be discussed later in further detail (see Figure 5.23). Therefore, the question as to whether DNA's magnetism is intrinsic or extrinsic remains a matter of some controversy.

The intrinsic aspect of DNA's magnetism is related to the charge transport mechanism established in recent years. The first suggestion in this regard was put forth by French workers who observed a ferromagnetic-like magnetization (M)-magnetic field (H) curve for B-DNA at liquid helium temperature (Figure 5.11), and proposed that the observed magnetic phenomenon was the result of a mesoscopic loop current flowing along the λ-DNA's circular structure, as shown in Figure 5.12.

This work is very important in the sense that the observed magnetic properties could be correlated with a possible charge transport mechanism (orbital magnetization). Although no microscopic evidence, such as ferromagnetic resonance, was noted, the idea was sufficient to inspire further and more advanced physical models of DNA magnetism (Figure 5.13).

5.2.2 DNA MAGNETISM

Soon afterward, we (Kwon et al. 2006; Lee, Do et al.; Kwon et al. 2007) also noted a very similar magnetic property for *salmon* sperm DNA in a dry state at room temperature and explained the observation using a model in which solenoid charge transport along helical structure of DNA occurs under a nonzero magnetic field.

According to this model, holes or electrons residing on the DNA helix can experience Lorentz forces, $F = qv \times B$ (q: electric charge, v: velocity, B: magnetic induction), exerted by a nonzero external magnetic field (Halliday, Resnick, and Walker 2008), resulting in their motion along the stacked π-network of DNA's base pairs. This will bring about a solenoid current aided by the cyclotron motion of charged particles, which is expected to transform the DNA helix into a magnetic solenoid. In particular, such a cyclotron motion can prove easier when the external magnetic field is parallel to the long axis of the DNA helix. As a point of evidence for such cyclotron motion, we demonstrated the suggested observation of an extremely broad electron magnetic resonance peak (linewidth = 1000 G) at 500 G, as demonstrated in Figure 5.14.

Furthermore, we (Lee, Kwon et al. 2006) determined that the extremely broad EMR signal is quite sensitive to water content in DNA, and that the S-shaped M–H

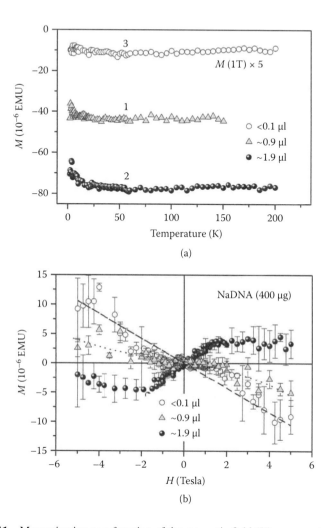

FIGURE 5.11 Magnetization as a function of the magnetic field (H) measured at 2 K without the contribution of water (-7.2×10^{-10} EMU·G^{-1}μL^{-1}). The straight dashed line represents the diamagnetic component of NaDNA measured at 150 K. The dotted curved line superimposed on the data points with H$_2$O = 0.9 μL is provided as a visual guide. Error bars represent the standard deviation in the measurements. Some error bars on low field data, $H \leq 1$T, were removed in order to avoid crowding the graph. A ferromagnetic component originating from impurities in the quartz tube was detected whose magnitude saturates at H~7000 G (M ~ 5 × 10^{-6} EMU for both NaDNA and MgDNA). This component was independent of temperature and water content, and has also been excised. (Reprinted with permission from the American Physical Society: Nakamae, S., M. Cazayous, A. Sacuto, P. Monod, and H. Bouchiat, *Phys. Rev. Lett.*, 94(24), 248102, 2005, Copyright (2005).)

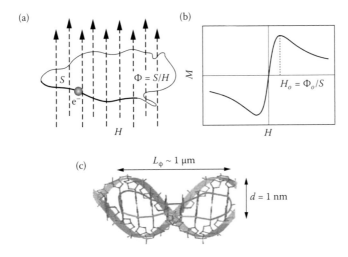

FIGURE 5.12 (a) Schematic view of a mesoscopic ring flux in the perpendicular magnetic field H. (b) The shape of orbital magnetization as a function of magnetic field in the nonlinear regime. (c) Simplified picture of a persistent current path inside and along the helical length of a B-DNA molecule involving interstrand charge transfer. (Reprinted with permission from the American Physical Society: Nakamae, S., M. Cazayous, A. Sacuto, P. Monod, and H. Bouchiat, *Phys. Rev. Lett.*, 94(24), 248102, 2005, Copyright (2005).)

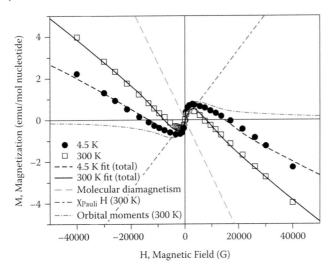

FIGURE 5.13 The M–H curves at room and liquid helium temperatures. Both of the M–H curves evidence positive slopes up to 2500 G, but, above this field, they exhibit negative slopes. The solid and dashed lines represent the curves fitted by the linear combination of the calculated molecular diamagnetism (-2.71×10^{-4} emu/G·mol nucleotide; blue dot), the Pauli paramagnetism (green-dashed dotted line), and the mesoscopic magnetism (orange solid line) due to the helical current loops induced by the applied magnetic field. (Reprinted with permission from the American Physical Society: Lee, C. H., Y.-W. Kwon, D.-D. Do, D. H. Choi, J.-I. Jin, D.-K. Oh, and J. Kim, *Phys. Rev. B*, 73, 224417, 2006, Copyright (2006).)

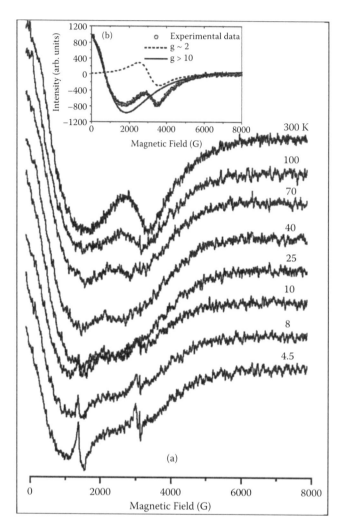

FIGURE 5.14   (a) The representative EMR signals of dry A-DNA obtained at various room temperatures. One can see that the EMR spectra consist of two distinct signals. One is quite broad, with a linewidth of more than 1000 G centered at approximately 500 G, and the other is relatively narrow, at approximately 3000 G. (b) Deconvolution of an EMR peak at 300 K by Lorentzian fitting. (Reprinted with permission from the American Physical Society: Lee, C. H., Y.-W. Kwon, D.-D. Do, D. H. Choi, J.-I. Jin, D.-K. Oh, and J. Kim, *Phys. Rev. B*, 73, 224417, 2006, Copyright (2006).)

curve in the superconducting quantum interference device (SQUID) measurement disappeared simultaneously with the extremely broad EMR signal when the sample became wet. It is even more surprising that the EMR signal noted at 3000 G originated from conductive spins with changing motional dimension from one to three dimensions as the water content decreases from 24 wpn (water molecules per nucleotide) to 0.5 wpn (Figures 5.15, 5.16, 5.17) (Kwon et al. 2008). This dimensional crossover demonstrated that, above 2 wpn, the charge dynamics are limited to the single DNA molecular level without helical motion, but below it the charge motion is delocalized gradually over intermolecules with helical motion, thus rendering them three-dimensional.

The molecular packing structure of DNA helix (Bloomfield 1996; Besteman et al. 2007; Cherstvy 2008; Keller et al. 1996; Podgornik et al. 1996) is believed to influence the three-dimensional delocalization of the conductive spins. We did indeed show, by cross-polarizing optical microscopy, that there exists a liquid crystalline order (Figure 5.18) in the fibrous A-DNA.

It was suggested that, in the nonzero magnetic field, DNA helices containing water molecules below 2 wpn can initially have a helical charge transport caused by Lorentz forces (Halliday, Resnick, and Walker 2008) forming a molecular solenoid, and can then couple coherently to one another in a closely packed hexagonal structure via a mutual induction mechanism (Faraday effect) (Halliday, Resnick, and Walker 2008), leading to the mesoscopic loop current required for the S-shaped M–H curve (Figure 5.19).

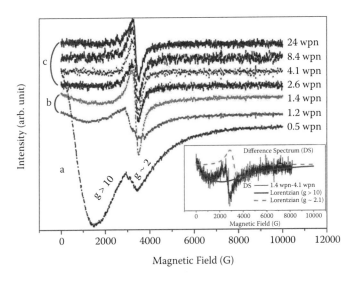

FIGURE 5.15 Schematic stack plots of the EMR signals acquired from dsDNA samples of varying water content. The inset shows the difference spectrum obtained from the spectra of DNA samples with 1.4 and 4.1 wpn. It also shows the deconvolution into two components. (Reprinted with permission from the Korean Chemical Society: Kwon, Y.-W., E. D. Do, D. H. Choi, J.-I. Jin, C. H. Lee, J. S. Kang, and E.-K. Koh. 2008. *Bull. Korean Chem. Soc.* 29 (6):1233–1242, Copyright (2008).)

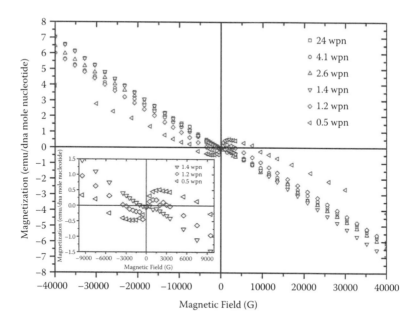

FIGURE 5.16 The magnetization–magnetic field strength (M–H) curves obtained at room temperature for the dsDNA samples with different water contents. The inset shows the magnified central S-shaped portions of the M–H curves. (Reprinted with permission from the Korean Chemical Society: Kwon, Y.-W., E. D. Do, D. H. Choi, J.-I. Jin, C. H. Lee, J. S. Kang, and E.-K. Koh. 2008. *Bull. Korean Chem. Soc.* 29 (6):1233–1242, Copyright (2008).)

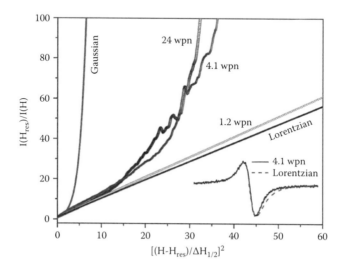

FIGURE 5.17 The motional dimension analyses of spins via $[(H\text{-}H_{res})/\Delta H_{1/2}]^2$ versus $I(H_{res})/I(H)$ plots. (Reprinted with permission from the Korean Chemical Society: Kwon, Y.-W., E. D. Do, D. H. Choi, J.-I. Jin, C. H. Lee, J. S. Kang, and E.-K. Koh. 2008. *Bull. Korean Chem. Soc.* 29 (6):1233–1242, Copyright (2008).)

FIGURE 5.18 Birefringence of a DNA fibril observed through a cross-polarizing optical microscope at room temperature (magnification 200X). (Reprinted with permission from the Korean Chemical Society: Kwon, Y.-W., E. D. Do, D. H. Choi, J.-I. Jin, C. H. Lee, J. S. Kang, and E.-K. Koh. 2008. *Bull. Korean Chem. Soc.* 29 (6):1233–1242, Copyright (2008).)

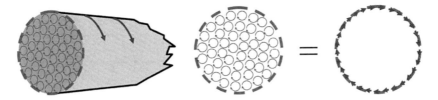

FIGURE 5.19 An idealized picture of the cyclotron motion of spin carriers in an elementary fibril of dsDNA. The inner small circles designate the solenoidal motions along the dsDNA. (Reprinted with permission from the Korean Chemical Society: Kwon, Y.-W., E. D. Do, D. H. Choi, J.-I. Jin, C. H. Lee, J. S. Kang, and E.-K. Koh. 2008. *Bull. Korean Chem. Soc.* 29 (6):1233–1242, Copyright (2008).)

This implies that every DNA evidencing an S-shaped M–H curve should harbor a water content below 2 wpn as well as at least a partly closely packed hexagonal structure. Here it is worth noting that the T1 and T2 phages evidencing an extremely broad EMR peak also harbor a DNA toroid morphology of closely packed hexagonal structures as demonstrated in Figure 5.20 (Hud and Downing 2001; Lepault et al. 1987).

Despite all these experimental results supporting the intrinsic orbital magnetization, the molecular solenoid model of the DNA charge transport may necessitate robust experimental methods. The simplest method of confirming a solenoid charge transport is to measure conductivity for a single-molecule DNA as a function of the external magnetic field (Minot et al. 2004). In order to accomplish such experiment, one must prepare an amine DNA-functionalized molecule that covalently contacts carboxyl SWNT-modified electrodes (two-probe method without any contact resistance). Additionally, the trenched-SiO_2/Si substrate should be utilized to avoid any possible interaction between the substrate and the DNA molecule. If one measures DC conductivity in a vacuum as a function of the external magnetic field, the maximum DC conductivity at a constant voltage should be detected due to the cyclotron

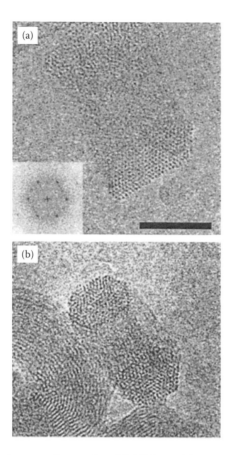

FIGURE 5.20 Cryoelectron micrographs of λ-DNA toroids with the plane of the toroid oriented ~90° with respect to the microscope image plane; edge-view toroid images. (a) A toroid in which the hexagonal packing of DNA helices is clearly apparent in the outer regions. (Inset) Fourier transform of image region containing the highly ordered DNA lattice. (b) A toroid for which the outer regions are well-defined hexagons. (Scale bar is 50 nm. All micrographs are shown at the same magnification.) (Reprinted with permission from the National Academy of Sciences U.S.A.: Hud, N. V., and K. H. Downing. 2001. *Proc. Natl. Acad. Sci. U.S.A.* 98 (26):14925–14930, Copyright (2001).)

motion of the charge carriers under the external magnetic field. Under an identical configuration, one may also measure AC conductivity as a function of the frequency and the external magnetic field. In particular, the microwave frequency of ~10 GHz is largely appropriate for the detection of cyclotron resonance absorption, which may be referred to as an electrically detected cyclotron resonance (EDCR).

Experimental evidence for the inductive coupling (Cai and Sevilla 2000) between the nearest DNA-neighbored molecules can be achieved as follows; at first, one configures two DNA molecules as shown in Figure 5.21. Then, if one switches DC voltage on the left circuit, a corresponding voltage variation should be induced on the opposite circuit via a Faraday induction mechanism.

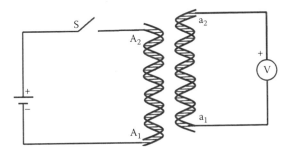

FIGURE 5.21 Schematic diagram for the DC test of DNA molecular transformer. One can measure the voltage difference induced between a_1 and a_2 in the secondary ds-DNA helix (right side) immediately after the switch S in the primary ds-DNA molecule on the left side is closed.

An AC test of Faraday induction (Halliday, Resnick, and Walker 2008) can also be conducted by first driving AC current on the left circuit and, subsequently, detecting the AC voltage induced on the right-side circuit as shown in Figure 5.21. Their phase difference should always be $\pi/2$.

The first direct evidence of the possible extrinsic origin of DNA's magnetism was reported by Schulman et al. (Walsh Jr., W. M., Shulman, and Heidenriech 1961) who examined every DNA sample emitting EMR signals using an electron microscope and succeeded in clearly showing inclusions of ferromagnetic Fe_2O_3 as shown in Figure 5.22 (W. M. Walsh Jr, Shulman, and Heidenriech 1961).

In fact, ferromagnetic inclusion has always been a vital problem in the study of the weak magnetic properties of organic materials (Garcia et al. 2009; Makarova 2009). Therefore, thorough sample preparation with extreme caution is the only way to ensure the exclusion of unwanted ferromagnetic impurities in sample specimens. DNA samples are no exceptions to this. However, it is also the case that the presence of magnetic impurities alone is insufficient to explain all of these diverse experimental results reported until now. For instance, some types of interactions between ferromagnetic inclusions and the DNA molecule may be considered in further efforts to understand the total observed magnetic properties in the specimens containing ferromagnetic contaminants.

Another external magnetic source is the presence of paramagnetic oxygen molecules whose contribution to DNA's magnetism was demonstrated by Mizoguchi et al. (2006). They prepared a quartz wool sample whose surface was contaminated with paramagnetic oxygen molecules and, by measuring its magnetic M–H behavior, succeeded in demonstrating the same ferromagnetic-like M–H saturation as reported earlier for B-DNA (Figure 5.23).

This tells us that the ferromagnetic-like M–H saturation in B-DNA did not derive from an intrinsic magnetism of B-DNA but rather from the paramagnetic oxygen molecules included in the sample. Indeed, it is not difficult to discern whether the magnetic behavior derives from the paramagnetic oxygen molecules or the DNA itself; the EMR signals of DNA (Kwon et al. 2008) and paramagnetic oxygen molecules are completely different from one another, as shown in Figure 5.24 (van Vleck

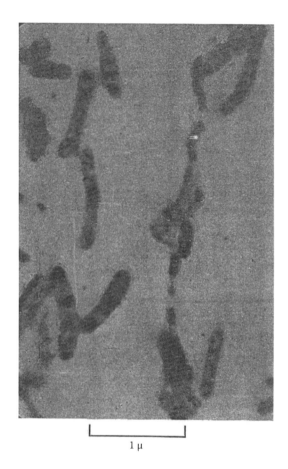

1 μ

FIGURE 5.22 Electron micrograph (80 KeV) of a nucleic acid sample (calf thymus ΣII) showing crystalline inorganic inclusions in groups of DNA molecules. (Reprinted with permission from Macmillan Publishers Ltd: [*Nature*] Walsh Jr., W. M., R. G. Shulman, and R. D. Heidenriech. 1961. *Nature* 16 (4807):1041–1043, Copyright (1961).)

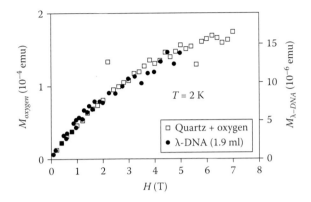

FIGURE 5.23 Comparison of M–H curves at 2K between the oxygen molecules on quartz and the reported data for the B-form of λ-DNA. (Reprinted with permission from the American Physical Society: Mizoguchi, K., S. Tanaka, and H. Sakamoto, *Phys. Rev. Lett.*, 96, 089801, 2006, Copyright (2006).)

1947; Tinkham and Strandberg 1955a, 1955b, 1955c; Defotis 1981; Altman et al. 2003; Kon 1973; Gardiner et al. 1981; Yagi et al. 2004; Kaneko et al. 1998).

DNA's magnetism may not only be caused by the accidental inclusion of unwanted ferromagnetic impurities but also may be intentionally induced via stoichiometric binding or the coordination of transition metal ions (M-DNA). The latter approach is a highly stoichiometric process and thus appropriate for studies of the magnetism of DNAs modified by various metal ions substituting for natural sodium ions (Wettig et al. 2003). In nature, the Na^+ cation of DNA is bonded in one-to-one fashion to the PO_4^- anion by Coulomb forces. In particular, under physiological conditions, the Na^+ cation is surrounded by H_2O, which reduces its cationic characteristics, and which in turn weakens Coulombic binding to the PO_4^- ion. Therefore, one can replace the Na^+ cation with other ones, including K^+, Li^+, Mg^{2+}, Ca^{2+}, Zn^{2+}, Sr^{2+}, Cu^{2+}, Mn^{2+}, Fe^{2+}, Co^{2+}, Ni^+, Hg^{2+}, and Ag^{2+} et al. (Duguid et al. 1993, 1995; Arakawa et al. 2001; Ahmad et al. 2003; Wettig et al. 2003, 2005; Muñoz et al. 2001; Aich et al. 1999). In the dication case, one dication can usually be bonded to two PO_4^- anions (Marincola et al. 2000; Kornilova et al. 1997; Andrushchenko et al. 1997; Hackl et al. 1997). Systematic studies conducted using a variety of cations have demonstrated that detailed binding sites differ from each other and depend on pH values (Wood and Lee 2005) and the types and concentration of cations. With increasing concentrations, cations incorporated into DNA are likely to evidence a higher probability to bind to the PO_4^- backbone, sugar moiety, and base pairs in that order. A schematic diagram is shown in Figure 5.25.

A surprising result associated with DNA's magnetism was very recently reported by a Slovenian group (Omerzu et al. 2010) studying freeze-dried Zn-DNA (calf thymus DNA). As has been well established, the Zn^{2+} cation is originally nonmagnetic. However, they clearly observed strong EMR peaks with Dysonian lineshapes (Dyson 1955) at the g value of 2.2, as demonstrated in Figure 5.26. In order to confirm the extrinsic origins, they prepared three control samples with the same starting chemicals

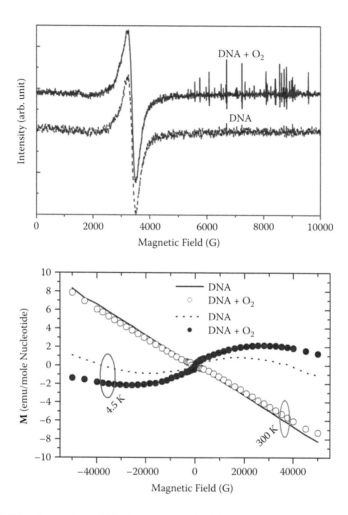

FIGURE 5.24 Comparison of EMR spectra acquired from a dsDNA sample in the absence or presence of oxygen molecules (upper figure). The EMR peaks induced by molecular oxygen are observed as very complex signals over the magnetic field range of 5,000–10,000 G. The lower figure provides the SQUID magnetization (M–H curve) measurements on the same sample specimens used in EMR spectroscopy. One can clearly observe that the effects of paramagnetic oxygen contamination are relatively minimal at room temperature but are apparently manifested at liquid helium temperature. (Reprinted with permission from the Korean Chemical Society: Kwon, Y.-W., E. D. Do, D. H. Choi, J.-I. Jin, C. H. Lee, J. S. Kang, and E.-K. Koh. 2008. *Bull. Korean Chem. Soc.* 29 (6):1233–1242, Copyright (2009).)

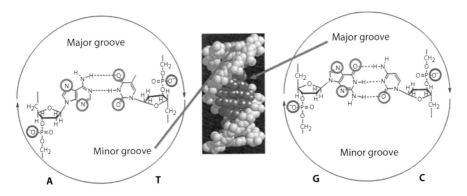

FIGURE 5.25 The available ligands of the major groove are the carbonyl group of guanine ($C_6=O_6$), the nitrogen (N_7C_8H) of adenine (N_7) and the carbonyl group ($C_4=O_4$) of thymine. The ligands of the minor groove are the carbonyl groups of thymine ($C_2=O_2$) and cytosine ($C_2=O_2$), and the nitrogens of the adenine (N_3) and guanine (N_3) moieties.

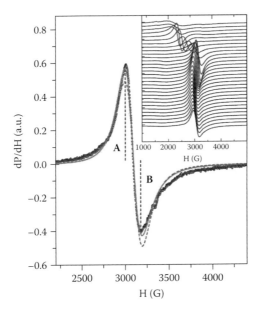

FIGURE 5.26 The room-temperature EMR spectrum of Zn-DNA (black circles) fitted with asymmetric Lorentzian (Dysonian) (solid red line) and Lorentzian (dashed blue line) curves. The inset shows a temperature development of the Zn-DNA EMR resonance measured on cooling from room temperature (the bottom curve) down to 10 K (the top curve) measured at every 10 K. (Reprinted with permission from the American Physical Society: Omerzu, A., B. Anzelak, I. Turel, J. Strancar, A. Potocnik, D. Arcon, I. Arcon, D. Mihailovic, and H. Matsui, *Phys. Rev. Lett.*, 104 (15), 156804, 2010, Copyright (2010).)

via the same freeze-drying method: (a) tris-HCl + $ZnCl_2$ + DNA at pH 7, (b) tris-HCl + $ZnCl_2$ at pH 9, and (c) tris-HCl + DNA at pH 9. None of them evidences EMR signals because none of them was capable of forming the Zn-DNA complex. X-ray absorption near edge structure (XANES) results demonstrated that a valence state of Zn was in Zn^{2+}, thereby implying that Zn^{2+} was not, in and of itself, paramagnetic. Meanwhile, the EMR signal detected was orders of magnitude larger than any signal that could have originated from paramagnetic impurities. As a result, they concluded that the EMR signals from Zn-DNA were attributable to the unpaired electrons residing on the Zn-DNA complex (Apalkov and Chakraborty 2008; Starikov 2003; Lee, Ahn et al. 2006).

The Dysonian lineshape is generally associated with a skin-depth phenomenon and thus is correlated strongly with the metallic conductivity of material at microwave frequencies, suggesting the possibility that a given Zn-DNA is metallic. The metallic nature of Zn-DNA was also demonstrated by the temperature dependence of EMR susceptibility as shown in Figure 5.27(A). As one can plainly see in the

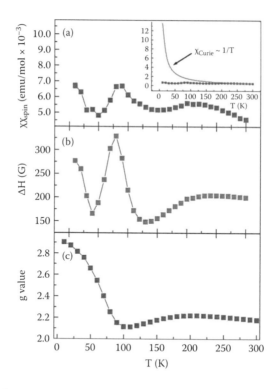

FIGURE 5.27 Temperature dependences of (a) a spin susceptibility χ_{spin}, (b) a resonance line width the ΔH, and (c) a g value. The inset of (a) shows the comparison of the temperature dependence of the spin susceptibility of Zn-DNA with the Curie law on an expanded scale. The susceptibility is expressed in mole of DNA base pairs. (Reprinted with permission from the American Physical Society: Omerzu, A., B., Anzelak, I. Turel, J. Strancar, A. Potocnik, D. Arcon, I. Arcon, D. Mihailovic, and H. Matsui, *Phys. Rev. Lett.*, 104 (15), 156804, 2010, Copyright (2010).)

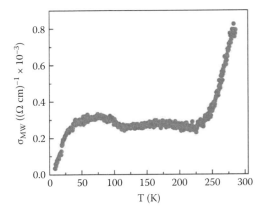

FIGURE 5.28 Temperature dependence of the microwave conductivity of Zn-DNA. (Reprinted with permission from the American Physical Society: Omerzu, A., B. Anzelak, I. Turel, J. Strancar, A. Potocnik, D. Arcon, I. Arcon, D. Mihailovic, and H. Matsui, *Phys. Rev. Lett.*, 104 (15), 156804, 2010, Copyright (2010).)

inset of Figure 5.27, the temperature dependence of EMR susceptibility deviates severely from Curie's law anticipated for a localized spin entity. Interestingly, the same metallic effect was reflected well in the temperature dependence of microwave conductivity measured via a cavity perturbation method (Figure 5.28).

Microwave conductivity maintains a finite value of 3×10^{-4} S/cm over the temperature range from 50 to 225 K, but both below 50 K and above 225 K, it increases with increasing temperature. The onset temperature (225 K) of the high temperature increase was associated with an activation of the rotational degree of freedom of water molecules remaining in Zn-DNA (Brovchenko et al. 2007; Sokolov et al. 1999; Banerjee and Bhat 2008), whereas the low temperature decreasing point at 50 K was correlated with a localization of the doped electron. All of the experimental data clearly support the conclusion that delocalized electrons exist. The question being raised is how such delocalized electrons are generated. The authors insisted that the Zn^{2+} cation replaced imino protons, H^+, residing at the hydrogen bonds between complementary nucleobases in the core of the double helix, thereby giving rise to a doping effect. Their fascinating investigation demonstrated that DNA becomes a good electronic and magnetic material as the result of doping under appropriate conditions. A similar EMR study was conducted by a Japanese group, but contradictory results were generated in that case (Figure 5.29). In their study, Zn-DNA was prepared at a pH of 8.5 and precipitated in an ethanol solution.

The EMR results demonstrated that approximately 20 ppm of Zn per base pair alone contributed to the production of the charge carriers in the base π band. In an effort to understand this discrepancy between the Slovenian and Japanese groups' results, one should note that the coordination of metal cations to the phosphate groups of a polynucleotide tends to stabilize the helix, whereas their coordination with the nitrogenous bases tends to destabilize the helix (Eichhorn and Shin, 1968). Therefore, it remains possible that the incorporation of metal ions at high concentrations disrupted or unzipped the helical conformation and simultaneously altered the

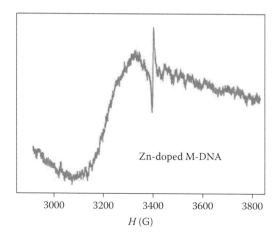

3000   3200   3400   3600   3800

H (G)

FIGURE 5.29 EMR spectrum of Zn-DNA observed by Japanese group. The sharp peak of g ~2.0022 is attributed to the π-radical. (Reprinted with permission from the American Physical Society: Mizoguchi, K., S. Tanaka, T. Ogawa, N. Shiobara, and H. Sakamoto, *Phys. Rev. B*, 72 (3), 033106, 2005, Copyright (2005).)

natural double helical morphology (amorphous or crystalline). Despite the importance of such parameters in understanding the electronic and magnetic properties of Zn-DNA, no studies have yet been conducted involving the control of those variables. Another important point to note is that no one has yet critically evaluated the influence of water concentration. Water molecules may function as scattering centers for electron motion as alluded earlier.

Another example can be found in Mn-DNA (Mizoguchi et al. 2005), which differs from Zn-DNA in that the coordinated metal cation is magnetic. Figure 5.30 shows the EMR spectra obtained for Mn-DNA (1 Mn/BP, BP: base pair) and $Ca_{0.9}Mn_{0.1}$-DNA.

The perfect single-line shape of the Mn-DNA was regarded as a contribution of the Heisenberg exchange interaction (Heisenberg 1926; Dirac 1926) between the neighboring Mn spins. For the inset figure, the sextet structure superimposed upon the single broad EMR line was interpreted as the result of a hyperfine splitting by the Mn nucleus with $I = 5/2$, thereby suggesting the presence of isolated Mn^{2+} ions. The results of the EDS (Energy Dispersive Spectrometry) (Goldstein 2003) study confirmed that one Mn^{2+} cation was bonded to two PO_4^-, reflecting the lack of a carrier doping effect. SQUID measurements (Figure 5.31) (Mizoguchi et al. 2005) of magnetization for Mn-DNA revealed a weak antiferromagnetic interaction between the neighboring Mn spins with Curie–Weiss temperature of $\Theta \approx -0.8$ K.

Based on these two experimental results, one can surmise that DNA molecules can become a magnetic carrier either via charge doping or via the introduction of magnetic cations. However, it should be noted that contradicting magnetic behaviors have been reported for the same Zn-DNA by Slovenian and Japanese groups. Therefore, additional consideration of other factors such as water concentration and the packing structures or morphology of DNA is necessary for a deeper understanding of the extrinsic magnetism of DNA. Needless to say, an integrated view encompassing

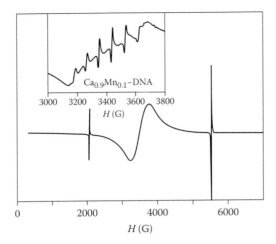

FIGURE 5.30 EMR spectra of Mn-DNA taken at room temperature at X-band. Two sharp peaks above and below the Mn spectrum are the signals from the ruby standard. The inset represents the six hyperfine split peaks in the 10% Mn-doped DNA. (Reprinted with permission from the American Physical Society: Mizoguchi, K., S. Tanaka, T. Ogawa, N. Shiobara, and H. Sakamoto, *Phys. Rev. B*, 72 (3), 033106, 2005, Copyright (2005).)

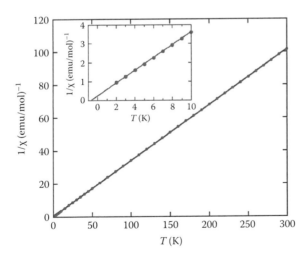

FIGURE 5.31 The magnetic susceptibility of Mn-DNA measured with a SQUID susceptometer at 1 T. The estimated spin concentration is approximately one $S = 5/2$ spin per base pair. The inset shows the expanded view at low temperatures, demonstrating a Curie–Weiss temperature of $\Theta \approx -0.8$ K. (Reprinted with permission from the American Physical Society: Mizoguchi, K., S. Tanaka, T. Ogawa, N. Shiobara, and H. Sakamoto, *Phys. Rev. B*, 72 (3), 033106, 2005, Copyright (2005).)

intrinsic and extrinsic magnetisms is very desirable to describe overall DNA magnetism. In this connection, it is necessary to perform an appropriate charge carrier doping (Aich et al. 1999, 2002; Skinner et al. 2005; Furukawa et al. 2007; Brancolini and Felice 2008; Nokhrin et al. 2007; Jian et al. 2008; Kino et al. 2004; Apalkov and Chakraborty 2008; Mac Naughton et al. 2006) without perturbing the helical structure of the DNA duplex, which would give rise to molecular solenoidal current in a controlled manner, inducing strong magnetism in the nonzero magnetic field.

5.2.3 DISCOTIC LIQUID CRYSTALS AS DNA-MIMICKING COMPOUNDS

The molecular solenoid concept for DNA magnetism is expected to bring about a new type of magnetic materials synthesizable via chemical routes. Indeed, Tagami et al. (2003) reported a theoretical model for the purely organic molecular magnetism caused by a helical current. Moreover, we reported very recently the room-temperature ferromagnetic hysteresis for a discotic liquid crystal (DLC) intercalated with Fe(III)–phthalocyanine (FePc) (Lee et al. 2010). Here, the DLC used was hydrophobic 2,9(10),16(17),23(24)-tetra(2-decyltetradecyloxy)-phthalocyanine (H2Pc), whose morphology was reported earlier to be of a columnar hexagonal structure at room temperature. Interestingly, the DLC and the Fe(III)Pc showed diamagnetic and paramagnetic behavior, respectively, while some of the DLC compositions containing proper levels of Fe(III)Pc revealed ferromagnetic hysteresis as shown in Figure 5.32.

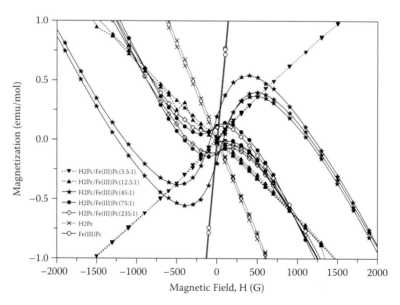

FIGURE 5.32 The magnetization M versus magnetic field H curves obtained at RT for H$_2$Pc doped with various concentrations of Fe(III)Pc. (Lee, C. H., Y.-W. Kwon, D. H. Choi, Y. H. Geerts, E. Koh, and J.-I. Jin.: High-temperature ferromagnetism of a discotic liquid crystal dilutely intercalated with iron(III) phthalocyanine. *Adv. Mater.* 2010. 22 (39):4405–4409, Copyright Wiley-VCH Verlag CmbH & Co., KGaA., Reproduced with permission.)

More interesting was that the ferromagnetic hysteresis observed only for specific ratios of DLC/Fe(III)Pc ranged from 3.5:1 to 235:1. Above 3.5:1, the hexagonal structure of DLC disappeared. This strongly suggests that the room-temperature ferromagnetism is correlated with DLC morphology. It is imaginable that the core aromatic part of the DLC molecule can flow a loop current (Walker 2003). Therefore, the DLC molecules forming the hexagonal structure with an interlayer distance of 3.4 Å can be regarded as a chemically synthesizable version of the DNA duplex. Such a possibility was also confirmed by the presence of the extremely broad EMR peak from DLC itself, although its intensity was very weak when compared with the corresponding peak from dry DNA, implying the presence of the similar motion of the charge carriers under the nonzero magnetic field for both cases.

5.3 CONCLUDING REMARKS

The basic question as to whether or not the DNA single molecule is intrinsically conductive is very important scientifically, as well as for applications of DNA as a new class of materials. It appears that DNA is a wide-bandgap semiconductor with a bandgap energy ranging from 2.5 to 4 eV. In particular, the results of recent nanotechnology-aided experiments support this conclusion.

Insofar as the magnetic properties of DNA are concerned, one learns that DNA molecules with a defined morphology can became fascinating magnetic carriers. However, two different and more detailed scenarios are suggested at present, namely, intrinsic and extrinsic origins. In the intrinsic origin hypothesis, the observed magnetic property was proposed to be attributable to the transport of helical charge along the DNA duplex. Such a possibility is expected to be of particular importance for dry A-DNA with <2 wpn in closely packed hexagonal morphology. This model, however, requires more definitive confirmation of helical charge transport, which is essential to understanding the intrinsic magnetism of the dry DNA duplex. On the contrary, the extrinsic model demonstrates that diverse magnetisms might be realized by varying the degree of coordination with and the doping effects of various metal cations (including nonmagnetic and magnetic) in the DNA duplex. However, it should be noted that contradictory magnetic behavior has been reported for the same Zn-DNA by the Slovenian and Japanese groups. Therefore, additional consideration of other factors, such as water concentration and the packing structures or morphology of DNA, is necessary for a deeper understanding of the extrinsic magnetism of DNA. Needless to say, an integrated view encompassing both intrinsic and extrinsic magnetisms is very desirable for the description of overall DNA magnetism. In regard to this connection, it is clearly desirable to perform appropriate charge carrier doping without perturbing the helical structure of the DNA duplex, which would give rise to a molecular solenoidal current in a controlled fashion, inducing strong magnetism in the nonzero magnetic field. Indeed, a theoretical model for the purely organic molecular magnetism induced by a helical current was proposed (Tagami et al. 2003).

Every electronic circuit consists of a resistor, a capacitor, and an inductor. The same is applied to molecular nanoelectronics. Therefore, a molecular resistor, capacitor, and inductor are all requirements for the completion of a molecular circuit. From this viewpoint, the molecular solenoid concept for DNA magnetism is expected to bring about a new type of molecular inductors that can be synthesized via chemical routes.

Synthetic mimic of DNA could be realized by a DLC compound, which is an exciting discovery demonstrating the possibility of opening up a new approach to easily attaining processable, high-temperature ferromagnetic organic compositions. It may not appear too optimistic to bring about new organic ferromagnetic materials technologies that are developed based on DNA science.

REFERENCES

Ahmad, R., H. Arakawa, and H. A. Tajmir-Riahi. 2003. A comparative study of DNA complexation with Mg(II) and Ca(II) in aqueous solution: Major and minor grooves bindings. *Biophys. J.* 84 (4):2460–2466.

Aich, P., S. L. Labiuk, L. W. Tari, L. J. T. Delbaere, W. J. Roesler, K. J. Falk, R. P. Steer, and J. S. Lee. 1999. M-DNA: A complex between divalent metal ions and DNA which behaves as a molecular wire. *J. Mol. Biol.* 294 (2):477–485.

Aich, P., R. J. S. Skinner, S. D. Wettig, R. P. Steer, and J. S. Lee. 2002. Long range molecular wire behaviour in a metal complex of DNA. *J. Biomol. Struct. Dyn.* 20 (1):93–98.

Albuquerque, E. L., M. S. Vasconcelos, M. L. Lyra, and F. A. B. F. de Moura. 2005. Nucleotide correlations and electronic transport of DNA sequences. *Phys. Rev. E* 71 (2):021910.

Altman, I. S., P. V. Pikhitsa, Y. J. Kim, and M. Choi. 2003. Magnetism of adsorbed oxygen at low coverage. *Phys. Rev. B* 67 (14):144410.

Andrushchenko, V. V., S. V. Kornilova, L. E. Kapinos, E. V. Hackl, V. L. Galkin, D. N. Grigoriev, and Y. P. Blagoi. 1997. IR-spectroscopic studies of divalent metal ion effects on DNA hydration. *J. Mol. Struct.* 408:225–228.

Apalkov, V., and T. Chakraborty. 2008. Influence of correlated electrons on the paramagnetism of DNA. *Phys. Rev. B* 78:104424.

Apalkov, V. M., and T. Chakraborty. 2005. Electron dynamics in a DNA molecule. *Phys. Rev. B* 71 (3):033102.

Arakawa, H., J. F. Neault, and H. A. Tajmir-Riahi. 2001. Silver(I) complexes with DNA and RNA studied by Fourier transform infrared spectroscopy and capillary electrophoresis. *Biophys. J.* 81 (3):1580–1587.

Banerjee, D., and S. V. Bhat. 2008. Spin probe ESR signature of freezing in water: Is it global or local?, arXiv:0810.4682.

Bansal, M. 2003. DNA structure: Revisiting the Watson-Crick double helix. *Curr. Sci.* 85 (11):1556–1563.

Barnett, R. N., C. L. Cleveland, A. Joy, U. Landman, and G. B. Schuster. 2001. Charge migration in DNA: Ion-gated transport. *Science* 294 (5542):567–571.

Berashevich, J., and T. Chakraborty. 2008. How the surrounding water changes the electronic and magnetic Properties of DNA. *J. Phys. Chem. B* 112 (44):14083–14089.

Besteman, K., K. van Eijk, I. D. Vilfan, U. Ziese, and S. G. Lemay. 2007. Influence of charged surfaces on the morphology of DNA condensed with multivalent ions. *Biopolymers* 87 (2–3):141–148.

Bhalla, V., R. P. Bajpai, and L. M. Bharadwaj. 2003. DNA electronics. *EMBO Reports* 4 (5):442–445.

Bliumenfeld, L. A., and V. A. Benderskii. 1960. Magnetic and dielectric properties of high-ordered macromolecular structures. *Dokl. Akad. Nauk SSSR* 133 (6):1451–1454.

Bloomfield, V. A. 1996. DNA condensation. *Curr. Opin. Struct. Biol.* 6 (3):334–341.

Bloomfield, V. A. 1999. Statistical thermodynamics of helix–coil transitions in biopolymers. *Am. J. Phys.* 67 (12):1212–1215.

Boal, A. K., E. Yavin, O. A. Lukianova, V. L. O'Shea, S. S. David, and J. K. Barton. 2005. DNA-bound redox activity of DNA repair glycosylases containing [4Fe-4S] clusters. *Biochemistry* 44 (23):8397–8407.

Bonincontro, A., G. Careri, A. Giansanti, and F. Pedone. 1988. Water-induced DC conductivity of DNA—a dielectric-gravimetric study. *Phys. Rev. A* 38 (12):6446–6447.

Boon, E. M., A. L. Livingston, N. H. Chmiel, S. S. David, and J. K. Barton. 2003. DNA-mediated charge transport for DNA repair. *Proc. Natl. Acad. Sci. USA* 100 (22):12543–12547.

Boon, E. M., J. E. Salas, and J. K. Barton. 2002. An electrical probe of protein–DNA interactions on DNA-modified surfaces. *Nat. Biotechnol.* 20 (3):282–286.

Brancolini, G., and R. D. Felice. 2008. Electronic properties of metal-modified DNA base pairs. *J. Phys. Chem. B* 112 (45):14281–14290.

Braun, E., Eichen, U. Sivan, and G. Ben-Yoseph. 1998. DNA-templated assembly and electrode attachment of a conducting silver wire. *Nature* 391 (6669):775–778.

Briman, M., N. P. Armitage, E. Helgren, and G. Gruner. 2004. Dipole relaxation losses in DNA. *Nano Lett.* 4 (4):733–736.

Brisker-Klaiman, D., and U. Peskin. 2010. Coherent elastic transport contribution to currents through ordered DNA molecular junctions. *J. Phys. Chem. C* 114 (44):19077–19082.

Brovchenko, I., A. Krukau, A. Oleinikova, and A. K. Mazur. 2007. Water clustering and percolation in low hydration DNA shells. *J. Phys. Chem. B* 111 (12):3258–3266.

Cai, L. T., H. Tabata, and T. Kawai. 2000. Self-assembled DNA networks and their electrical conductivity. *Appl. Phys. Lett.* 77 (19):3105–3106.

Cai, Z., and M. D. Sevilla. 2000. Electron spin resonance study of electron transfer in DNA: Inter-double-strand tunneling processes. *J. Phys. Chem. B* 104 (29):6942–6949.

Cherstvy, A. G. 2008. DNA cholesteric phases: The role of DNA molecular chirality and DNA–DNA electrostatic interactions. *J. Phys. Chem. B* 112 (40):12585–12595.

Cohen, H., C. Nogues, R. Naaman, and D. Porath. 2005. Direct measurement of electrical transport through single DNA molecules of complex sequence. *Proc. Natl. Acad. Sci. USA* 102 (33):11589–11593.

Conwell, E. M. 2005. Charge transport in DNA in solution: The role of polarons. *Proc. Natl. Acad. USA* 102 (25):8795–8799.

Conwell, E. M., and D. M. Basko. 2006. Effect of water drag on diffusion of drifting polarons in DNA. *J. Phys. Chem. B* 110 (46):23603–23606.

Conwell, E. M., and S. V. Rakhmanova. 2000. Polarons in DNA. *Proc. Natl. Acad. Sci. USA* 97 (9):4556–4560.

Cuniberti, G., G. Fagas, and K. Richter. 2001. Conductance of a molecular wire attached to mesoscopic leads: Contact effects. *Acta Phys. Pol. B* 32 (2):437–442.

de Pablo, P. J., F. Moreno-Herrero, J. Colchero, J. Gómez Herrero, P. Herrero, A. M. Baró, Pablo Ordejón, José M. Soler, and Emilio Artacho. 2000. Absence of DC-conductivity in λ-DNA. *Phys. Rev. Lett.* 85 (23):4992–4995.

Defotis, G. C. 1981. Magnetism of solid oxygen. *Phys. Rev. B* 23 (9):4714–4740.

Delhaes, P., ed. 2000. *Graphite and Precursors*. Amsterdam: Gordon and Breach Science Publishers.

Dirac, P. A. M. 1926. On the theory of quantum mechanics. *Proc. Roy. Soc. London A* 112 (762):661–677.

Drummond, T. G., M. G. Hill, and J. K. Barton. 2003. Electrochemical DNA sensors. *Nat. Biotechnol.* 21 (10):1192–1199.

Duguid, J., V. A. Bloomfield, J. Benevides, and G. J. Thomas. 1993. Raman spectroscopy of DNA-metal complexes. I. Interactions and conformational effects of the divalent cations: Mg, Ca, Sr, Ba, Mn, Co, Ni, Cu, Pd, and Cd. *Biophys. J.* 65 (5):1916–1928.

Duguid, J. G., V. A. Bloomfield, J. M. Benevides, and G. J. Thomas. 1995. Raman spectroscopy of DNA-metal complexes. II. The thermal denaturation of DNA in the presence of Sr^{2+}, Ba^{2+}, Mg^{2+}, Ca^{2+}, Mn^{2+}, Co^{2+}, Ni^{2+}, and Cd^{2+}. *Biophys. J.* 69 (6):2623–2641.

Dyson, F. J. 1955. Electron Spin Resonance absorption in metals. 2. Theory of electron diffusion and the skin effect. *Phys. Rev.* 98 (2):349–359.

Eichhorn, G. L. and A. E. Shin 1968. Interaction of metal ions with polynucleotides and related compounds. XII. The relative effect of various metal ions on DNA helicity. *J. Am. Chem. Soc.* 90(26): 7223–7328.

Eley, D. D., and D. I. Spivey. 1962. Semiconductivity of organic substances .9. Nucleic acid in dry state. *Trans. Faraday Soc.* 58 (470):411–417.

Endres, R. G., D. L. Cox, and R. R. P. Singh. 2004. Colloquium: The quest for high-conductance DNA. *Rev. Modern Phys.* 76 (1):195–214.

Feng, J. F., X. S. Wu, S. J. Xiong, and S. S. Jiang. 2006. Temperature dependence of transport behavior of a short DNA molecule. *Solid State Commun.* 139 (9):452–455.

Fink, H.-W., and C. Schönenberger. 1999. Electrical conduction through DNA molecules. *Nature* 398 (6726):407.

Fink, H. W. 2001. DNA and conducting electrons. *Cell. Mol. Life Sci.* 58 (1):1–3.

Fukada, E. 1995. Piezoelectricity of biopolymers. *Biorheology* 32 (6):593–609.

Fukada, E., and Y. Ando. 1972. Piezoelectricity in oriented DNA films. *J. Polym. Sci. A-2: Polym. Phys.* 10 (3):565–567.

Fukada, E., and Y. Ando. 1972. Piezoelectricity in oriented DNA films. *J. Polym. Sci. A-2: Polym. Phys.* 10 (3):565–567.

Fuller, W., T. Forsyth, and A. Mahendrasingam. 2004; Water-DNA interactions as studied by X-ray and neutron fibre diffraction. *Phil. Trans. R. Co. Lond. B* 359 (1448):1237–1248.

Furukawa, M., H. S. Kato, M. Taniguchi, T. Kawai, T. Hatsui, N. Kosugi, T. Yoshida, M. Aida, and M. Kawai. 2007. Electronic states of the DNA polynucleotides poly(dG)-poly(dC) in the presence of iodine. *Phys. Rev. B* 87 (4):045119.

Garcia, M. A., E. F. Pinel, J. de la Venta, A. Quesada, V. Bouzas, J. F. Fernandez, J. J. Romero, M. S. Martin-Gonzalez, and J. L. Costa-Kramer. 2009. Sources of experimental errors in the observation of nanoscale magnetism. *J. Appl. Phys.* 105 (1):013925.

Gardiner, W. C., H. M. Pickett, and M. H. Proffitt. 1981. Collisional linewidths of the Epr-spectrum of molecular-oxygen. *J. Chem. Phys.* 74 (11):6037–6043.

Genereux, J. C., and J. K. Barton. 2010. Mechanisms for DNA charge transport. *Chem. Rev.* 110 (3):1642–1662.

Goldstein, J. 2003. *Scanning electron microscopy and x-ray microanalysis.* 3rd ed. New York: Kluwer Academic/Plenum Publishers.

Guo, X., A. A. Gorodetsky, J. Hone, J. K. Barton, and C. Nuckolls. 2008. Conductivity of a single DNA duplex bridging a carbon nanotube gap. *Nat. Nanotech.* 3:163–167.

Gutierrez, R., S. Mandal, and G. Cuniberti. 2005. Dissipative effects in the electronic transport through DNA molecular wires. *Phys. Rev. B* 71 (23):235116.

Ha, D. H., H. Nham, K. H. Yoo, H. M. So, H. Y. Lee, and T. Kawai. 2002. Humidity effects on the conductance of the assembly of DNA molecules. *Chem. Phys. Lett.* 355 (5–6):405–409.

Hackl, E. V., S. V. Kornilova, L. E. Kapinos, V. V. Andrushchenko, V. L. Galkin, D. N. Grigoriev, and Y. P. Blagoi. 1997. Study of Ca^{2+}, Mn^{2+} and Cu^{2+} binding to DNA in solution by means of IR spectroscopy. *J. Mol. Struct.* 408:229–232.

Halliday, D., R. Resnick, and J. Walker. 2008. *Fundamentals of physics.* 8th ed. Hoboken, NJ: Wiley.

Hatcher, E., A. Balaeff, S. Keinan, R. Venkatramani, and D. N. Beratan. 2008. PNA versus DNA: Effects of structural fluctuations on electronic structure and hole-transport mechanisms. *J. Am. Chem. Soc.* 130 (35):11752–11761.

Hayakawa, R., and Y. Wada. 1973. Piezoelectricity and related properties of polymer films. *Adv. Polym. Sci.* 11:1–55.

Heisenberg, W. 1926. Multi-body problem and resonance in the quantum mechanics. *Zeitschrift Fur Physik* 38 (6/7):411–426.

Henderson, P. T., D. Jones, G. Hampikian, Y. Kan, and G. B. Schuster. 1999. Long-distance charge transport in duplex DNA: The phonon-assisted polaron-like hopping mechanism. *Proc. Natl. Acad. Sci. USA* 96 (15):8353–8358.

Hennig, D., E. B. Starikov, J. F. R. Archilla, and F. Palmero. 2004. Charge transport in poly(dG)-poly(dC) and poly(dA)-poly(dT) DNA polymers. *J. Biol. Phys.* 30 (3):227–238.

Hua, W. J., B. Gao, S. H. Li, H. Agren, and Y. Luo. 2010. Refinement of DNA structures through near-edge x-ray absorption fine structure analysis: Applications on guanine and cytosine nucleobases, nucleosides, and nucleotides. *J. Phys. Chem. B* 114 (41):13214–13222.

Hua, W. J., H. Yamane, B. Gao, J. Jiang, S. H. Li, H. S. Kato, M. Kawai, T. Hatsui, Y. Luo, N. Kosugi, and H. Agren. 2010. Systematic study of soft x-ray spectra of Poly(Dg)center dot Poly(Dc) and Poly(Da)center dot Poly(Dt) DNA duplexes. *J. Phys. Chem. B* 114 (20):7016–7021.

Hud, N. V., and K. H. Downing. 2001. Cryoelectron microscopy of lambda phage DNA condensates in vitreous ice: The fine structure of DNA toroids. *Proc. Natl. Acad. Sci. USA* 98 (26):14925–14930.

Hwang, J. S., K. J. Kong, D. Ahn, G. S. Lee, D. J. Ahn, and S. W. Hwang. 2002. Electrical transport through 60 base pairs of poly(dG)–poly(dC) DNA molecules. *Appl. Phys. Lett.* 81 (6):1134.

Ito, Y., and E. Fukusaki. 2004. DNA as a "nanomaterial." *J. Mol. Catal. B-Enzymatic* 28 (4–6):155–166.

Jian, P.-C. Jang, T.-F. Liu, C.-M. Tsai, M.-S. Tsai, and C.-C. Chang. 2008. Ni^{2+} doping DNA: A semiconducting biopolymer. *Nanotechnology* 19 (35):355703.

Jo, Y.-S., Y. Lee, and Y. Roh. 2003. Current–voltage characteristics of λ- and poly-DNA. *Mater. Sci. Engin. C* 23 (6–8):841–846.

Jurchescu, O. D., J. Baas, and T. T. M. Palstra. 2005. Electronic transport properties of pentacene single crystals upon exposure to air. *Appl. Phys. Lett.* 87:052102.

Kaneko, K., J. Suzuki, and C. Ishii. 1998. Low temperature magnetic properties of an O_2 and H_2O mixed molecular assembly, confined in a graphitic nanospace. *Chem. Phys. Lett.* 282 (2):176–180.

Kanvah, S., J. Joseph, G. B. Schuster, R. N. Barnett, C. L. Cleveland, and U. Landman. 2010. Oxidation of DNA: Damage to nucleobases. *Acc. Chem. Res.* 43 (2):280–287.

Kasumov, A. Y., D. V. Klinov, P.-E. Roche, S. Gueron, and H. Bouchiat. 2004. Thickness and low-temperature conductivity of DNA molecules *Appl. Phys. Lett.* 84 (6):1007.

Kats, E. I., and V. V. Lebedev. 2002. Base pair dynamic assisted charge transport in DNA. *JETP Lett.* 75 (1):37–40.

Keller, S. L., H. H. Strey, R. Podgornik, D. C. Rau, and V. A. Parsegian. 1996. Single DNA mesophases observed by electron and polarization microscopy. *Biophys. J.* 70 (2):Wp378–Wp378.

Kino, H., M. Tateno, M. Boero, J. A. Torres, T. Ohno, K. Terakura, and H. Fukuyama. 2004. Possible origin of carrier doping into DNA. *J. Phys. Soc. Jpn.* 73 (8):2089–2092.

Kleine-Ostmann, T., C. Jordens, K. Baaske, T. Weimann, M. H. de Angelis, and M. Koch. 2006. Conductivity of single-stranded and double-stranded deoxyribose nucleic acid under ambient conditions: The dominance of water. *Appl. Phys. Lett.* 88 (10):102102.

Klotsa, D., R. A. Romer, and M. S. Turner. 2005. Electronic transport in DNA. *Biophys. J.* 89 (4):2187–2198.

Kneuer, C., M. Sameti, U. Bakowsky, T. Schiestel, H. Schirra, H. Schmidt, and C.-M. Lehr. 2000. A nonviral DNA delivery system based on surface modified silica-nanoparticles can efficiently transfect cells in vitro. *Bioconj. Chem.* 11 (6):926–932.

Kon, H. 1973. Paramagnetic-resonance of molecular-oxygen in condensed phases. *J. Am. Chem. Soc.* 95 (4):1045–1049.

Kornilova, S. V., P. Miskovsky, A. Tomkova, L. E. Kapinos, E. V. Hackl, V. V. Andrushchenko, D. N. Grigoriev, and Y. P. Blagoi. 1997. Vibrational spectroscopic studies of the divalent metal ion effect on DNA structural transitions. *J. Mol. Struct.* 408:219–223.

Kummer, K., D. V. Vyalikh, G. Gavrila, A. B. Preobrajenski, A. Kick, M. Bonsch, M. Mertig, and S. L. Molodtsov. 2010. Electronic structure of genomic DNA: A photoemission and x-ray absorption study. *J. Phys. Chem. B* 114 (29):9645–9652.

Kutnjak, Z., G. Lahajnar, C. Filipic, R. Podgornik, L. Nordenskiold, N. Korolev, and A. Rupprecht. 2005. Electrical conduction in macroscopically oriented deoxyribonucleic and hyaluronic acid samples. *Phys. Rev. E* 71 (4):041901.

Kwon, Y.-W., E. D. Do, D. H. Choi, J.-I. Jin, C. H. Lee, J. S. Kang, and E.-K. Koh. 2008. Hydration effect on the intrinsic magnetism of natural deoxyribonucleic acid as studied by EMR spectroscopy and SQUID measurements. *Bull. Korean Chem. Soc.* 29 (6):1233–1242.

Kwon, Y.-W., C. H. Lee, D. H. Choi, and J.-I. Jin. 2009. Materials science of DNA. *J. Mater. Chem.* 19:1353–1380.

Kwon, Y.-W., C. H. Lee, E. D. Do, K. M. Jung, D. H. Choi, J.-I. Jin, and D. K. Oh. 2007. Photomagnetism of A-DNAs intercalated with photoresponsive molecules. *Mol. Cryst. Liq. Cryst.* 472:727–732.

Laudát, J., and F. Laudát. 1992. Dielectric study of charge motion in DNA. *Eur. Biophys. J. Biophy.* 21 (3):233–239.

Lee, C. H., E. D. Do, Y.-W. Kwon, D. H. Choi, J.-I. Jin, D. K. Oh, H. Nishide, and T. Kurata. 2006. Magnetic properties of natural and modified DNAs. *Nonlinear Opt. Quantum Opt.* 35 (1–3):165–174.

Lee, C. H., Y.-W. Kwon, D. H. Choi, Y. H. Geerts, E. Koh, and J.-I. Jin. 2010. High-temperature ferromagnetism of a discotic liquid crystal dilutely intercalated with iron(III) phthalocyanine. *Adv. Mater.* 22 (39):4405–4409.

Lee, C. H., Y.-W. Kwon, E.-D. Do, D. H. Choi, J.-I. Jin, D.-K. Oh, and J. Kim. 2006. Electron magnetic resonance and SQUID measurement study of natural A-DNA in dry state. *Phys. Rev. B* 73:224417.

Lee, H.-Y., H. Tanaka, Y. Otsuka, K.-H. Yoo, J.-O. Lee, and T. Kawai. 2002. Control of electrical conduction in DNA using oxygen hole doping. *Appl. Phys. Lett.* 80 (9):1670.

Lee, H. K., and M. H. C. Jin. 2010. Negative differential resistance in hydrated deoxyribonucleic acid thin films mediated by diffusion-limited water redox reactions. *Appl. Phys. Lett.* 97 (1):013306.

Lee, J. M., S. K. Ahn, K. S. Kim, Y. Lee, and Y. Roh. 2006. Comparison of electrical properties of M- and λ-DNA attached on the Au metal electrodes with nanogap. *Thin Solid Films* 515 (2):818–821.

Lith, D. van, J. M. Warman, M. P. de Haas, and A. Hummel. 1986. Electron migration in hydrated DNA and collagen at low temperatures. Part 1.—Effect of water concentration. *J. Chem. Soc. Farad. T. I* 82 (9):2933–2943.

Lepault, J., J. Dubochet, W. Baschong, and E. Kellenberger. 1987. Organization of double-stranded DNA in bacteriophages—a study by cryoelectron microscopy of vitrified samples. *Embo J.* 6 (5):1507–1512.

Long, Y.-T., C.-Z. Li, H.-B. Kraatz, and J. S. Lee. 2003. AC impedance spectroscopy of native DNA and M-DNA *Biophys. J.* 84:3218–3225.

MacNaughton, J. B., M. V. Yablonskikh, A. H. Hunt, E. Z. Kurmaev, J. S. Lee, S. D. Wettig, and A. Moewes. 2006. Solid versus solution: Examining the electronic structure of metallic DNA with soft x-ray spectroscopy. *Phys. Rev. B* 74 (12):125101.

MacNaughton, J., A. Moewes, and E. Z. Kurmaev. 2005. Electronic structure of the nucleobases. *J. Phys. Chem. B* 109 (16):7749–7757.

Maiya, B. G., and T. Ramasarma. 2001. DNA, a molecular wire or not—The debate continues. *Curr. Sci.* 80 (12):1523–1530.

Makarova, T. L. 2009. Nanomagnetism in otherwise nonmagnetic materials. arXiv:0904.1550.

Marincola, F. C., M. Casu, G. Saba, C. Manetti, and A. Lai. 2000. Interaction of divalent metal ions with DNA investigated by 23Na NMR relaxation. *Phys. Chem. Chem. Phys.* 2 (10):2425–2428.

Matsuo, Y., K. Sugita, and S. Ikehata. 2005. Doping effect for ionic conductivity in DNA film. *Synth. Met.* 154 (1–3):13–16.

Matta, C. F., N. Castillo, and R. J. Boyd. 2006. Extended weak bonding interactions in DNA: Pi-stacking (base-base), base-backbone, and backbone-backbone interactions. *J. Phys. Chem. B* 110 (1):563–578.

Mehrez, H., and M. P. Anantram. 2005. Interbase electronic coupling for transport through DNA. *Phys. Rev. B* 71 (11):115405.

Mehrez, H., S. Walch, and M. P. Anantram. 2005. Electronic properties of O_2-doped DNA. *Phys. Rev. B* 72 (3):035441.

Minot, E. D., Y. Yaish, V. Sazonova, and P. L. McEuen. 2004. Determination of electron orbital magnetic moments in carbon nanotubes. *Nature* 428 (6982):536–539.

Mizogushi, K., S. Tanaka, T. Ogawa, N. Shiobara, and H. Sakamoto. 2005. Magnetic study of the electronic states of B-DNA and M-DNA doped with metal ions. *Phys. Rev. B* 72 (3):033106.

Mizoguchi, K., S. Tanaka, and H. Sakamoto. 2006. Comment on "Intrinsic low temperature paramagnetism in B-DNA." *Phys. Rev. Lett.* 96:089801.

Muñoz, J., J Sponer, P. Hobza, M. Orozco, and F. J. Luque. 2001. Interactions of hydrated Mg^{2+} cation with bases, base pairs, and nucleotides. electron topology, natural bond orbital, electrostatic, and vibrational study. *J. Phys. Chem. B* 105 (25):6051–6060.

Muller, A., G. Hotz, and K. G. Zimmer. 1961. Electron spin resonances in bacteriophage: Alive, dead and irradiated. *Biochem. Biophys. Res. Commun.* 4 (3):214–217.

Nakamae, S., M. Cazayous, A. Sacuto, P. Monod, and H. Bouchiat. 2005. Intrinsic low temperature paramagnetism in B-DNA. *Phys. Rev. Lett.* 94 (24):248102.

Neubert, M., R. Bakule, and J. Nedbal. 1985. Temperature and humidity dependence of the dielectric spectrum of NaDNA in solid state. Paper read at 5th International Symposium on Electrets, at Heidelberg, Germany.

Nokhrin, S., M. Baru, and J. S. Lee. 2007. A field-effect transistor from M-DNA. *Nanotechnology* 18 (9):095205.

O'Neill, M. A., and J. K. Barton. 2004. DNA charge transport: Conformationally gated hopping through stacked domains. *J. Am. Chem. Soc.* 126 (37):11471–11483.

Odenthal, K. J., and J. J. Gooding. 2007. An introduction to electrochemical DNA biosensors. *Analyst* 132:603–610.

Okahata, Y., T. Kobayashi, K. Tanaka, and M. Shimomura. 1998. Anisotropic electric conductivity in an aligned DNA cast film. *J. Am. Chem. Soc.* 120 (24):6165–6166.

Omerzu, A., B. Anzelak, I. Turel, J. Strancar, A. Potocnik, D. Arcon, I. Arcon, D. Mihailovic, and H. Matsui. 2010. Strong correlations in highly electron-doped Zn(II)-DNA complexes. *Phys. Rev. Lett.* 104 (15):156804.

Otsuka, Y., H. Y. Lee, J. H. Gu, J. O. Lee, K. H. Yoo, H. Tanaka, H. Tabata, and T. Kawai. 2002. Influence of humidity on the electrical conductivity of synthesized DNA film on nanogap electrode. *Jpn. J. Appl. Phys.* 41 (2A):891–894.

Pablo, P. J. de, F. Moreno-Herrero, J. Colchero, J. Gómez Herrero, P. Herrero, A. M. Baró, Pablo Ordejón, José M. Soler, and Emilio Artacho. 2000. Absence of dc-conductivity in λ-DNA. *Phys. Rev. Lett.* 85 (23):4992–4995.

Pierre, D., ed. 2000. *Graphite and Precursors*. Amsterdam: Gordon and Breach Science Publishers.

Podgornik, R., H. H. Strey, K. Gawrisch, D. C. Rau, A. Rupprecht, and V. A. Parsegian. 1996. Bond orientational order, molecular motion, and free energy of high-density DNA mesophases. *Proc. Natl. Acad. Sci. USA* 93 (9):4261–4266.

Porath, D., A. Bezryadin, S. de Vries, and C. Dekker. 2000. Direct measurement of electrical transport through DNA molecules. *Nature* 403 (6770):635–638.

Priyadarshy, S., S. M. Risser, and D. N. Beratan. 1996. DNA is not a molecular wire: Protein-like electron-transfer predicted for an extended π-electron system. *J. Phys. Chem.* 100 (44):17678–17682.

Rajski, S. R., B. A. Jackson, and J. K. Barton. 2000. DNA repair: Models for damage and mismatch recognition. *Mutat. Res. Fundam. Mol. Mech. Mugag.* 447 (1):49–72.

Rakhmanova, S. V., and E. M. Conwell. 2001. Polaron motion in DNA. *J. Phys. Chem. B* 105 (10):2056–2061.

Rakitin, A., P. Aich, C. Papadopoulos, Y. Kobzar, A. S. Vedeneev, J. S. Lee, and J. M. Xu. 2001. Metallic conduction through engineered DNA: DNA nanoelectronic building blocks. *Phys. Rev. Lett.* 86 (16):3670–3673.

Ronne, C., L. Thrane, P. O. Astrand, A. Wallqvist, K. V. Mikkelsen, and S. R. Keiding. 1997. Investigation of the temperature dependence of dielectric relaxation in liquid water by THz reflection spectroscopy and molecular dynamics simulation. *J. Chem. Phys.* 107 (14):5319–5331.

Roy, S., H. Vedala, A. D. Roy, D. H. Kim, M. Doud, K. Mathee, H. K. Shin, N. Shimamoto, V. Prasad, and W. B. Choi. 2008. Direct electrical measurements on single-molecule genomic DNA using single-walled carbon nanotubes. *Nano Letters* 8 (1):26–30.

Schroder, D. K. 1998. *Semiconductor Material and Device Characterization.* 2nd ed. New York: Wiley.

Shapir, E., H. Cohen, A. Calzolari, C. Cavazzoni, D. A. Ryndyk, G. Cuniberti, A. Kotlyar, R. Di Felice, and D. Porath. 2008. Electronic structure of single DNA molecules resolved by transverse scanning tunnelling spectroscopy. *Nat. Mater.* 7 (1):68–74.

Shinwari, M. W., M. J. Deen, E. B. Starikov, and G. Cuniberti. 2010. Electrical conductance in biological molecules. *Adv. Funct. Mater.* 20 (12):1865–1883.

Shukla, L. I., A. Adhikary, R. Pazdro, D. Becker, and M. D. Sevilla. 2004. Formation of 8-oxo-7,8-dihydroguanine-radicals in γ-irradiated DNA by multiple one-electron oxidations. *Nucl. Acids Res.* 32 (22):6565–6574.

Skinner, R. J. S., J. S. Lee, Y. F. Hu, D. T. Jiang, P. Aich, S. Wettig, J. Maley, and R. Sammynaiken. 2005. Local structure of M-DNA at the nitrogen K-edge: Evidence towards a metal ion induced conduction band in DNA. *J. Nanosci. Nanotech.* 5 (9):1557–1560.

Sokolov, A. P., H. Grimm, and R. Kahn. 1999. Glassy dynamics in DNA: Ruled by water of hydration? *J. Chem. Phys.* 110 (14):7053–7057.

Starikov, E. B. 2003. Role of electron correlations in deoxyribonucleic acid duplexes: Is an extended Hubbard Hamiltonian a good model in this case? *Phil. Mag. Lett.* 83 (11):699–708.

Storm, A. J., J. van Noort, S. de Vries, and C. Dekker. 2001. Insulating behavior for DNA molecules between nanoelectrodes at the 100 nm length scale. *Appl. Phys. Lett.* 79 (23):3881–3883.

Tabata, H., L. T. Cai, J.-H. Gu, S. Tanaka, Y. Otsuka, Y. Sacho, M. Taniguchi, and T. Kawai. 2003. Toward the DNA electronics. *Synth. Met.* 133:469–472.

Tagami, K., M. Tsukada, Y. Wada, T. Iwasaki, and H. Nishide. 2003. Electronic transport of benzothiophene-based chiral molecular solenoids studied by theoretical simulations. *J. Chem. Phys.* 119 (14):7491–7497.

Takada, T., K. Kawai, M. Fujitsuka, and T. Majima. 2004. Direct observation of hole transfer through double-helical DNA over 100 Å. *Proc. Natl. Acad. Sci. USA* 101 (39):14002–14006.

Taniguchi, M., and T. Kawai. 2006. DNA electronics. *Physica E* 33 (1):1–12.

Taniguchi, M., Y. Otsuka, H. Tabata, and T. Kawai. 2003. Humidity dependence of electrical resistivity in Poly(dG)center dot Poly(dC) DNA thin film. *Jpn. J. Appl. Phys.* 42 (10):6629–6630.

Terawaki, A., Y. Otsuka, H. Y. Lee, T. Matsumoto, H. Tanaka, and T. Kawai. 2005. Conductance measurement of a DNA network in nanoscale by point contact current imaging atomic force microscopy. *Appl. Phys. Lett.* 86 (11):113901.

Tinkham, M., and M. W. P. Strandberg. 1955a. Interaction of molecular oxygen with a magnetic field. *Phys. Rev.* 97 (4):951–965.

Tinkham, M., and M. W. P. Strandberg. 1955b. Line breadths in the microwave magnetic resonance spectrum of oxygen. *Phys. Rev.* 99 (2):537–539.

Tinkham, M., and M. W. P. Strandberg. 1955c. Theory of the fine structure of the molecular oxygen ground state. *Phys. Rev.* 97 (4):937–951.

Tran, P., B. Alavi, and G. Gruner. 2000. Charge transport along the lambda-DNA double helix. *Phys. Rev. Lett.* 85 (7):1564–1567.

van Lith, D., J. M. Warman, P. P. de Haas, and A. Hummel. 1986. Electron migration in hydrated DNA and collagen at low temperatures. Part 1.—Effect of water concentration. *J. Chem. Soc. Farad. T. 1* 82 (9):2933–2943.

van Vleck, J. H. 1947. The absorption of microwaves by oxygen. *Phys. Rev.* 71 (7):413–424.

Ventra, M. D., and M. Zwolak. 2004. DNA electronics. In *Encyclopedia of Nanoscience and Nanotechnology*, edited by H. S. Nalwa. Los Angeles: American Scientific Publishers.

Voityuk, A. A., J. Jortner, M. Bixon, and N. Rösch. 2000. Energetics of hole transfer in DNA *Chem. Phys. Lett.* 324 (5–6):430–434.

Walsh Jr., W. M., R. G. Shulman, and R. D. Heidenriech. 1961. Ferromagnetic inclusions in nucleic acid samples. *Nature* 16 (4807):1041–1043.

Walker, F. A. 2003. Pulsed EPR and NMR Spectroscopy of paramagnetic iron porphyrinates and related iron macrocycles: How to understand patterns of spin delocalization and recognize macrocycle radicals. *Inorg. Chem.* 42 (15):4526–4544.

Wan, C., T. Fiebig, S. O. Kelley, C. R. Treadway, J. K. Barton, and A. H. Zewail. 1999. Femtosecond dynamics of DNA-mediated electron transfer. *Proc. Natl. Acad. Sci. USA* 96 (11):6014–6019.

Wang, J. 2008. Electrical conductivity of double stranded DNA measured with AC impedance spectroscopy. *Phys. Rev. B* 78 (24):245304.

Wang, X. F., and T. Chakraborty. 2006. Charge transfer via a two-strand superexchange bridge in DNA. *Phys. Rev. Lett.* 97 (10):106602.

Watson, J. D., and F. H. Crick. 1953. Molecular structure of nucleic acids; a structure for deoxyribose nucleic acid. *Nature* 171 (4356):737–738.

Wei, J. H., L. X. Wang, K. S. Chan, and Y. Yan. 2005. Trapping and hopping of bipolarons in DNA: Su-Schrieffer-Heeger model calculations. *Phys. Rev. B* 72 (6):064304.

Wettig, S. D., C.-Z. Li, Y.-T. Long, H.-B. Kraatz, and J. S. Lee. 2003a. M-DNA: A self-assembling molecular wire for nanoelectronics and biosensing. *Analyst. Sci.* 19 (1):23–26.

Wettig, S. D., C. Z. Li, Y. T. Long, H. B. Kraatz, and J. S. Lee. 2003b. M-DNA: A self-assembling molecular wire for nanoelectronics and biosensing. *Analyst. Sci.* 19 (1):23–26.

Wettig, S. D., D. O. Wood, P. Aich, and J. S. Lee. 2005. M-DNA: A novel metal ion complex of DNA studied by fluorescence techniques. *J. Inorg. Biochem.* 99 (11):2093–2101.

Wettig, S. D., D. O. Wood, and J. S. Lee. 2003b. Thermodynamic investigation of M-DNA: A novel metal ion–DNA complex. *J. Inorg. Biochem.* 94 (1–2):94–99.

Wood, D. O., and J. S. Lee. 2005. Investigation of pH-dependent DNA–metal ion interactions by surface plasmon resonance. *J. Inorg. Biochem.* 99 (2):566–574.

Xu, M. S., R. G. Endres, S. Tsukamoto, M. Kitamura, S. Ishida, and Y. Arakawa. 2005. Conformation and local environment dependent conductance of DNA molecules. *Small* 1 (12):1168–1172.

Xu, M. S., S. Tsukamoto, S. Ishida, M. Kitamura, Y. Arakawa, R. G. Endres, and M. Shimoda. 2005. Conductance of single thiolated poly(GC)-poly(GC) DNA molecules. *Appl. Phys. Lett.* 87:083902.

Yagi, M., S. Takemoto, and R. Sasase. 2004. Measurement of concentration of singlet molecular oxygen in the gas phase by electron paramagnetic resonance. *Chem. Lett.* 33 (2):152–153.

Yavin, E., A. K. Boal, E. D. A. Stemp, E. M. Boon, A. L. Livingston, V. L. O'Shea, S. S. David, and J. K. Barton. 2005. Protein-DNA charge transport: Redox activation of a DNA repair protein by guanine radical. *Proc. Natl. Acad. Sci. USA* 102 (10):3546–3551.

Yoo, K. H., D. H. Ha, J. O. Lee, J. W. Park, J. Kim, J. J. Kim, H. Y. Lee, T. Kawai, and H. Y. Choi. 2001. Electrical conduction through poly(dA)-poly(dT) and poly(dG)-poly(dC) DNA molecules. *Phys. Rev. Lett.* 87 (19):198102.

Zhang, Y., R. H. Austin, J. Kraeft, E. C. Cox, and N. P. Ong. 2002. Insulating behavior of λ-DNA on the micron scale. *Phys. Rev. Lett.* 89 (19):198102.

Zheng, L., J. P. Brody, and P. J. Burke. 2004. Electronic manipulation of DNA, proteins, and nanoparticles for potential circuit assembly. *Biosens. Bioelectron.* 20 (3):413–664.

Zheng, L. F., S. D. Li, P. J. Burke, and J. P. Brody. 2003. Towards single molecule manipulation with dielectrophoresis using nanoelectrodes. Paper read at 2003 Third IEEE Conference on Nanotechnology, IEEE-NANO 2003, at San Francisco, CA, U. S. A.

Zimmerman, R. L. 1976. Induced piezoelectricity in isotropic biomaterial. *Biophys. J.* 16 (12):1341–1348.

Zwolak, M., and M. Di Ventra. 2002. DNA spintronics. *Appl. Phys. Lett.* 81 (5):925–927.

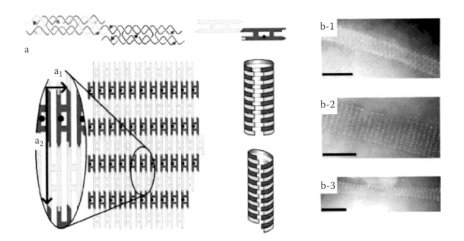

FIGURE 2.3 Assembly of chiral DNA nanotubes from two DAE-O tiles. Samples were protein labeled and negatively stained for an easy observation under TEM. (Reprinted with permission from Mitchell, J. C. et al. 2004. *J. Am. Chem. Soc.* 126: 16342–16343. Copyright 2004 American Chemical Society.)

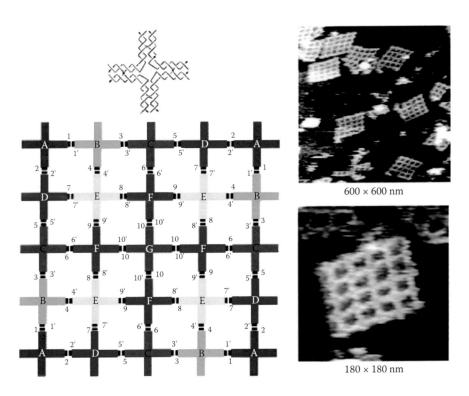

FIGURE 2.9 A 5×5 DNA array assembled from seven unique tiles (only sticky ends were unique) based on a fourfold rotational symmetry of the lattice. (Reprinted with permission from Liu, Y. et al. 2005. *J. Am. Chem. Soc.* 127: 17140–17141. Copyright 2005 American Chemical Society.)

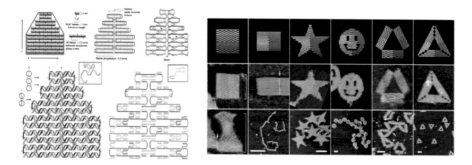

FIGURE 2.14 Scaffolded DNA origami capable of making accurate geometric shapes in two dimensions. Left panel: schematic drawings showing how a long DNA strand is folded into a defined shape with the help of short-fixing strands. Right panel: theoretical models and experimentally achieved geometrical shapes. (Reprinted by permission from Macmillan Publishers Ltd: [*Nature*] Rothemund, P. W. K. 2006. *Nature* 440: 297–302. Copyright 2006.)

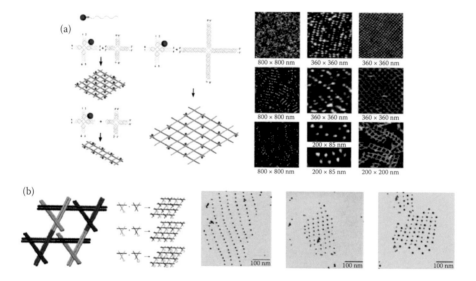

FIGURE 2.18 (a,b) DNA-programmed two-dimensional assembly of gold nanoparticles into various predetermined arrays; (b) enhanced structural control on a nanoparticle array was achieved by the use of a rigid triangular DNA motif with DX edges for the assembly. ([a] Reproduced with permission from Sharma, J. et al. 2006. *Angew. Chem. Int. Ed.* 45: 730–735. Copyright Wiley-VCH Verlag GmbH & Co, KGaA.. [b] Reprinted with permission from Zheng, J. W. et al. 2006. *Nano Lett.* 6: 1502–1504. Copyright 2006 American Chemical Society.)

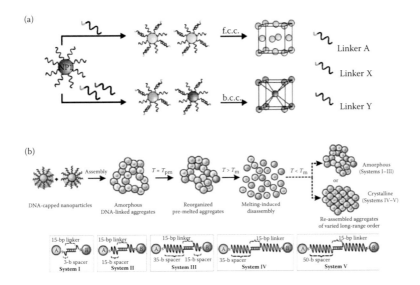

FIGURE 2.19 Schematic drawings showing the formation of three-dimensional gold nanoparticle crystals in two different crystalline states based on a single- and a binary-component system. (Copyright request pending, credit line to be added) ([a] Reprinted by permission from Macmillan Publishers Ltd: [*Nature*] Park, S. Y. et al. 2008. *Nature* 451: 553–556. Copyright 2008. [b] Reprinted by permission from Macmillan Publishers Ltd: [*Nature*] Nykypanchuk, D. et al. 2008. *Nature* 451: 549–552. Copyright 2008.)

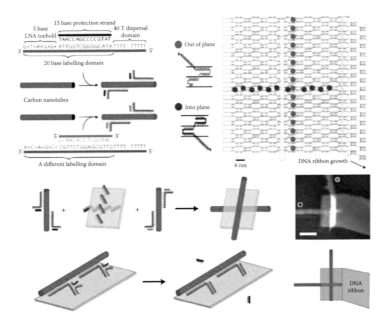

FIGURE 2.23 DNA origami-based self-assembly of two single-walled carbon nanotubes into a cross junction with a 90° orientation related to each other. Note that the two nanotubes were assembled on opposite faces of a DNA origami. (Reprinted by permission from Macmillan Publishers Ltd: [*Nature Nanotechnology*], Maune, H. T. et al. 2010. *Nat. Nanotechnol.* 5: 61–66. Copyright 2010.)

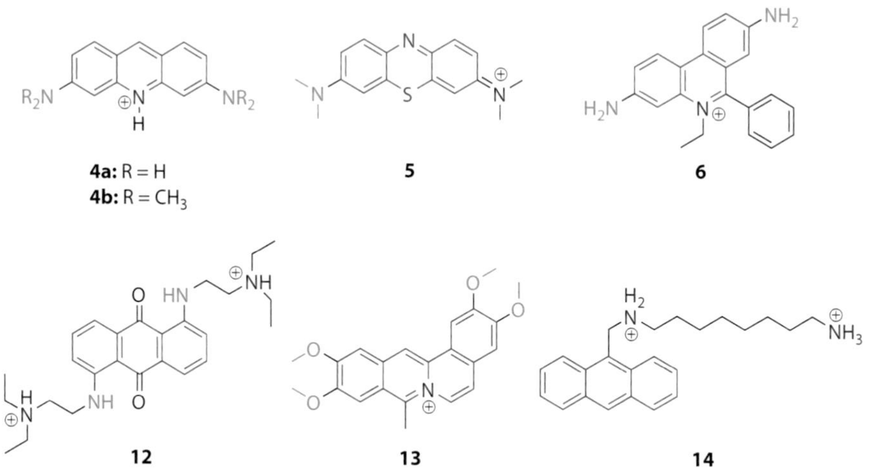

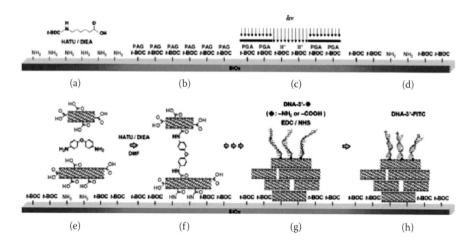

FIGURE 3.4 Structure of representative DNA intercalators.

FIGURE 4.5 Overall scheme for the fabrication of patterned SWNT multilayer films and the selective immobilization of DNA: (a) surface treatment with *t*-BOC (*tert*-butyloxycarbonyl) protecting group, (b) deposition of a layer of a PAG (photoacid generator), (c) UV exposure (405 nm, 25 mW) using a photomask for patterning, (d) development process, (e) selective immobilization of SWNTs onto the aminated regions of the substrate, (f) chemical attachment of additional SWNT layers using an ODA (4,4′-oxydianiline) linker molecule and a condensation agent (HATU, *O*-(7-azabenzotriazol-1-yl)-*N,N,N′,N′*-tetramethyluronium hexafluorophosphate), (g) covalent immobilization of oligonucleotide probes onto the patterned SWNT multilayer regions [●, functional groups (NH₂ or COOH) terminated to the oligonucleotide], and (h) hybridization of the FITC (fluorescein isothiocyanate)-labeled complementary oligonucleotide. (From Jung, D.–H., *Langmuir* 20, 8886, 2004. With permission.)

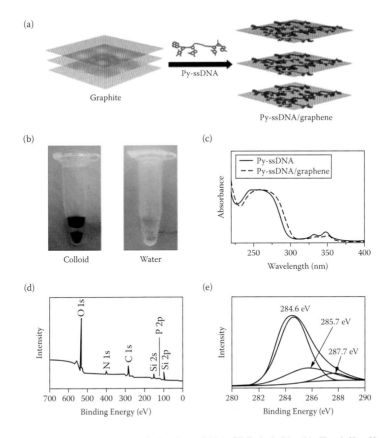

FIGURE 4.20 (a) Schematics of the Py-ssDNA-GRP hybrid. (b) Tyndall effect of a Py-ssDNA-GRP colloid. (c) UV-vis absorption spectra of Py-ssDNA and Py-ssDNA-GRP. XPS data for (d) Py-ssDNA-GRP hybrids and (e) core C 1s level. (From Liu, F., *Chem. Commun.*, 46, 2844, 2010. With permission.)

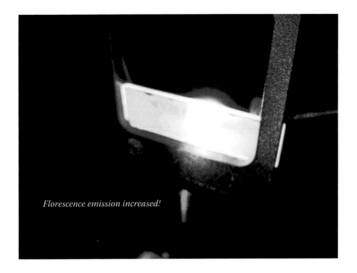

FIGURE 8.8 Fluorescence emission of DNA–CTMA film doped with Eu–FOD.

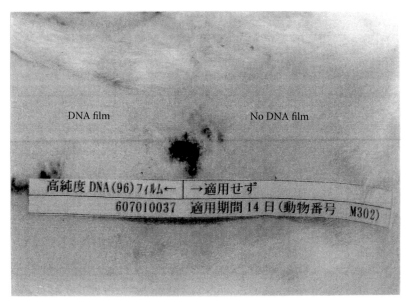

DNA film   No DNA film

高純度 DNA(96)フィルム←   │ →適用せず
          607010037   適用期間 14 日（動物番号   M302)

Wound of a rat skin after 14 days, patched by DNA film

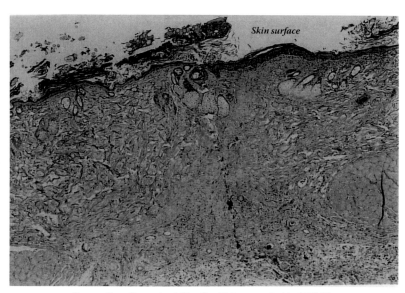

Skin surface

Cross section of wound rat skin after 14 days, patched by DNA film

FIGURE 8.22 Comparison of DNA-film-attached wounded skin and non-DNA-film-attached skin.

FIGURE 9.3 Two beams propagating through three-layer poled DNA-based waveguide. One beam is propagating under the electrode, and the other is next to the electrode. A second beam between the electrodes is also shown. (Reprinted with permission from E. Heckman, Ph.D Dissertation 2006.)

FIGURE 9.6 The probe station setup used for measuring the DNA-based transistors. (Reprinted with permission from C. Bartsch, Ph.D Dissertation 2007.)

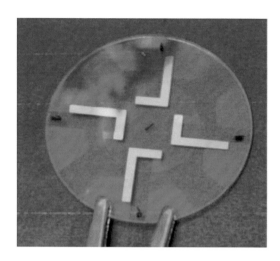

FIGURE 9.16 Photograph of a finished device. (Reprinted with permission from J. Hagen, Ph.D Dissertation 2006.)

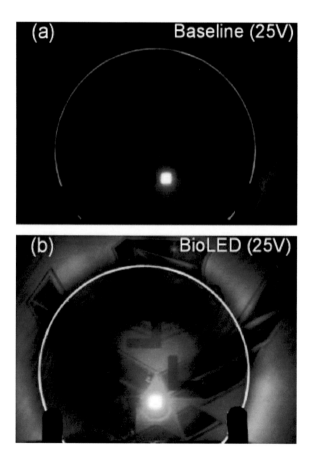

FIGURE 9.20 Photographs of green-emitting BioLED and baseline devices in operation. (Reprinted with permission from J. Hagen, Ph.D Dissertation 2006.)

6 DNA Ionic Liquid

Naomi Nishimura and Hiroyuki Ohno

CONTENTS

6.1 Introduction 163
6.2 DNA and Ionic Liquid Mixture 164
6.3 Ionic liquidized DNA-Inner Column.. 164
    6.3.1 Low-Molecular-Weight Model Compounds 165
    6.3.2 Ionic liquidized Bases in DNA 167
6.4 Ionic liquidized DNA-Outer Column 170
6.5 Conclusion 176
Acknowledgments.. 176
References.. 176

6.1 INTRODUCTION

Ionic liquids are composed only of ions, and they exist in liquid from over a wide temperature range with no vapor pressure. Ionic liquids, formerly called room temperature organic molten salts, behave as liquids in the absence of an organic solvent. Lowering the melting temperature of salts is not so difficult. For example, NaCl is in a molten state above 801°C, but when heteroaromatic ammonium cations are used, chloride salts show a melting temperature below 100°C (Bonhôte et al. 1996). After 2000, an increasing number of studies on the ionic liquids has been found in several scientific fields. Ion-conductive material is a typical example.

In the 1990s, ionic liquids had been recognized as excellent ion-conductive materials (Wilkes et al. 1992). Since these salts are liquid with a high ionic concentration at room temperature, they are exceptionally ion conductive (Ohno, 2005). Examples have recently been found in which ionic conductivity is as high as that of aqueous salt solutions (Xu et al. 2003). Ionic liquid mixed with a polymer matrix is one realization of a good solid polymer electrolyte (Ueki et al. 2008). There are many studies on this kind of composite with polar polymers such as Nafion®, PMMA, polypyrrole, and polyvinyl alcohol. To promote salt dissociation, polymers should be chosen to have high dielectric constants. Fluorination of polymers is also an effective method to induce microphase separation of aqueous-type ion-conductive fluid. Although mixtures of fluorinated polymer and ionic liquid show higher ionic conductivity, there is little interaction between these components and, accordingly, their phases are not so stable. Furthermore, the ionic conductivity of the polymer/ionic liquid mixed system was one order lower than that of pure ionic liquid.

DNA has been used here as a host polymer. DNA is composed of four nucleic acid bases, and these bases are heteroaromatic rings. Based on the ionic liquid formation

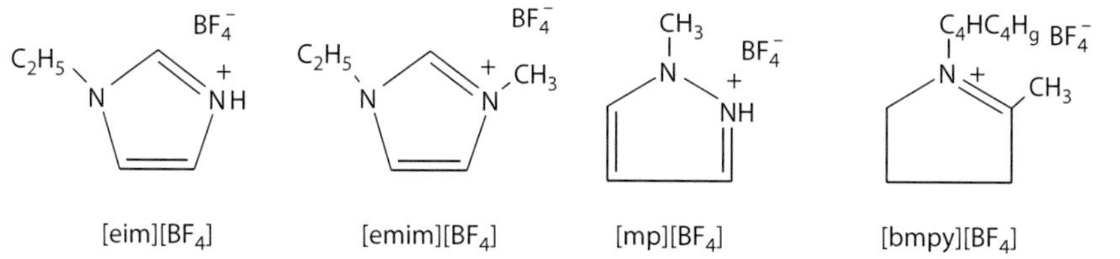

$[eim][BF_4]$ $[emim][BF_4]$ $[mp][BF_4]$ $[bmpy][BF_4]$

SCHEME 6.1 Structure of ionic liquids used in this study.

composed of these nucleic acid bases, after explaining a simple mixture of DNA and ionic liquids, direct ionic liquidization of DNA will be mentioned. DNA was ionic liquidized to prepare new ion-conductive materials. In the present chapter, the very unique application of DNAs for ion-conductive materials will be mentioned.

6.2   DNA AND IONIC LIQUID MIXTURE

DNA was used here as a host polymer. Since DNA has a number of heteroaromatic rings on the chain, it should have an excellent affinity with ionic liquids containing aromatic cations. Here, typical ionic liquids as seen in Scheme 6.1 have been used. First, preparation of a highly ion-conductive matrix by mixing ionic liquid with DNA is examined. Starting materials were amines having similar structure to nucleic acid bases or expected to have a high affinity with DNA. The mixtures of DNA and ionic liquid are evaluated here as ion conductors. Scheme 6.1 shows the structure of four kinds of ionic liquids.

The synthesis has been mentioned in a study (Ohno et al., 2001). All salts are obtained as liquids, and these are known to show a high ionic conductivity (Hirao et al., 2000). DNA was dissolved in 25 to 85 wt% of the aforementioned four kinds of ionic liquids in an aqueous solution. These solutions were then cast on a Teflon® plate and dried in vacuo. DNA containing up to 88.5 wt% ionic liquids formed films.

The ionic conductivity measurement has been carried out for these DNA containing 88.5 wt% ionic liquids and pure ionic liquids as references (Figure 6.1).

Though the ionic conductivity of ionic liquids is reduced considerably after mixing with DNA, those mixtures gave good mechanical properties as flexible films. To study the dependence of ionic conductivity on salt concentration, four kinds of ionic liquids were mixed with DNA in different proportions from 5 to 95 wt%. Ionic conductivity changes smoothly with the ionic liquid content. The salt concentration corresponding to the maximum ionic conductivity is different for each DNA/salt system. 1-Ethylimidazolium tetrafluoroborate ($[eim][BF_4]$), which is one of the best salts for high ionic conductivity and forming a flexible film, was added to DNA in a different mixing ratio. The highest ionic conductivity, 4.0×10^{-3} S cm^{-1} at 50°C, was observed when the DNA film was prepared containing 93 wt% $[eim][BF_4]$.

6.3   IONIC LIQUIDIZED DNA-INNER COLUMN

The typical cation structure of ionic liquids is a heteroaromatic ring such as imidazolium cations or pyridinium cations. Molecular orbital calculations suggested that

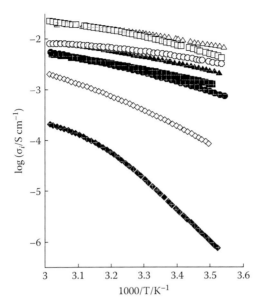

FIGURE 6.1 Effect of added ionic liquid species on the ionic conductivity of DNA films. ●: [emim][BF₄], ■: [eim][BF₄], ◆: [bmpy][BF₄], and ▲: [mp][BF₄]. Ionic liquid content: 88.5 wt%. Data for pure ionic liquids without DNA were also shown as a reference (corresponding open plots).

imidazole is a good starting material to prepare onium cations for excellent ionic liquids. Biosystems also contain many kinds of amines. In particular, nucleic acid bases composed of heteroaromatic rings promise excellent ion-conductive materials. In this section, nucleic acid bases are examined as starting materials for ionic liquids by means of neutralization or alkylation. The resulting ionic liquid can be regarded as a model for ionic liquidized DNA. DNA is then converted into ionic liquid. A methodology is set out for preparing flexible DNA films having a high ionic conductivity.

6.3.1 LOW-MOLECULAR-WEIGHT MODEL COMPOUNDS

There are two methods for preparing ionic liquids: (1) quaternization of the tertiary amines with alkyl halide followed by anion exchange, and (2) neutralization of bases with acids in pure water (Hirao et al., 2000). It is not easy to put a positive charge on nucleic acid bases by simple quaternization with an alkylating agent because the bases are heteroaromatic rings with a few labile hydrogen atoms. One alkylation method was used in a certain nitrogen atom of the nucleic acid base, and it was immediately followed by the release of proton to keep it neutral in spite of increase in unit volume and hydrophobicity. Therefore, we used the second method, neutralization.

Since DNA is soluble in an aqueous medium, ionic liquidized DNA can easily be prepared by the neutralization method in pure water. To select a suitable acid for the formation of ionic liquid, cytosine was neutralized as a model base with 11 different acids (HBF₄, HTFSI (bis(trifluoromethane)sulfonylimide), CF₃SO₃H, HBr, HClO₄,

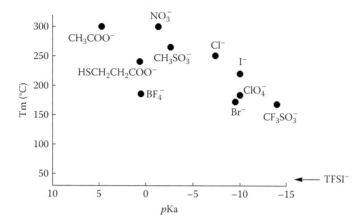

FIGURE 6.2 Relationship between T_m of cytosine after neutralization as a function of the pKa of the acids used.

HI, HCl CH$_3$SO$_3$H, HNO$_3$, CH$_3$COOH, and HSCH$_2$CH$_2$COOH). The obtained salts were analyzed with ^{1}H-NMR spectroscopy. Protons at the 5 and 6 positions of cytosine shifted to the lower magnetic field side depending on the degree of salt formation (Bonhôte et al. 1996). Though pure cytosine decomposed above 360°C without melt, the obtained salts showed the melting point (T_m). The T_m of cytosine salts thus prepared depends on the acid species used. For example, [C][TFSI] showed the lowest T_m, at 32.3°C. No other acid proved more effective in making the ionic liquid. Figure 6.2 summarizes the relation between T_m of the resulting salts and pKa of the acid used for neutralization. Since the pKa of HTFSI is not determined yet, the T_m of this salt was indicated with an arrow in Figure 6.2.

This figure clearly shows that strong acids such as HClO$_4$ and CF$_3$SO$_3$H lowered the T_m of these salts. The tendency seen in Figure 6.2 was similar to that for imidazolium-type ionic liquids, that is, stronger acid made salt with a lower T_m. However, ionic liquids from corresponding bases displayed a much higher T_m than imidazolium salts.

Four kinds of nucleic acid bases were neutralized by HBF$_4$ and HTFSI in pure water. TFSI$^-$ is an excellent anion for polymer electrolyte preparation (Martinelli et al., 2009), and it also yielded ionic liquids with a low T_m. All of these bases before neutralization are white powders, and they decompose above 320°C without melting (Budavari, 1996). Salts prepared by neutralization with HBF$_4$ were also obtained as white powder, but their T_m varied from 317.9°C to 185.9°C, depending strongly on the base species. Adenine neutralized with HTFSI ([A][TFSI]) had no melting point but showed glass transition temperature (T_g) at –13.2°C. Furthermore, the [C][TFSI] salt obtained by the neutralization of cytosine with HTFSI became an ionic liquid with T_m = 32.3°C. These should be classified as a new kind of ionic liquid prepared from the corresponding bases of the DNA. [A][TFSI] and [C][TFSI] were only slightly soluble in water due to hydrophobic property of the TFSI anion. The pK$_b$ of A, C, T, and G was 9.8, 9.4, 4.1, and 3.2, respectively (Budavari, 1996; Dawson et al., 1986), and these agree well with the experimental data such that both A and C can be neutralized with acid. No reaction occurred when thymine was mixed with

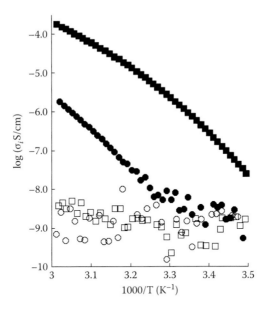

FIGURE 6.3 Temperature dependence of the ionic conductivity of neutralized bases. ●: [A][TFSI], ■: [C][TFSI], ○: [A][BF₄], and □: [C][BF₄].

acid. Guanine was gradually degraded by the acid, yielding a product having a lower T_m at 50°C. FT-IR and ¹H-NMR measurements suggested cleavage of purine rings. These results show that both adenine and cytosine are suitable for forming ionic liquids. Figure 6.3 shows the temperature dependence of the ionic conductivity of the corresponding bases after neutralization with HBF₄ or HTFSI. Both adenine and cytosine neutralized with HBF₄ show low ionic conductivity, around 10^{-9} S cm⁻¹ (in Figure 6.3; ○ and □, respectively). In contrast, [A][TFSI] and [C][TFSI] showed relatively high ionic conductivity; [C][TFSI], especially, showed 6.85×10^{-5} S cm⁻¹ at 50°C (Figure 6.3; ■). This high ionic conductivity is related to their low T_g, since [A][TFSI] and [C][TFSI] had T_g of –13.2°C and –30.8°C, respectively. These remarkable characteristics of the corresponding bases are the results of ionic liquid formation through neutralization with HTFSI.

6.3.2 Ionic Liquidized Bases in DNA

Then DNA was used to prepare DNA ionic liquid (Nishimura and Ohno, 2002). (DNA sodium salt isolated from salmon milt was a gift from Daiwa Kasei Co.) The molecular weight distribution was about 60,000 to 300,000. DNA 1.00 g was dissolved in 50 mL pure water and slowly stirred. Each acid (HBF₄, HTFSI, CF₃SO₃H, HBr, HClO₄, HI or HCl), of 50 mol% relative to the total amount of the bases, was added dropwise to the solution, then stirred at 0°C for 24 h. The samples were phase separated and collected by filtration. After washing with pure water, they were dried in vacuo for 4 d.

Since adenine and cytosine formed ionic liquids after neutralization, these bases in DNA should also form ionic liquids after acid treatment. DNA has many bases aligned in the double helix. If all of these bases are converted into ionic liquids, successive ionic liquid domains should be created along the chain. According to the results in Figure 6.2, HBF_4, HTFSI, CF_3SO_3H, HBr, $HClO_4$, HI, or HCl are strongly suggested to be effective acids for this purpose. DNA was then treated with these acids. The four kinds of bases were assumed to be contained equally in the DNA. Acids of 50 mol% to the total bases were mixed with DNA to neutralize all adenines and cytosines. After neutralization with the stated acids individually in pure water, all products were obtained as precipitates. It is known that purine rings are dissociated from DNA by strong acid treatment. There are some fears of dissociation of the purine rings in the present treatments; still we expected to prepare ionic liquid domain of adenine salts in DNA. The resulting samples were washed with water, removing unreacted acid, dissociated purine (salts), etc. In the case of the model reaction, degradation of guanine was detected after acid treatment. Guanosine-5′-phosphate disodium salt (GMP) was also neutralized with HTFSI. Since the guanine ring was confirmed to be unchanged by ^{1}H-NMR, guanine rings in the DNA were considered not to be cleaved by neutralization. The hydrogen bonding between complementary base pairs was broken upon neutralization and the hydrophobic bases turned outside the helix to make the whole chain more hydrophobic. CD and IR spectra strongly suggest that these DNAs lose their double-strand helical structure (Voet et al., 1963).

The ionic conductivity of the neutralized DNA was about 1×10^{-9} S cm^{-1} at room temperature. This can be explained by insufficient ionic liquid fraction. The weight fraction of all the bases in solid DNA is about 40 wt%. However, if all of the adenines and cytosines were supposed to yield ionic liquids, the corresponding domain fraction should be only about 20 wt%. A closely packed ion conduction path could not be formed with this low fraction.

The [C][TFSI], which had the lowest T_m of the neutralized bases, was then added to the neutralized DNA to assist in the construction of the continuous ion conduction path. [DNA][BF_4] 10 mg was mixed with 5 to 80 wt% of [C][TFSI] in 0.5 mL pure water to give a homogeneous mixture. All samples were stirred for 24 h at room temperature. [DNA][BF_4] 10 mg was dissolved in 1 mL pure water, and 5 to 96 wt% [eim][BF_4] was added to the solution. The same procedure was used to prepare the sample for conductivity and thermal response measurements.

The neutralized DNA and [C][TFSI] were mixed and then cast to prepare films. [DNA][TFSI] is an excellent matrix from the viewpoint of ionic conductivity, but it cannot be dissolved in any solvent, and we have no clear way to process it. Since [DNA][BF_4] was soluble in excess of water, it was used in further experiments. When the [C][TFSI] content in [DNA][BF_4] was less than 50 wt%, the ionic conductivity of the mixture was very low, around 1×10^{-8} S cm^{-1} at 50°C (Figure 6.4). It abruptly increased when more than 70 wt% [C][TFSI] was added, reaching 4.76×10^{-5} S cm^{-1} at 50°C when 80 wt% of [C][TFSI] was added to the [DNA][BF_4].

The mixture was, however, obtained as a flexible film even with 80 wt% [C][TFSI]. Since the ionic conductivity of [C][TFSI] in the bulk was only 6.85×10^{-5} S cm^{-1}, the [DNA][BF_4] and [C][TFSI] mixed film showed reasonable conductivities, suggesting the formation of successive ionic liquid phases. To improve the ionic conductivity of [DNA][BF_4] film, [DNA][BF_4] should be mixed with more conductive ionic liquid. For this purpose, ethylimidazolium tetrafluoroborate ([eim][BF_4])

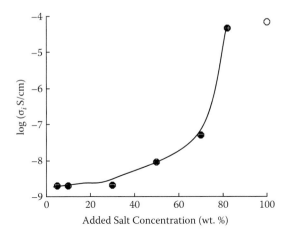

FIGURE 6.4   Effect of added ionic liquid concentration on the ionic conductivity of [DNA][BF$_4$]/ [C][TFSI] (●) and pure [C][TFSI] (○) at 50°C.

was added to the [DNA][BF$_4$]. The [eim][BF$_4$] shows very high ionic conductivity of around 10^{-2} S cm^{-1} at 50°C. Since both [DNA][BF$_4$] and [eim][BF$_4$] have the same BF$_4^-$ anion and are soluble in pure water, the [DNA][BF$_4$]/[eim][BF$_4$] mixture was conveniently prepared as a homogeneous film by casting. As shown in Figure 6.5, when the amount of the added [eim][BF$_4$] was up to 10 wt%, the ionic conductivity of the mixture was the same as that for pure [DNA][BF$_4$]. However, the ionic conductivity of the film containing 15 wt% [eim][BF$_4$] was about 4.62×10^{-7} S cm^{-1}, and that of the film containing 23.7 wt% [eim][BF$_4$] was 1.74×10^{-4} S cm^{-1} at 50°C. This excellent ionic conductivity was maintained up to 85 wt% addition of [eim][BF$_4$]. The highest ionic conductivity, 5.05×10^{-3} S cm^{-1} at 50°C, was observed when the film was prepared with 93 wt% [eim][BF$_4$]. Further addition of [eim][BF$_4$] maintained

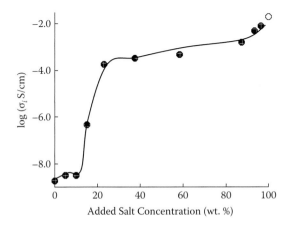

FIGURE 6.5   Effect of added ionic liquid concentration on the ionic conductivity of [DNA][BF$_4$]/[eim][BF$_4$] (●), and pure [eim][BF$_4$] (○) at 50°C.

high ionic conductivity, but stable film was no longer obtained. It is surprising that only 7 wt% DNA forms film to solidify 93 wt% ionic liquid. This is because of the high molecular weight of DNA, and the strong affinity of [eim][BF$_4$] to [DNA][BF$_4$]. We have already reported the preparation of excellent ion-conductive films from the mixture of native DNA and [eim][BF$_4$] (Ohno et al., 2001). There is no difference in the ionic conductivity between [DNA][BF$_4$]/[eim][BF$_4$] film and DNA/[eim][BF$_4$] film at high ionic liquid content (>70 wt%). However, [DNA][BF$_4$]/[eim][BF$_4$] film displayed much higher ionic conductivity even at low [eim][BF$_4$] content. This can be explained by the formation of an effective ionic liquid pathway in the DNA matrix due to neutralization of the bases. Further, [DNA][BF$_4$]/[eim][BF$_4$] film showed high ionic conductivity and excellent flexibility over a wide ionic liquid content.

[DNA][BF$_4$] was mixed with 40 wt% [eim][BF$_4$], and the ionic conductivity of the film was found to be 1.32×10^{-3} S cm^{-1} at room temperature. In other words, flexible film having a high ionic conductivity was obtained when [DNA][BF$_4$] was mixed with only a small amount of [eim][BF$_4$]. The experiments were carried out in a dry nitrogen atmosphere at room temperature.

6.4   IONIC LIQUIDIZED DNA-OUTER COLUMN

In the previous section, nucleic acid bases composed of heteroaromatic rings were used in the preparation of ionic liquids. Since the bases are located inside the double-stranded DNA, the DNA did not keep the double-stranded structure after converting the bases into an ionic liquid. In this chapter, the double-strand structure of DNA is not cleaved, yet an ionic liquid has been prepared with it. The phosphate groups on the outside helix are carrying negative charges. Ionic liquids are obtained by the reaction of the phosphate groups to form a continuous ion-conductive domain along with the DNA helix (Nishimura and Ohno, 2005).

We preliminarily confirmed that salts composed of phosphoric acid di-*n*-butyl ester (PDE) and alkylmethylimidazolium cation (c$_n$mim) showed a low T_g and high ionic conductivity. Ionic liquids were then successfully prepared with phosphate groups in DNA. (DNA sodium salt isolated from salmon milt was a gift from Daiwa Kasei Co.) The average molar weight was about 300,000. Imidazolium-type ionic liquids with an alkyl chain length (N1) of 2, 4, 8, and 12 were synthesized according to the method briefly mentioned in the following text.

c$_2$mim-DNA, c$_4$mim-DNA, and c$_8$mim-DNA: H-DNA (all sodium cations were neutralized to protons on the phosphate residues of DNA) was obtained from Na-DNA by cation exchange using Amberlite IR-120B H AG. The degree of proton exchange was determined by the titration to be 98%. The H-DNA was then neutralized with [c$_n$mim][OH] (n = 2, 4, or 8) to prepare c$_n$mim-DNA aqueous solution. The c$_n$mim-DNA powder was obtained from this solution after freeze drying.

c$_{12}$mim-DNA: Na-DNA and [c$_{12}$mim][OH] were slowly mixed in an aqueous solution. The precipitate was collected and washed with water. No precipitation of AgBr was detected when AgNO$_3$ was added in an aqueous solution, and the solution was then freeze dried.

The countercations of the phosphate group in DNA are exchanged into an imidazolium cation. c$_n$mim-DNA (n = 2, 4, and 8) was prepared from H-DNA. Since dodecyl imidazolium cation has a longer alkyl chain, c_{12}mim-DNA was insoluble in water. When Na-DNA was mixed with dodecylbromide in an aqueous solution, the resulting c_{12}mim-DNA was precipitated, and they were collected and dried. All imidazolium-DNA ($c_n$mim-DNA) was obtained as a white powder. A flexible film was obtained by casting a methanol solution of c_{12}mim-DNA (Figure 6.6 a). The film prepared by casting an aqueous solution of Na-DNA alone was transparent, but it was brittle.

When imidazolium cation was the countercation, flexible film was obtained. Surprisingly, DNA film became flexible only after the countercation was changed to imidazolium cation due to its low T_g. With FT-IR spectroscopy, a band was observed at 2360 cm^{-1}, which is characteristic of the C-H stretching vibrations. Furthermore,

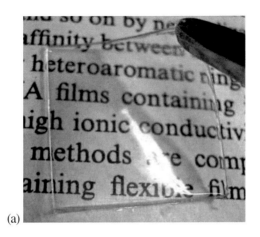

(a)

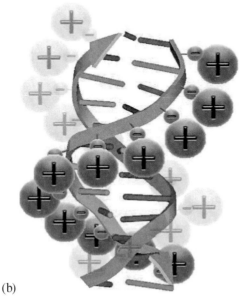

(b)

FIGURE 6.6 (a) Photograph of flexible and transparent c_{12}mim-DNA film. (b) Scheme of ionic liquidized DNA ($c_n$mim-DNA) keeping double-stranded structure.

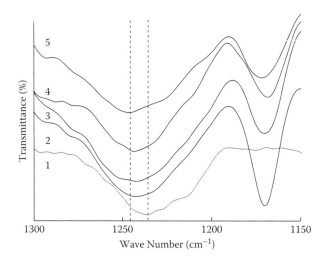

FIGURE 6.7 FT-IR spectra for $c_n$mim-DNA. **1**: Na-DNA, **2**: n = 2, **3**: n = 4, **4**: n = 8, and **5**: n = 12.

the band due to P=O stretching vibrations at 1230 cm^{-1} shifted to higher frequencies by increasing the cation size as shown in Figure 6.7. In addition, cation exchange was confirmed that $c_n$mim-DNA was soluble in methanol.

To confirm the double helical structure after exchange of countercations, Raman spectroscopy of $c_n$mim-DNA was performed. Generally, the single- or double-stranded helix of DNA is confirmed from the molar ellipticities of CD spectroscopy, supported also by UV spectroscopy. Since, in this study, dry DNA was used and discussed, the CD spectroscopy measurement in a solution does not serve the purpose. However, there is interest in $c_n$mim-DNA, which we synthesized here. The CD spectroscopy in an aqueous solution (Figure 6.8a), and in a methanol solution (Figure 6.8 b) are shown as a reference. The type-B DNA structure was, of course, seen in water, and type-A DNA structure was found in methanol the same as mentioned by Hanlon et al. (1975).

The confirmation analysis of DNA in a dry state has been carried out using SQUID (Lee et al., 2006) or FT-IR (Zhu et al., 2007). We chose Raman spectra for the dry-state conformational analysis of DNA. The three-dimensional structure of DNA was analyzed here by Raman spectroscopy as shown in Figure 6.9. The three-dimensional structure was confirmed from the intensity ratio (I_{1240}/I_{1094}) of the band at 1240 cm^{-1} attributed to thymine to the band at 1094 cm^{-1} attributed to the phosphate group (Erfurth et al., 1975). The double-stranded structure of DNA is implied by the I_{1240}/I_{1094} ratio of 0.3. If DNA becomes partly single stranded, the I_{1240}/I_{1094} ratio should be more than 0.3. As seen in Figure 6.9, all of $c_n$mim-DNAs show the I_{1240}/I_{1094} ratio of 0.3, indicating that all $c_n$mim-DNAs was confirmed to keep the double-stranded structure.

To find the suitable alkyl chain length for the highest ionic conductivity, the latter was measured for a series of $c_n$mim-DNAs. The ionic conductivity in the bulk was small at 10^{-9} S cm^{-1}. Although more than 98% of the phosphate groups of DNA were confirmed to be converted into an ionic liquid by titration, the observed ionic

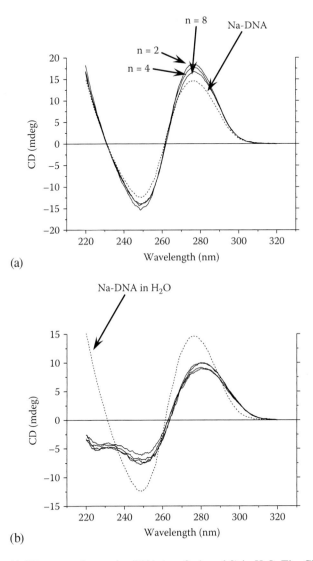

(a)

(b)

FIGURE 6.8 (a) CD spectra for $c_n$mim-DNA (n = 2, 4, and 8) in H_2O. The CD spectra for Na-DNA is also shown (dotted line) as a reference. (b) CD spectra for $c_n$mim-DNA (n = 2, 4, 8, and 12) in methanol. That for Na-DNA in H_2O is also shown as a reference.

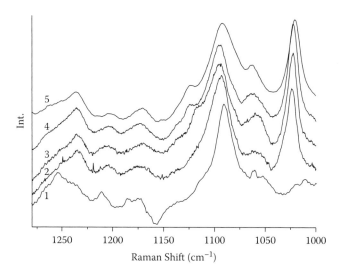

FIGURE 6.9   Raman spectra of $c_n$mim-DNA. 1: Na-DNA, 2: n = 2, 3: n = 4, 4: n = 8, and 5: n = 12.

conductivity was not high. The distance between phosphate groups between base pairs should be more than 0.34 nm considering the radius of the strand (1.0 nm) (Saenger, 1984). It seems to be difficult to conduct ions along with the DNA after considering the following few points such as that the diameter of ethylimidazolium cation is about 0.3 nm (Ue et al., 2002), the diameter of Na cation is 0.1 nm (Lide, 2009), and it is known that there are hydrogen bonds between the imidazolium cation and the anion (Gao et al., 2010). Imidazolium cations should be oriented outside the DNA. The bulk ionic conductivity of $c_n$mim-DNAs was very low, possibly because effective ion conduction did not take place owing to the larger distance between cations.

To improve ionic conductivity, 15 wt% of [eim][BF_4] was added to each $c_n$mim-DNA. These samples were obtained as transparent films. Ionic conductivity was increased up to 10^{-6} S cm^{-1} at 30°C as shown in Figure 6.10. The highest ionic conductivity was observed for $c_2$mim-DNA. It was found that the shorter the alkyl chain length, the higher the ionic conductivity. These are similar results in the case of low molecular-weight model compounds.

To find the salt concentration that optimizes the ionic conductivity of $c_2$mim-DNA, the latter was measured for the mixture of $c_2$mim-DNA and various concentrations of [eim][BF_4]. When [eim][BF_4] was mixed with $c_2$mim-DNA up to 15 wt%, the ionic conductivity of the mixture was similar to that of pure $c_2$mim-DNA. However, the ionic conductivity of film containing 20 wt% [eim][BF_4] was about 1.9×10^{-6} S cm^{-1}. Further addition of [eim][BF_4] continued the dramatic improvement in ionic conductivity. The observed ionic conductivity at about 20 wt% [eim][BF_4] added system was comparable to that of pure [eim][BF_4] as shown in Figure 6.11. Even after the 60 wt% [eim][BF_4] addition, the mixture was obtained as a flexible film.

When ethylimidazolium cation was the countercation of the phosphate anion, no continuous ion-conductive domain was formed because of the larger distance

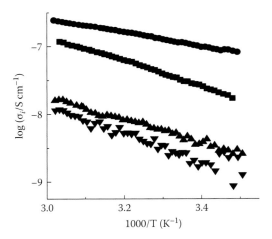

FIGURE 6.10 Temperature dependence on the ionic conductivity for $c_2$mim-DNA; ($\bullet$), $c_4$mim-DNA; ($\blacksquare$), $c_8$mim-DNA; ($\blacktriangle$), and c_{12}mim-DNA; ($\blacktriangledown$).

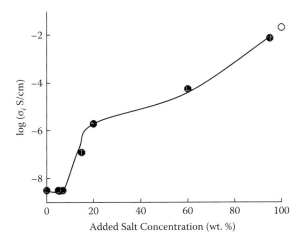

FIGURE 6.11 Effect of added ionic liquid concentration on the ionic conductivity of $c_2$mim-DNA / [eim][BF$_4$] ($\bullet$) at 50°C.

between cations. The [eim][BF$_4$] was added to $c_2$mim-DNA to generate continuous ion-conduction paths. When 20 wt% of [eim][BF$_4$] was added to $c_2$mim-DNA, the ionic conductivity, of the mixture certainly improved. The film consisting of $c_2$mim-DNA and [eim][BF$_4$] showed reasonable ionic conductivity implying the formation of the successive ionic liquid phase. Further addition of salt continued to improve the ionic conductivity. When 95 wt% of [eim][BF$_4$] was mixed with $c_2$mim-DNA, the mixture showed a phase separation into DNA and [eim][BF$_4$]. A $c_2$mim-DNA/[eim][BF$_4$] (less than 90 wt%) film showed excellent stability, and no leakage of ionic liquid was detected. No phase separation was observed during storage for over several months at room temperature.

6.5   CONCLUSION

From the viewpoint of ionic liquid science, DNA contains two interesting units such as nucleic acid bases and phosphate anions. Additionally, these are located along with the double-stranded structure. Both of these structures are useful in the preparation of ionic liquids. The random coiled ionic liquid polymer system is easily prepared by ionic liquidization of nucleic acid bases of DNA. They are effective in the preparation of flexible (and transparent) ion-conductive films. Phosphate anions were ionic liquidized to keep the double-helix structure of DNA. In both cases, sufficient ionic conductivity was not observed in spite of low glass transition temperature derived from the characteristic properties of ionic liquids. Ionic conductivity was, however, considerably improved by the addition of ionic liquids to these ionic-liquidized DNAs.

ACKNOWLEDGMENTS

The study was supported by a Grant-in-Aid for Scientific Research from the Japan Society of Promotion of Science. The authors also acknowledge Daiwa Kasei Co. for their kind donation of DNA samples.

REFERENCES

Bonhôte, P. et al. 1996. Hydrophobic, highly conductive ambient-temperature molten salts. *Inorg Chem* 35: 1168.

Budavari, S. Ed., 1996. *The Merck Index, 12th ed.* Whitehouse Station, NJ: Merck and Co.

Dawson, R. N. et al.1986. *Data for Biochemical Research, 3rd ed.* Clarendon Press, Oxford.

Erfurth, S. C. et al. 1975. Melting and premelting phenomenon in DNA by laser Raman scattering. *Biopolymers* 14: 247.

Gao, Y. et al. 2010. Probing electron density of H-bonding between cation–anion of imidazolium-based ionic liquids with different anions by vibrational spectroscopy. *J Phys Chem B* 114: 2828.

Hanlon S. et al. 1975. Structural transitions of deoxyribonucleic acid in aqueous electrolyte solutions. I. Reference spectra of conformational limits. *Biochemistry* 14: 1648.

Hirao, M., H. Sugimoto, and H. Ohno. 2000. Preparation of novel room temperature molten salts by neutralization of amines. *J Electrochem Soc* 147: 4168.

Lee, C. H. et al. 2006. Electron magnetic resonance and SQUID measurement study of natural A-DNA in dry state. *Physical Rev B* 73: 224417.

Lide, D. R. Ed., 2009. *CRC Handbook of Chemistry and Physics, 90th ed.* CRC Press, Boca Raton, FL.

Martinelli, A. et al. 2009. Phase behavior and ionic conductivity in lithium bis(trifluoromethanesulfonyl) imide-doped ionic liquids of the pyrrolidinium cation and bis(trifluoromethanesulfonyl) imide anion. *J Phys Chem B* 113: 11247.

Nishimura, N. and H. Ohno 2002. Design of successive ion conduction paths in DNA films with ionic liquids. *J Mater Chem* 12: 2299.

Nishimura, N. and H. Ohno 2005. DNA strands robed with ionic liquid moiety. *Biomaterials* 26: 5558.

Ohno, H. et al. 2001. Ion conductive characteristics of DNA film containing ionic liquids. *J Electrochem Soc* 148: E168.

Ohno, H. Ed., 2005. *Electrochemical Aspects of Ionic Liquids.* John Wiley & Sons, New York.

Saenger, W. Eds, 1984. *Principles of Nucleic Acid Structure*. Springer-Verlag, Berlin.

Ue, M. et al., 2002. A convenient method to estimate ion size for electrolyte materials design. *J Electrochem Soc* 149: A1385.

Ueki, T. and M. Watanabe. 2008. Macromolecules in ionic liquids: progress, challenges, and opportunities. *Macromolecules* 41: 3739.

Voet, D. et al., 1963. Absorption spectra of nucleotides, polynucleotides, and nucleic acids in the far ultraviolet. *Biopolymer* 1: 193.

Wilkes, J. S. et al. 1992. Air and water stable 1-ethyl-3-methylimidazolium based ionic liquids. *J Chem Soc Chem Commun* 966.

Xu, W. and C. A. Angell. 2003. Solvent-free electrolytes with aqueous solution-like conductivities. *Science* 302: 422.

Zhu, B. et al., 2007. Natural DNA mixed with Trehalose persists in B-form double-stranding even in the dry state. *J Phys Chem B* 111: 5542.

7 DNA-Surfactant Thin-Film Processing and Characterization

Emily M. Heckman, Carrie M. Bartsch,
Perry P. Yaney, Guru Subramanyam,
Fahima Ouchen, and James G. Grote

CONTENTS

7.1 Introduction 180
7.2 DNA Processing 180
    7.2.1 Molecular Weight 180
    7.2.2 Precipitation with CTMA Surfactant 181
    7.2.3 Preparation of DNA–CTMA Films.. 182
        7.2.3.1 Non-Cross-Linked DNA–CTMA Films............................ 182
        7.2.3.2 Cross-Linked DNA–CTMA Films.................................... 183
        7.2.3.3 DNA–CTMA–Chromophore Films 184
        7.2.3.4 DNA:PEDOT:CTMA.. 185
7.3 Material Characterization.. 185
    7.3.1 DNA–CTMA Structure 185
    7.3.2 Index of Refraction 187
    7.3.3 Optical Loss... 187
        7.3.3.1 Absorption Loss 187
        7.3.3.2 Propagation Loss.. 189
    7.3.4 Thermal Properties 189
7.4 RF Electrical Characterization.. 190
    7.4.1 Capacitive Test Structure 190
        7.4.1.1 Experimental Procedure 194
        7.4.1.2 Results and Analysis.. 196
        7.4.1.3 Capacitance Measurements 200
    7.4.2 Electric Force Microscopy... 201
7.5 DC Resistivity Studies..203
    7.5.1 Introduction 203
    7.5.2 Measurement Technique 205
    7.5.3 Data Analysis...208
    7.5.4 DATA..212
        7.5.4.1 DNA Compared to Nonbiopolymers 212

            7.5.4.2   Silk 215
            7.5.4.3   Effect of Humidity and Measurement Accuracy 216
            7.5.4.4   DNA with Conductive Dopants 218
            7.5.4.5   Discussion 220
      7.5.5   Summary 224
Acknowledgments 226
References 226

7.1   INTRODUCTION

The use of biopolymer-based materials, such as deoxyribonucleic acid (DNA), in organic electronic and photonic devices is rapidly becoming an area of interest in the photonics community. Compared to conventional polymer materials, biopolymers, either naturally occurring or artificially produced, can provide additional degrees of freedom in device design and produce enhancements in device performance. This is due to their unique electronic and optical properties. Additionally, biopolymers derived from DNA are a renewable resource that is non-fossil-fuel based and inherently biodegradable.

The DNA-surfactant biopolymer is one of the most studied biopolymers currently being used in photonic devices. However, the process required to transform raw, genomic DNA from a natural, biological source into an optical quality material is nontrivial. This chapter will focus on the processing that has been developed in the past several years to form the optical-quality DNA biopolymer from genomic salmon DNA, the various dopants integrated into the DNA biopolymer, and the relevant optical, electronic, and material characterization of the biopolymer for electronic and photonic devices (Grote et al. 2003, 2004a, 2004b, 2005, 2006; Heckman et al. 2004; Heckman et al. 2005; Heckman et al. 2005; Heckman et al. 2006; Heckman et al. 2006; Heckman 2006; Bartsch et al. 2007, 2006; Hagen et al. 2006; Ouchen et al. 2009; Yaney et al. 2007, 2008, 2009).

7.2   DNA PROCESSING

The DNA used in this research was purified by the Chitose Institute of Science and Technology (CIST) in Hokkaido, Japan, from salmon roe and milt sacs, waste products of the Japanese fishing industry, using an enzyme isolation process. The use of salmon spermatozoa as a rich source of DNA is not new; what is new is the ability to extract and purify the DNA on a scale of mass production (Zamenhof 1957). Although the exact purification details are proprietary, the general steps of the purification process are known, as this process is simply a variation on time-proven DNA extraction techniques that have been in use for decades. The DNA received from CIST is reported to have a purity of ~96% with a protein content of ~2%.

7.2.1   MOLECULAR WEIGHT

The molecular weight of the DNA provided by CIST was measured to be greater than 8000 kDa using agarose gel electrophoresis (Ausubel et al. 1999). This high-molecular-weight DNA is difficult to process into an optical waveguide quality film

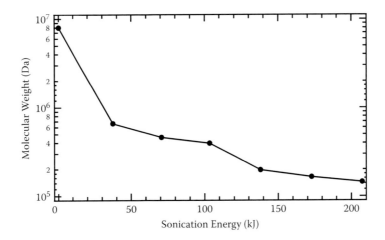

FIGURE 7.1   Molecular weight of DNA as a function of total sonication energy. The DNA was sonicated on ice in 10-s-long pulses with a 20 s rest period between pulses to prevent overheating of the sample. (Reprinted with permission from E. Heckman, Ph.D. Dissertation 2006.)

due to its high viscosity in solution. In addition, it is necessary to reduce the molecular weight of the DNA to at least 1000 kDa to achieve an appreciable EO coefficient through electrode poling. To reduce the molecular weight, an ultrasonic processor is used (Heckman et al. 2005, 2006). The Sonics & Materials ultrasonic processor model VC-750 with a 19-mm-diameter high-gain solid probe can process up to 500 mL of a solution and can reduce the mean molecular weight of the CIST DNA to as low as 200 kDa.

To begin with, 2 g of DNA is dissolved in 500 mL of 14 MΩ·cm distilled/deionized water at room temperature. The DNA is then sonicated on ice in 10 s pulses with a 20 s rest period between each pulse. Figure 7.1 shows the molecular weight as a function of total sonication energy. Sonication does not uniformly reduce the size of the DNA strands; rather, it randomly breaks the DNA into a Gaussian-like distribution of molecular weights. The molecular weight reported is the mean of this distribution of molecular weights with a variation of approximately ±50%. The molecular weight is measured using agarose gel electrophoresis with a 0.8% agarose gel. After sonication, the aqueous DNA solution is filtered through a nylon filter with a 0.45 µm pore size to remove non-DNA particles (such as carbon black) created during sonication.

7.2.2 Precipitation with CTMA Surfactant

The DNA is initially soluble only in aqueous solutions and does not dissolve in any organic solvent. It is precipitated with the cationic surfactant complex hexadecytl-trimethylammonium-chloride (CTMA) to make it water insoluble and to provide increased molecular stability (Zhang 2001, 2002). This is done through an ion exchange reaction. The CTMA surfactant was chosen over other surfactants for two reasons. First, the alkyl chain length of 16 for CTMA was ideal. Cationic surfactants with alkyl chains shorter than 16 might induce poor mechanical properties of the material, while surfactants with longer alkyl chains are water insoluble (which complicates the precipitation process, since it cannot be initially dissolved in water) and pose the risk of damaging the DNA double-helical structure (Zhang 2001). Second, the CTMA surfactant was commercially available. It was obtained from Fisher Scientific and used without further purification.

To precipitate the DNA with the CTMA cationic surfactant, the DNA is first dissolved at room temperature in 14 MΩ·cm distilled/deionized water at a ratio of 4 g/L using a magnetic stirrer. If necessary, the molecular weight of the DNA is reduced with the ultrasonic processor as previously described. An equal amount by weight of CTMA is likewise dissolved in 14 MΩ·cm distilled/deionized water, also at a concentration of 4 g/L. Using equal amounts of DNA and CTMA by weight ensures that there is one CTMA molecule for each DNA base pair.

The DNA solution is added dropwise to the CTMA solution with a burette. A white DNA–CTMA precipitate forms as the DNA is added to the CTMA. The solution is mixed for an additional 4 h at room temperature. The precipitate is removed by filtering the solution using a TX 609 Technicloth low-lint clean-room paper as a filter. Using this type of filter allows for the precipitate to be periodically squeezed within the filter paper to remove any unprecipitated CTMA. It also provides a more complete rinsing of the DNA–CTMA precipitate. During the filtering process, an additional 3 to 4 L of 14 MΩ·cm distilled/deionized water is poured through the filter to rinse the precipitate and to ensure that any CTMA that did not bind to the DNA is thoroughly rinsed away. When the water running through the filter comes out clear and there is no evidence of surfactant bubbles, it is assumed that the precipitate has been thoroughly rinsed. The precipitate is then collected, placed in a Teflon beaker, and dried in a vacuum oven overnight at 40°C. This excessive rinsing and periodic squeezing of the precipitate within the filter is essential to obtaining a pure DNA–CTMA sample with no unprecipitated CTMA mixed in. This method yields a white, powdery DNA–CTMA precipitate. Although squeezing the DNA–CTMA precipitate in the filter paper is the most effective way we have found to remove the excess CTMA, one drawback of this method is that filter fibers tend to stick to the precipitate. These fibers are removed by using a 0.2 μm syringe filter.

7.2.3 Preparation of DNA–CTMA Films

7.2.3.1 Non-Cross-Linked DNA–CTMA Films

The resulting DNA–CTMA compound is soluble in many of the alcohols, including butanol, methanol, ethanol, and isopropanol as well as an alcohol–chloroform blend. To spin-coat a thin film for photonics applications, butanol is the solvent of choice because its slow evaporation due to a high boiling point (116°C–118°C) ensures a smooth, uniform film during the spin-coating process. Butanol has a density of 0.81 g/mL and was purchased from Sigma-Aldrich and used without further purification. The amount of solvent used in dissolving the DNA–CTMA is dependent on the molecular weight of the starting DNA. Higher-molecular-weight (>1000 kDa) DNA forms a more viscous solution than lower-molecular-weight (<500 kDa) DNA. A greater amount of solvent is, therefore, required for higher-molecular-weight DNA.

TABLE 7.1

Film Thickness of Spin-Coated DNA-CTMA Films at 800 rpm for 10s

Molecular Weight of DNA (kDa)[a]	DNA–CTMA Concentration in Butanol (mM bp^{-1})	Film Thickness (μm)[b]
8000	54.2	5.0
8000	39.8	2.4
1300	61.6	3.0
1300	46.9	1.3
500	85.0	3.0
500	69.3	1.7
300	101.5	2.0
300	85.0	1.5
200	85.0	2.6

[a] Estimated uncertainty ±50%.
[b] Estimated uncertainty ±0.1 μm.

To make a thin film, DNA–CTMA is dissolved in butanol at a concentration appropriate for its molecular weight. The most common molecular weights used are 500 and 200 kDA. For these lower molecular weights, a solution of 110 mM bp^{-1} DNA–CTMA in butanol yields a film thickness of between 2 and 3 μm for spinning parameters of 800 rpm for 10 s. The solution is mixed in a sealed glass bottle, in a 60°C oven, using an ATR Rotamix for 6 h. Once completely dissolved, the solution is filtered through a 0.2 μm pore size syringe filter. Because DNA–CTMA solutions are more viscous at room temperature, the filtering takes place inside a 60°C oven using a New Era pump systems motorized syringe pump. The solution is left to sit overnight in a tightly capped container in the 60°C oven to allow any microbubbles induced by filtering to dissipate. It is then spin-coated on to a substrate at a speed of 800–1000 rpm for 10 s with a 5 s ramp. After spinning, the sample is cured in an 80°C oven for 1 h. Table 7.1 provides examples of measured film thickness as a function of DNA molecular weight and DNA–CTMA concentration in a solution. These data show that the film thickness is quite sensitive to the DNA–CTMA concentration for a given molecular weight and, as might be expected, higher concentrations are possible with the lower molecular weights.

7.2.3.2 Cross-Linked DNA–CTMA Films

The non-cross-linked DNA–CTMA films, while of high optical quality, are soft and scratch easily (a visible mark is left on the film when scratched by a fingernail). This makes them incompatible with more aggressive processing techniques, such as sawing, that are often required to fabricate a photonic device. One solution to this problem is to cross-link the DNA–CTMA films. The cross-linker used in this work is poly(phenyl isocyanate)-co-formaldehyde (PPIF) with a formula weight of 400 g/mol. It was purchased from Sigma-Aldrich and used without further purification. The resulting cross-linked DNA–CTMA films are significantly harder (the films show no mark when scratched by a fingernail) than the non-cross-linked films, allowing them to withstand more demanding fabrication procedures than non-cross-linked films due to their increased hardness. Cross-linked films are also resistant to a wider range of solvents, including butanol and other alcohols. This allows multilayer DNA-based structures to be fabricated.

It was found through systematic study that the lowest amount of PPIF that can be used and still cross-link the DNA–CTMA film is 10 wt% (36.1 mol%) PPIF with respect to DNA–CTMA. However, concentrations of up to 20 wt% (81.2 mol%) can be used. Beyond 20 wt%, optical losses are too high for photonics applications (>1 dB/cm). The addition of the cross-linker significantly increases the refractive index, and this must be taken into consideration when choosing the amount of cross-linker to use.

To prepare a cross-linked DNA–CTMA film, the DNA–CTMA is dissolved in butanol at a concentration appropriate for its molecular weight. The PPIF cross-linker is dissolved in butanol at a ratio of 1:4, PPIF:Butanol by weight. The DNA–CTMA–butanol solution is mixed in a 60°C oven for at least 6 h. The PPIF-butanol solution is not mixed using the Rotamix, but rather is left standing in the 60°C oven. After a few hours of standing in the oven, the PPIF is fully dissolved in the butanol. Once fully dissolved, the DNA–CTMA-butanol solution is added to the PPIF-butanol solution, and the resulting solution is mixed for an additional 2 h in the 60°C oven. The DNA–CTMA–PPIF solution is filtered through a 0.2 μm pore size syringe filter to remove any impurities and left to sit overnight in a 60°C oven to allow any microbubbles induced by filtering to dissipate. It is spin-coated onto a substrate using the same spin parameters as the non-cross-linked DNA–CTMA films. The substrates are baked in an 80°C oven for 5 min and then cured in a vacuum oven at 175°C for 15 min. After curing, the cross-linked films are resistant to butanol.

7.2.3.3 DNA–CTMA–Chromophore Films

For the core waveguide layer, a chromophore dye DR1 is added to the DNA–CTMA–PPIF solution. DR1 has a formula weight of 314.34 g/mol and was purchased from Sigma-Aldrich and used without further purification. Both cross-linked and non-cross-linked DNA–CTMA–chromophore films were made. For the core layer, an amount between 5 wt% (21.8 mol%) and 10 wt% (45.9 mol%) DR1 with respect to DNA–CTMA was used. The dye was dissolved separately in dioxane at a concentration of 24.6 mM. Dioxane has a density of 1.034 g/mL and was purchased from Sigma-Aldrich and used without further purification. For the lower concentration of chromophore dye, the lower amount of PPIF, 10 wt%, is suitable to achieve cross-linking; however, a longer curing time of 20 min at 175°C is required. For the higher concentration of chromophore dye, a higher amount of PPIF, 20 wt%, is required to achieve cross-linking.

To fabricate a DNA–CTMA–chromophore film, DNA–CTMA is dissolved in butanol as previously discussed and an appropriate amount of PPIF is dissolved separately in butanol at a concentration of 1:4, PPIF:Butanol by weight, by being allowed to stand in a 60°C oven for several hours. Once dissolved, the DNA–CTMA-butanol solution is added to the PPIF-butanol solution and mixed for an additional 2 h. The chromophore dye is dissolved separately in dioxane at a concentration of 24.6 mM.

The chromophore dye-dioxane solution is then added to the DNA–CTMA–PPIF solution and mixed for an additional hour. The solution is filtered through a 0.2 μm syringe filter and allowed to stand overnight. It is spin-coated and cured in a manner appropriate for the dye concentration, as previously discussed.

7.2.3.4 DNA:PEDOT:CTMA

A processing technique similar to that used to create DNA–CTMA is used to obtain DNA:PEDOT:CTMA (Hagen et al. 2007; Bartsch 2007). This biopolymer is formed from the raw DNA obtained from CIST, poly(3,4-ethylenedioxythiophene) poly(styrenesulfonate) (PEDOT:PSS), and CTMA. The aqueous dispersion of PEDOT:PSS is commercially known as Baytron P. Two grams of DNA are dissolved in 500 mL of deionized water to produce a concentration of 4 g per liter for sonication. This beaker of dissolved DNA is then placed into a bucket, and the beaker is surrounded by ice to prevent overheating. Then a Sonics & Materials ultrasonic processor model VC-750 with a 19-mm-diameter high gate solid probe is used in pulse mode for ten 9 min cycles to reduce the molecular weight of the DNA from 6000–8000 kDa to an average molecular weight of about 300 kDa. Each pulse consists of sonicating for 10 s and resting for 20 s. Each 9 min sonication cycle releases around 20 kJ of energy into the dissolved DNA. After sonication is complete, the dissolved DNA is filtered through a 0.65 μm nylon filter. Baytron P is purified before titration by filtering through a 0.45 μm polytetrafluoroethylene (PTFE) ringe filter. Then, 17.0 g is added dropwise to the dissolved DNA at a ratio of approximately 1:80 PEDOT molecules to DNA base pairs. This is equivalent to adding 36.0 wt% of PEDOT molecules into the DNA, or 89.9 wt% Baytron P into the DNA. The DNA:PEDOT solution is mixed with a magnetic stirrer at room temperature for 4 h. An excess amount of CTMA is dissolved in deionized water. The CTMA solution is then slowly poured into the DNA:PEDOT solution as it is stirred, keeping the stream of CTMA solution as small as possible. This results in the DNA:PEDOT:CTMA complex. The DNA:PEDOT:CTMA solution stirs for 2 h, and afterward the precipitate settles out of the water. Finally, the precipitate is thoroughly rinsed and dried overnight in a vacuum oven. This renders an organically soluble, semiconducting complex of DNA:PEDOT:CTMA.

7.3 MATERIAL CHARACTERIZATION

7.3.1 DNA–CTMA Structure

Circular dichroism (CD) and absorption measurements at 260 nm were used to determine if the double-helical structure of DNA is preserved in the complexed DNA–CTMA material. CD measurements confirmed that a DNA–CTMA solution does retain its double-helical structure at room temperature. Absorption measurements were used to determine the denaturation temperature.

The CD measurements were taken using a Jasco model J-720 spectropolarimeter. The DNA–CTMA is dissolved in butanol at a concentration of 26.0 mM. The absorption measurements are taken using a Cary 100 Bio UV-Visible Spectrophotometer. The liquid samples are placed into a quartz cuvette, and the temperature is controlled

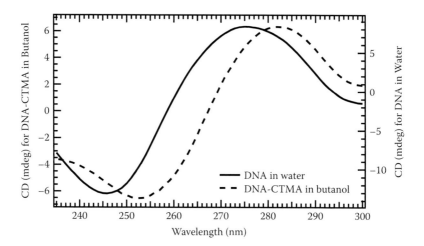

FIGURE 7.2 CD spectra of DNA in water and DNA–CTMA in butanol at room temperature. DNA–CTMA shows the peaks characteristic of DNA, although they are shifted slightly due to the addition of CTMA to the DNA chain. This suggests that the helix is intact in the DNA–CTMA butanol solution. The CD signal strength is different for each sample due to differences in concentration. (Reprinted with permission from E. Heckman, Ph.D. Dissertation 2006.)

with a Peltier device connected to Varian Bio-Melt software. The temperature is controlled by the software and is increased 5°C every 2 min, with a 30 s wait time at each temperature before each measurement. The sample concentrations were 50 μg/mL for aqueous DNA measurements and 34 mg/mL for DNA–CTMA in butanol.

A CD spectrum taken at room temperature of DNA–CTMA in butanol was shown to exhibit the same characteristics as a CD spectrum of aqueous DNA, although it is shifted slightly (Figure 7.2) (Rodger and Norden 1997; Tinoco et al. 1995). This shift is most likely due to the addition of the CTMA molecules on the DNA chain. This spectrum implies that DNA–CTMA retains the double-helical structure of DNA at room temperature.

An absorption scan at 260 nm as a function of temperature was used to determine the denaturation temperature of the complexed DNA–CTMA. In aqueous DNA, denaturation occurs at ~90°C–95°C (Kuchel and Ralston 1998). Figure 7.3 shows the melting curve at 260 nm for the DNA received from CIST for both aqueous DNA and complexed DNA–CTMA in butanol. The absorption spectrum of the aqueous DNA received from CIST shows the expected increase in absorption centered around 90°C, confirming its expected denaturation temperature. The absorption spectrum for DNA–CTMA dissolved in butanol does not, however, show a substantial increase in absorption even up to 100°C (the high-temperature limit for the instrument). This implies that the denaturing temperature for complexed DNA–CTMA is greater than 100°C.

The CD data confirm that the helix is preserved for the DNA–CTMA solution at room temperature, and the 260 nm absorption data indicate that it remains preserved for temperatures >100°C. It is unknown if the helix is intact for a cured DNA–CTMA

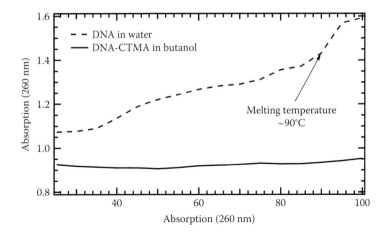

FIGURE 7.3 Melting curves of DNA in water and complexed DNA–CTMA in butanol. The denaturation temperature of DNA in water is ~90°C, and the denaturation temperature of DNA–CTMA is >100°C. (Reprinted with permission from E. Heckman, Ph.D. Dissertation 2006.)

film. However, because the non-cross-linked DNA–CTMA films are fully cured at 80°C, well below 100°C, it can be concluded that the double helix also remains intact in the cured, non-cross-linked DNA–CTMA films. Even though the cross-linked films are fully cured at temperatures greater than 100°C, the solvent is completely removed at 80°C. It is therefore reasonable to conclude that the double helix is also intact for cross-linked DNA–CTMA films. A more detailed structural analysis is needed to fully confirm this supposition.

7.3.2 INDEX OF REFRACTION

The index of refraction was measured with the Metricon 2010 Prism Coupler at three discrete wavelengths (632.8, 1152, and 1523 nm). A Cauchy fit to these data was used to plot the dispersion. The indices of refraction were measured for non-cross-linked DNA–CTMA for three different molecular weights (200, 500, and 5,000 kDa) and are shown with a Cauchy fit in Figure 7.4. Within experimental uncertainty, there is no appreciable difference in the indices as a function of molecular weight. The addition of the PPIF cross-linker raises the index of refraction as does the addition of a chromophore dye.

7.3.3 OPTICAL LOSS

7.3.3.1 Absorption Loss

A Hitachi model 4001 UV-VIS spectrophotometer was used to measure the optical absorption loss for a thick (>100 μm) non-cross-linked DNA–CTMA film over a wavelength range of 400–1600 nm. The thick DNA–CTMA free-standing film used for the transmission measurement was made by slowly evaporating a DNA–CTMA-

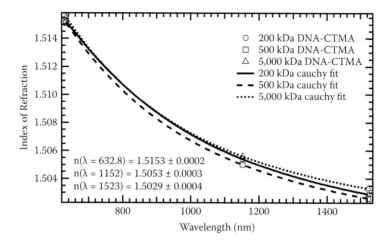

FIGURE 7.4 Index of refraction with Cauchy fit of DNA–CTMA for three different molecular weights. Within experimental uncertainty, there is no appreciable change in an index with molecular weight. (Reprinted with permission from E. Heckman, Ph.D. Dissertation 2006.)

butanol solution in a Teflon® mold in a 40°C vacuum oven. This technique could not be used to measure the absorption loss for a cross-linked DNA–CTMA film due to excessive cracking of the thick cross-linked films.

The absorption loss for a non-cross-linked DNA–CTMA film is shown in Figure 7.5. The film was measured to be 356 μm thick. The spectrum was shifted slightly to zero loss at 790 nm. The loss was found to be <1 dB/cm at the communications wavelengths: 0.1 dB/cm at 800 nm, 0.2 dB/cm at 1300 nm, and 0.7 dB/cm at 1550 nm. This shows that the conventional criterion that the loss for waveguide applications be ≤1 dB/cm at the communications wavelengths is met. The loss was

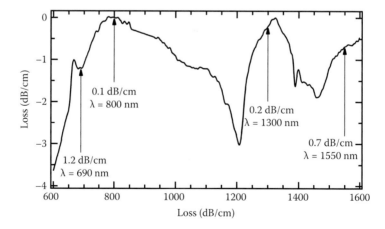

FIGURE 7.5 Absorption loss spectrum of a 356 μm thick film of non-cross-linked DNA–CTMA. The losses at the communications wavelengths are shown to be <1 dB/cm. (Reprinted with permission from E. Heckman, Ph.D. Dissertation 2006.)

TABLE 7.2

Waveguide Loss for Varying Concentrations of PPIF Cross-Linker in DNA–CTMA Films

Amount of PPIF Cross-Linker (wt%)	Amount of PPIF Cross-Linker (mol%)	Waveguide Loss (dB/cm)
0	0	1.25 ± 0.6%
10	36.1	1.48 ± 1.6%
15	57.3	1.79 ± 0.8%
20	81.2	3.08 ± 0.6%

also calculated at 690 nm because of the many measurements taken at this wavelength and was found to be 1.2 dB/cm.

7.3.3.2 Propagation Loss

Optical waveguide losses were measured at 690 nm for single-layer DNA–CTMA and DNA–CTMA–PPIF by coupling light into the films using the prism coupling method (Pollock 1994). The films act as the core layers of a one-dimensional waveguide with SiO_2 ($n = 1.46$ at 690 nm) and air ($n = 1$) acting as the two cladding layers. The streak propagating through the waveguide is captured with a Sony Cybershot DSC-P93 5.1M pixel digital camera. The length of the streak is calibrated by placing a ruler directly over the streak and recording a photo of the ruler with the camera in the exact position where the photo of the streak was recorded.

The optical waveguide loss was calculated for a non-cross-linked DNA–CTMA film and for three cross-linked DNA–CTMA films with varying concentrations of cross-linker: 10 wt%, 15 wt%, and 20 wt% PPIF. The losses for these films at 690 nm are shown in Table 7.2. The loss increases with increasing concentration of cross-linker. Although the losses reported are >1 dB/cm, these values are for measurements at 690 nm, where a higher loss is expected. At 690 nm, the cross-linked films show a loss increase of ~20% and ~40%, respectively, for the 10 wt% and 15 wt% PPIF films compared to the non-cross-linked film. Assuming the same percentage increase of loss at the communications wavelengths (where for the non-cross-linked film the loss was 0.2 dB/cm at 1300 nm and 0.7 dB/cm at 1550), the absorption and scattering losses in these materials should be sufficiently low at the communications wavelengths.

7.3.4 THERMAL PROPERTIES

Thermogravimetric analysis (TGA) and differential scanning calorimetry (DSC) were used to characterize the thermal properties of DNA in its various forms. The materials analyzed were (1) unprecipitated DNA (the raw material received from CIST), (2) DNA–CTMA powder, (3) a DNA–CTMA film, and (4) a cross-linked DNA–CTMA film. TGA was used to determine the thermal stability of the materials, and DSC was used to determine the glass-transition temperature T_g of the materials. The TGA curves for the four DNA materials are shown in Figure 7.6.

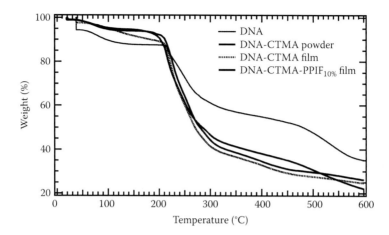

FIGURE 7.6 TGA curves for DNA specimens. There is no appreciable weight loss up to 200°C, indicating these materials are thermally stable up to this temperature. (Reprinted with permission from E. Heckman, Ph.D. Dissertation 2006.)

The materials show no sharp change in weight up to 200°C, suggesting that they are thermally stable and suitable for device operation up to this temperature. The DSC results are shown in Figure 7.7. A glass transition temperature of 148.17°C was detected for the DNA–CTMA powder; however, a clear T_g could not be detected for the DNA–CTMA film or for the other samples.

7.4 RF ELECTRICAL CHARACTERIZATION

7.4.1 CAPACITIVE TEST STRUCTURE

A new capacitive test structure is used to characterize the electrical properties of polymer and biopolymer thin films (Bartsch et al. 2006, 2007b). Specifically, this capacitive test structure provides a mechanism for determining the dielectric properties of polymer and biopolymer materials in an easily fabricated device. The dielectric properties obtained using the capacitive test structure are determined from microwave measurements.

The capacitive test structures used in these studies are fabricated on a 2-in. wafer of high resistivity (>10 kΩ) silicon using shadow masks and spin-coating. The capacitive test structure is shown in Figure 7.8. This structure is made up of four distinct layers on the silicon wafer. The adhesion layer consists of 100 Å of chromium sputtered through a shadow mask directly onto the silicon wafer. The bottom metal layer, which is formed from 7500 Å of gold deposited by electron beam (e-beam) deposition through the same shadow mask onto the adhesion layer, consists of two ground lines shunted together at their midpoints by a 100 μm wide conductor in the shape of an H. The next layer is the polymer under study, which is deposited as a thin film by spin-coating on top of the gold. The top layer is 3500 Å of gold, deposited by e-beam deposition through a second shadow mask. This layer forms a ground-signal-ground coplanar waveguide (CPW) transmission line. The CPW line consists of two ground

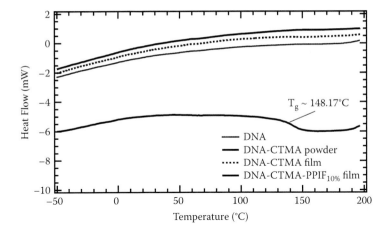

FIGURE 7.7 DSC curves for DNA specimens. The DNA–CTMA powder has a T_g of 148.17°C; however, a clear T_g could not be detected for the other materials. (Reprinted with permission from E. Heckman, Ph.D. Dissertation 2006.)

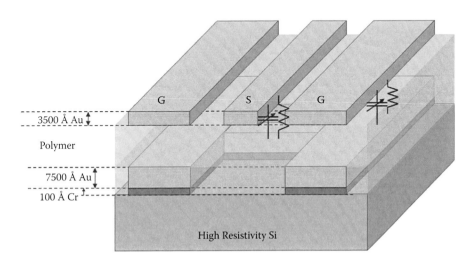

FIGURE 7.8 A three-dimensional representation of the capacitive test structure, showing the overlap of the signal conductor in the top metal and the shunt line in the bottom metal, which form the test capacitor. Note that the large ground pad capacitor is in a series with the test capacitor, resulting in the equivalent capacitance of the test capacitor. (Reprinted with permission from C. Bartsch, Ph.D. Dissertation 2007.)

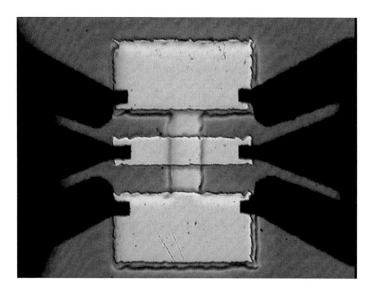

FIGURE 7.9 A photograph of a capacitive test structure with the CPW probes in place to make microwave measurements. (Reprinted with permission from C. Bartsch, Ph.D. Dissertation 2007.)

lines, positioned directly above the ground lines in the bottom metal layer, and a signal line, centered between the ground lines and perpendicular to the shunt line in the bottom metal layer. The active region of this test structure is the area where the signal line in the top metal layer overlaps the shunt line in the bottom metal layer. The active region forms a test capacitor, which has an area of approximately 100 μm by 100 μm. The test structure has a large ground pad capacitor that results from the overlap of the ground lines in the top and bottom metal. Since the test capacitor is in a series with the much larger ground pad capacitor, the effective capacitance is that of the test capacitor. When a dc bias is applied, the dc current passes through the leakage conductance of the test capacitor, the shunt line in the bottom metal, and the leakage conductance of the ground pad capacitor. This eliminates the need for via holes to ground the bottom conductor. Since the large ground pad capacitor has a much higher leakage conductance, the dc bias applied to the signal conductor drops almost entirely across the test capacitor. Figure 7.9 shows a photograph of the capacitive test structure being measured.

An electrical model of the capacitive test structure is shown in Figure 7.10. In this model, $C(V)$ is the test capacitance of the polymer in the active region, and $R(V)$ is the shunt resistance modeling the leakage conductance of the test capacitor. R_S and L are the parasitic series resistance and inductance, respectively, for the test capacitor. Using this electrical model and the scattering parameters (S parameters), the relative dielectric constant (ε_r) and loss tangent (tan δ) of the polymer can be derived as follows.

Since this system is a two-port reciprocal network, deriving the S parameters amounts to deriving S_{11} and S_{21}. For any reciprocal two-port network, S_{11} is identical to S_{22}, and S_{21} is identical to S_{12}. Any single S parameter, S_{ij}, is defined as

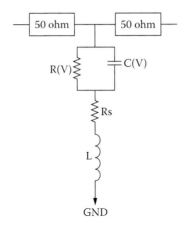

FIGURE 7.10 The electrical model for the capacitive test structure. (Reprinted with permission from C. Bartsch, Ph.D. Dissertation 2007.)

$$S_{ij} = \frac{V_i^-}{V_j^+}$$ (7.1)

where V_j^+ is the amplitude of the wave incident on port j and V_i^- is the amplitude of the wave reflected out of port i (Pozar 1998). This is determined when the only incident wave in the system is that incident upon the input port, port j, and all other ports are terminated with matched loads to avoid reflections. For the two-port network that describes the capacitive test structure, with port 1 as the input port and port 2 as the output port, it can be shown that

$$S_{11} = \frac{z_{in}^1 - z_0}{z_{in}^1 + z_0}$$ (7.2)

and

$$S_{21} = \frac{V_o}{V_{in}}$$ (7.3)

where Z_{in}^1 is the impedance seen looking into port 1, and Z_0 is the characteristic impedance of the transmission line. To algebraically calculate the S parameters for the capacitive test structure, one must account for the loss associated with the real transmission line. In this case, the loss associated with each 50 Ω transmission line is assumed to be 1 Ω. Using Figure 7.10 and the characteristic impedance $Z_0 = 50$ Ω, Z_{in} and the S parameters for the capacitive test structure can be computed as

$$S_{11} = \frac{2R + wj\omega L + 2\left(\dfrac{1}{R} + j\omega C\right)^{-1} - 2499}{102R + 102j\omega L + 102\left(\dfrac{1}{R} + j\omega C\right)^{-1} + 2601} \tag{7.4}$$

and

$$S_{21} = \frac{50\left[R + j\omega L + \left(\dfrac{1}{R} + j\omega C\right)^{-1}\right]}{51\left[51 + R + j\omega L + \left(\dfrac{1}{R} + j\omega C\right)^{-1}\right]} \tag{7.5}$$

The relative dielectric constant, ε_r, is determined from the capacitance in the electrical model. The capacitance is that of the parallel plate capacitor consisting of the polymer film in the active region. Manipulating the equation for the capacitance of a parallel plate capacitor, the relative dielectric constant can be computed as

$$\varepsilon_r = \frac{C(V)t}{\varepsilon_0 A} \tag{7.6}$$

where $C(V)$ is the capacitance, ε_0 is the permittivity of free space, A is the overlap area of the active region, and t is the thickness of the polymer. The loss tangent (Pozar 1998) is found from the electrical model of the capacitive test structure, using the equation for the shunt resistance such that (Ahamed and Subramanyam 2004)

$$\tan(\delta) = \frac{1}{\omega R(V)C(V)} \tag{7.7}$$

where ω is the angular frequency, $R(V)$ is the shunt resistance, and $C(V)$ is the capacitance shown in Figure 7.10 at a single bias voltage (Bartsch 2007).

7.4.1.1 Experimental Procedure

The specific values for the parameters in the electrical model are determined experimentally by fitting the circuit model to the experimentally obtained S parameters using the simulation package in Applied Wave Research's (AWR) Microwave Office® tools. A Hewlett-Packard 8720B Microwave Network Analyzer and an on-wafer microwave probe station, shown in Figure 7.11, are used to measure the S parameters of the capacitive test structure. The setup used to measure the S parameters is the same two-port network, with one port at each end of the signal line, as is described earlier for the theoretical calculations. The procedure for experimentally determining the S parameters in this study uses the following steps: First, the network analyzer and probe station are calibrated to the device over the frequency range of interest

FIGURE 7.11   The microwave probe station pictured with the right probe on a capacitive test structure containing BSA-PVA and the left probe hovering above the sample. (Reprinted with permission from C. Bartsch, Ph.D. Dissertation 2007.)

using a Line-Reflect-Reflect-Match (LRRM) calibration (Purroy and Pradell 2001). The frequency range of interest for this work is 5 to 20 GHz, when the data is taken at the Sensors Directorate of the Air Force Research Laboratory, since conductive thickness effects cause skin depth issues that limit the low frequency, and equipment limitations restrict the high frequency, and 10 to 18 GHz, when the data is taken at the University of Dayton, also due to equipment limitations. Then, the silicon wafer is raised to the temperature of interest and the appropriate dc bias voltage is applied to the signal lead of the probe. Finally, the S parameters are recorded.

The measured S parameters are then imported into the AWR Microwave Office® simulation package. Using the electrical model for the test structure, shown in Figure 7.10, the electrical parameters are determined. First, an initial estimate of the values in the electrical model of the capacitive test structure is made. Then, the electrical model is manually tuned to match the experimental results for both S11 and S21, simultaneously, across the frequency range. This procedure is repeated for each capacitive test structure at every bias voltage and temperature yielding the specific values for the electrical model. Since there are four parameters ($C(V)$, $R(V)$, R_S, and L) in the electrical model that are being varied to match the experimental data, using just the S_{21} or S_{12} curve can result in several matches that appear to be good. However, using both the S_{21} (or S_{12}) curve and the S_{11} (or S_{22}) curve generally produces a unique electrical model that matches both curves well. Figure 7.12 shows an example of the AWR Microwave Office screen while matching the electrical model to the experimental data. The electrical model is seen on the top right, with the variable tuner below it. The S_{12} graph containing both the experimental data and the theoretical data is seen on the top left and the S_{11} graph is visible on the bottom left (Bartsch et al. 2006, 2007).

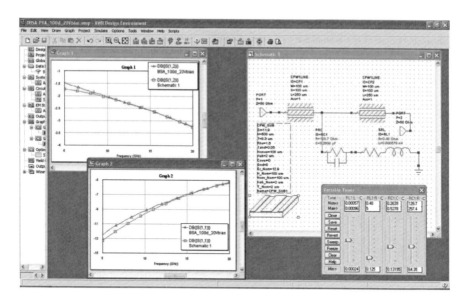

FIGURE 7.12 An example of the AWR Microwave Office screen while matching circuit parameters with the measured S parameters. The graphs of S_{11} and S_{12} with both experimental and theoretical data from the model are seen on the left side of this figure. The electrical model and the variable tuner are seen on the right side of this figure. (Reprinted with permission from C. Bartsch, Ph.D. Dissertation 2007.)

7.4.1.2 Results and Analysis

The polymer poly(bisphenol A carbonate-co-4,4'-(3,3,5-trimethylcyclohexylidene) diphenol carbonate), commonly known as APC, is used as a reference polymer for comparison to the DNA-biopolymer, DNA–CTMA. Variations in the S21 measurements are found to occur for different capacitive test structures on the same substrate. This is presumably due to thickness variations in the thin films, which cause different thicknesses in the active region of the various capacitive test structures containing the thin film. Since it is not possible to measure the thickness of the film for a particular test structure, the average film thickness is used to obtain the dielectric properties. The average thickness of the APC film is 2.03 μm and that of the DNA–CTMA film is 1.78 μm. The following subsections focus on identifying and analyzing the response of APC, DNA–CTMA, and a known ferroelectric material, barium strontium titanate (BST), under the various conditions measured.

7.4.1.2.1 APC

A typical example of S21 data obtained for the reference polymer, APC, is plotted in Figure 7.13. This figure shows that at room temperature the value of S21 decreases with increasing frequency, and at 100°C the value of S21 decreases with increasing frequency for frequencies above 6.4 GHz, while at frequencies below 6.4 GHz there is a brief decrease in S21 and then an increase. These lower-frequency, high-temperature effects on S21 are probably caused by the conductor thickness and skin depth effects occurring at slightly higher frequencies than expected, and therefore

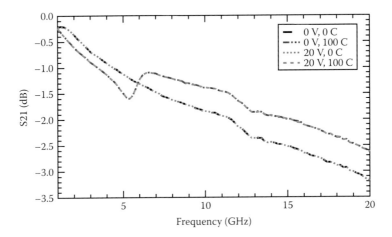

FIGURE 7.13 The frequency response of S21 for a capacitive test structure of APC. (Reprinted with permission from C. Bartsch, Ph.D. Dissertation 2007.)

the brief increase in S21 at 100°C should be neglected. From Figure 7.13, it is evident that changing the bias voltage applied to APC in the capacitive test structure does not affect its frequency response. Also, note that increasing the temperature of the polymer increases the transmittance of the signal for frequencies above 7 GHz.

7.4.1.2.2   DNA–CTMA

The frequency dependence of S21 for a capacitive test structure containing DNA–CTMA is plotted at room temperature with various applied bias voltages in Figure 7.14. This figure shows that for all applied biases measured, the transmittance decreases monotonically with increasing frequency. This means that more of

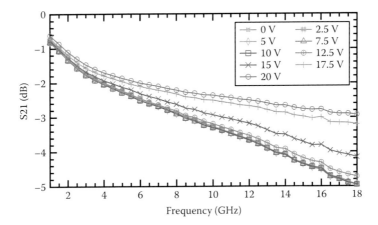

FIGURE 7.14 The frequency response of S21 for a capacitive test structure containing DNA–CTMA at room temperature for various applied biases from 0 to 20 V. (Reprinted with permission from C. Bartsch, Ph.D. Dissertation 2007.)

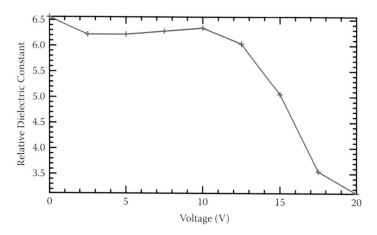

FIGURE 7.15   The calculated relative dielectric constant as a function of applied voltage for DNA–CTMA using the measured S parameters from a capacitive test structure. (Reprinted with permission from C. Bartsch, Ph.D. Dissertation 2007.)

the input signal put into port 1 of the capacitive test structure is measured at port 2 when the frequency is lower than is measured at port 2 when the frequency is higher. Figure 7.14 also shows that the transmittance of the capacitive test structure containing DNA–CTMA changes with applied bias. The transmittance is constant for applied biases up to 10.0 V. At 12.5 V, the transmittance increases slightly, with a greater increase taking place at 15.0 V. The largest increase in transmittance occurs between 15.0 and 17.5 V, and the increase in transmittance is again very slight between 17.5 and 20.0 V.

This change in transmittance corresponds to a decrease in the equivalent circuit capacitance and also in the relative dielectric constant of the DNA–CTMA. Figure 7.15 shows the calculated relative dielectric constant of DNA–CTMA as a function of applied voltage obtained from analyzing the S parameters in Figure 7.14 measured on a capacitive test structure containing DNA–CTMA. This figure shows that the relative dielectric constant decreases with increasing applied bias. When the slope is zero, the relative dielectric constant is a constant value. When the slope is nonzero, the relative dielectric constant is changing with applied voltage. The change in the relative dielectric constant is greatest when the slope is steepest. The relative dielectric constant of the DNA–CTMA film is 6.56 at 0.0 V and drops to a value of 3.12 at 20.0 V. This corresponds to 52% dielectric tuning as the applied electric field changes from 0 to 113 kV/cm.

To quantify if the measured tuning is statistically significant, a two-sided paired t-test is used to determine if there is a difference in the relative dielectric constant values for the DNA–CTMA film at 0.0 and 20.0 V. Using the paired t-test, $|t| = 2.15$, which is greater than $t_{0.05} = 1.80$. Therefore, it can be concluded at the 90% confidence level that there is a difference in the relative dielectric constants of the DNA–CTMA film at 0.0 and 20.0 V. These results from the two-sided paired t-test provide confidence in the analysis that the DNA–CTMA film tunes with applied voltage.

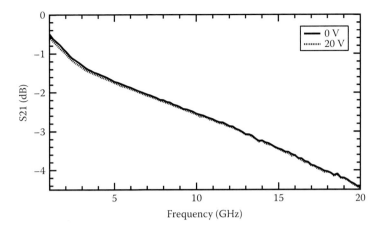

FIGURE 7.16 The frequency response of S21 for a capacitive test structure containing DNA–CTMA at 100°C for various applied biases from 0 to 20 V. (Reprinted with permission from C. Bartsch, Ph.D. Dissertation 2007.)

Figure 7.16 shows the frequency response of S21 for a capacitive test structure containing DNA–CTMA at 100°C with various applied biases. This figure shows that the transmittance decreases with increasing frequency for both 0 and 20 V at 100°C. Figure 7.16 also shows that S21 does not change with applied bias for DNA–CTMA at 100°C. This means that while the relative dielectric constant of DNA–CTMA tunes at room temperature, as the temperature increases it does not continue to tune. The reason why the dielectric tunability occurs at room temperature but not at 100°C is unknown. One possible explanation, however, is that the thin film of DNA–CTMA has a glass transition temperature, T_g, between room temperature and 100°C. The T_g of DNA–CTMA thin films is unknown because it is undetectable by differential scanning calorimetry, although the T_g of the DNA–CTMA powder has been measured to be 148°C (Heckman 2006). If the DNA–CTMA thin film has passed the transition temperature by 100°C, then the DNA–CTMA thin film is softer and more flexible at 100°C than it is at room temperature. This change might explain the differences in the transmittance measurements at room temperature and 100°C for this biopolymer (Bartsch 2007a; Bartsch et al. 2007b).

7.4.1.2.3 Barium Strontium Titanate (BST)

The dielectric properties of the tunable biopolymers described earlier should be compared to a known inorganic material with voltage tunable dielectric properties, such as barium strontium titanate (BST). The frequency response of S21 for a capacitive test structure containing 300 nm of BST is plotted in Figure 7.17 with applied biases varying from 0.0 to 3.0 V. This figure shows that the transmittance decreases with increasing frequency at each of these applied biases. The figure also shows that the transmittance increases with increasing applied bias, and confirms that the dielectric properties of BST tune with applied bias. The increasing transmittance corresponds to a decreasing capacitance and, therefore, a decreasing relative dielectric constant. The relative dielectric constant tunes by 47% as the applied voltage changes from

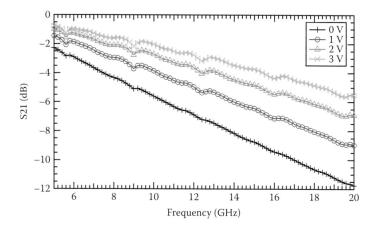

FIGURE 7.17 The frequency response of S21 for a capacitive test structure containing BST for applied biases from 0 to 3 V. (Reprinted with permission from C. Bartsch, Ph.D. Dissertation 2007.)

0.0 to 3.0 V, which for the 300-nm-thick film of BST corresponds to a change in the applied electric field from 0 to 100 kV/cm at room temperature.

It is useful to note that the DNA–CTMA film tunes more than the BST film at room temperature. The DNA–CTMA films are unoptimized and are already showing tuning on the order of that seen in the BST film. This suggests that DNA biopolymers can be used in applications that utilize dielectric tunability, such as sensors and antennas.

7.4.1.3 Capacitance Measurements

Capacitance is measured as a function of frequency to confirm that the electrical properties obtained from the microwave measurements are those of the polymer. This can be done by showing that the capacitance in the electrical model is that of the polymer and is not due to interface charges. An on-wafer probe station with micromanipulators and a HP 4284A Precision LCR (inductance-capacitance-resistance) meter are used to perform low-frequency characterization on thin films using the capacitive test structure. The LCR meter has the ability to vary the voltage from −40 V to 40 V, and the frequency from 20 Hz to 1 MHz, and to measure current. Using this equipment, the capacitance and dissipation factors can be obtained as a function of either frequency or bias voltage.

One example of the low-frequency capacitance measurements obtained as a function of frequency for a DNA–CTMA capacitive test structure is plotted in Figure 7.18. This figure shows that the capacitance is very high at low frequencies and decreases to the bulk capacitance with increasing frequency. The high capacitance present at very low frequencies is due to interface charges that are quickly depleted as the frequency increases. The rate of depletion slows down as the frequency increases and therefore equilibrium is approached at the higher-frequency values. The equilibrium capacitance value is that of the bulk capacitance and is the same as that obtained from the microwave frequency measurements. Other capacitive test structures have

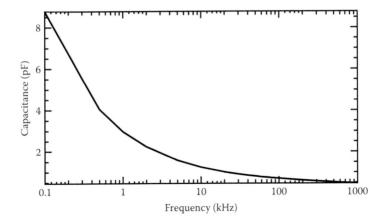

FIGURE 7.18 Capacitance as a function of frequency on a log scale for one capacitive test structure made with a DNA–CTMA thin film. (Reprinted with permission from C. Bartsch, Ph.D. Dissertation 2007.)

been measured in this same manner and produce similar results. This verifies that the capacitance values, and therefore the relative dielectric constant and loss tangents obtained from the microwave characterization measurements described earlier, are for the bulk polymer and are not due to the polymer–metal interfaces in the capacitive test structure (Bartsch 2007).

7.4.2 ELECTRIC FORCE MICROSCOPY

Electric force microscopy (EFM) images are taken to examine the mechanisms of the dielectric tunability observed for the capacitive test structure containing DNA–CTMA (Subramanyam et al. 2009; Bartsch 2007). These measurements map the surface topography of the test structure and show the effect of a constant dc bias applied to the DNA–CTMA film by measuring the change in the electric field gradient across the region of interest. Changing the applied dc bias will confirm the dielectric tunability if large changes in the EFM images are observed.

Atomic force microscopy (AFM) produces high-resolution images of the surface topography. To obtain these topography images, a sharp tip on the edge of a flexible cantilever is scanned across the surface of the sample while the system maintains a small and constant force between the sample and the tip. The tip scans the surface in a raster pattern. The deflection of the tip is detected by a laser beam reflecting off the back of the cantilever into a photodetector, where differences in the output voltage correspond to changes in the cantilever deflection (Russell et al. 2004).

In this research, TappingMode AFM is used, in conjunction with EFM and Phase Imaging, using a Veeco Instruments MultiMode Scanning Probe Microscope equipped with a NanoScope IIIa Controller and Quadrex Phase-Detection (Prater et al. 2004).

TappingMode AFM refers to the mode of operation in which the AFM tip alternately touches the surface to make high-resolution images and is lifted off the surface to move between imaging locations, to avoid damaging the surface of the sample. In this mode, the AFM cantilever oscillates at its resonance frequency to lightly tap the surface with the AFM tip while imaging. The deflection of the laser beam is used to keep the tip a constant distance from the sample by moving the scanner vertically at every data point. The topographical image of the surface is formed from a record of the scanner's movement.

EFM maps the gradient of the electric field present between the tip and the sample at every point. The EFM images are generated by plotting the change in the phase, frequency, or amplitude of the cantilever oscillation as it encounters vertical repulsive and attractive electric field forces above the sample surface at each point (Serry et al. 2004). The change in phase or frequency detection is more commonly used than amplitude detection, as these methods result in better images. A dc bias voltage is applied to the sample through a conductive platinum-iridium coated AFM tip. The field variations due to trapped charges are mapped using EFM. This study images the electric field present between the sample and AFM tip by phase detection, and relies on the vertical force gradients from the sample, causing changes in the resonance frequency of the cantilever. Attractive forces pull the AFM tip closer to the sample, thereby reducing the resonance frequency and increasing the phase component of the oscillating cantilever EFM.

Phase Imaging maps the phase of the cantilever oscillations during TappingMode AFM scans. The phase lag of the cantilever, with respect to the drive signal, is monitored for each point during the scan. This imaging highlights edges and provides more information about fine features that can be hidden by a rough surface (Babcock and Prater 2004).

The scans are performed on a 5 μm^2 area of the center of a capacitive test structure containing DNA–CTMA in air at ambient temperatures. A scan rate of 2 Hz is used with cantilever resonance frequencies of 65 to 68 kHz. Silver paint is applied from the steel magnetic AFM sample stub to the top of the grounding pads to provide a consistent ground. LiftMode is used to obtain the EFM maps of electric field gradient and distribution as a function of applied voltage, ranging from 0.0 to 12.0 V. LiftMode separately measures topography and electric force in a two-pass technique, using topographic information to track the probe tip at 41 nm above the surface.

Figures 7.19 and 7.20 show the TappingMode topography, EFM, and phase images from left to right, as measured on the active region of the capacitive test structure, with applied dc biases of 5.5, 6.0, 6.5, 7.0, and 7.5 V. The phase component of the electric field is mapped in the center images and shows a dramatic increase in the phase of the electric field from 5.5 to 7.0 V, and then decreases from 7.0 to 7.5 V. As the bias applied to the AFM tip increases from 5.5 to 7.0 V, the EFM images show that the phase increases. This increase in phase is due to an increase in the attractive force between the sample and the AFM tip that causes the resonant frequency to decrease. As the voltage on the AFM tip increases from 7.0 to 7.5 V, the EFM images show that the phase decreases, which corresponds to an increase in the resonant frequency of the AFM tip. This change in the direction of the phase is caused by a change in the effective dipole moment in the DNA–CTMA layer, which is seen as a decrease in the relative dielectric constant in the original S parameter measurements on the capacitive test structure made with the DNA–CTMA thin film.

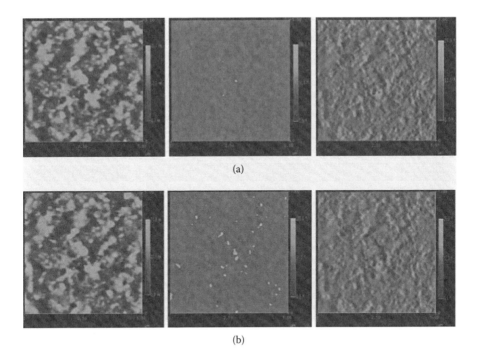

FIGURE 7.19 TappingMode topography (left), electric force microscopy (center), and phase images (right) with dc biases of (a) 5.5 V and (b) 6.0 V applied to the AFM tip. The phase component of the electric field is observed over a 5×5 μm region in the center of the image. (Reprinted with permission from C. Bartsch, Ph.D. Dissertation 2007.)

7.5 DC RESISTIVITY STUDIES

7.5.1 INTRODUCTION

The application of the biopolymer deoxyribonucleic acid (DNA) derived from salmon waste products (Zhang et al. 2001) has spawned research investigations and proof-of-principle development of DNA-based biotronic specimens (Grote et al. 2004). This molecule, with its helical symmetric strands, huge range of possible base pair sequences, and the large range of possible molecular weights (from <100 kDa to >8000 kDa, where Da = Dalton, the unit of molecular weight in grams/mol), has challenged efforts to use it in designing and fabricating electro-optic (Heckman 2006), photonic (Hagen et al. 2006; Zhou et al. 2007), and electronic specimens (Bartsch 2007; Ouchen et al. 2008). The key to success in these efforts is the availability of detailed data on the electrical (Yaney et al. 2009; Subramanyam et al. 2005), electro-optical, and optical properties (Yaney et al. 2006; Samoc et al. 2006) of this material. This section reports studies on the volume dc electrical resistivities of DNA–CTMA in comparison to common nonbiopolymers (Yaney et al. 2009). The dc resistivity behaviors of DNA–CTMA films doped with conductive additives and of two silk fibroin biopolymers prepared by different processing procedures are

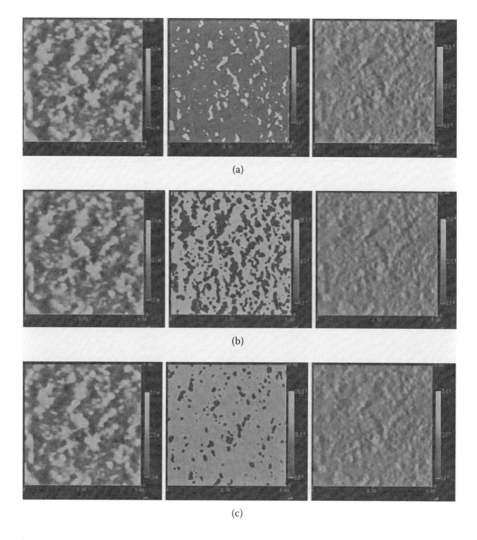

(a)

(b)

(c)

FIGURE 7.20 TappingMode topography (left), electric force microscopy (center), and phase images (right) with dc biases of (a) 6.5 V, (b) 7.0 V, and (c) 7.5 V applied to the AFM tip. The phase component of the electric field is observed over a 5 micron × 5 micron region in the center of the image. (Reprinted with permission from C. Bartsch, Ph.D. Dissertation 2007.)

included. Additionally, the quantitative dependence of DNA–CTMA conductivity on molecular weight is presented and discussed in detail.

7.5.2 Measurement Technique

Most of the DNA used in this work (see Section 7.2) was processed with high-energy sonication to obtain molecular weights from 200 to 1000 kDa from the as-received level of >8000 kDa (see Section 7.2.1 and Heckman et al. 2005). The DNA was then made water insoluble by treating it with the cationic surfactant hexacetyltrimethl-ammonium chloride (CTMA) that attaches to the helical strands in place of the Na^+ counter cations, which remain after the various processing steps that extract the DNA from the starting waste material. The resulting material, DNA–CTMA, is dissolved in butanol and spin-coated typically on half glass-slide substrates wherein the solution viscosity, solution and substrate temperatures, and spin parameters determine the film thickness and uniformity (see Section 7.2.3). Uniform films of thickness ranging from ~100 nm to over 10 μm have been made. Typical resistivity measurements use film thickness, t, between 1.5 and 5 μm with measurements made at ~1.5 V/μm. DNA–CTMA films with $t \leq 1$ μm often result in electrical shorts between the gold electrodes used for the measurements. Film thickness is measured using a Veeco 6M Dektak profiler.

The two electrode configurations shown in Figure 7.21 are used to measure the volume resistivity. The initial design had an electrode diameter, $D = 12$ mm, and included a guard electrode, which later was found unnecessary. (*Note:* The

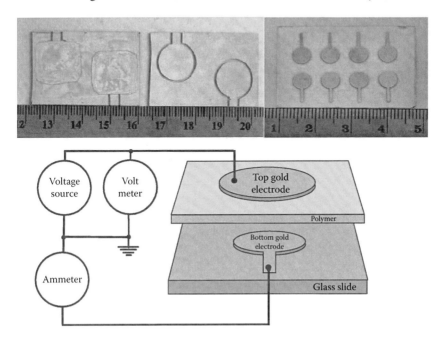

FIGURE 7.21 Electrode patterns of 12 and 5 mm diameter, and the schematic of the circuit used for volume resistivity measurements. (Reprinted with permission from Yaney, P.P. et al., *Proc. SPIE* 7403: 74030M-1–9, 2009.)

12-mm-diameter design with guard electrode greatly increased the number of sub-strates with shorted electrodes, apparently due to the large electrode areas. The guard electrode was found unnecessary, especially since the diameter-to-thickness ratio is typically >1600. Because the spin-coated polymer fills the gap between the guarded and guard electrodes, surface resistivity measurements were not possible as originally intended.) The current design has eight $D = 5$ mm electrode pairs or "devices," which increases the likelihood of having a device that is not shorted for a given spin-coated substrate. The 12 mm specimen is preferred for specimens hav-ing very high room temperature resistivity (>10^{15} ohm-cm), because of the increased electrode area, which reduces the measured resistance.

We use gold electrodes (50–80 nm thick) as shown in Figure 7.21 in order to mini-mize any chemical bonding at the electrode–polymer interface, which are usually applied by electron-gun-plasma deposition. A 10 nm chromium layer is deposited first for the bottom electrode followed by a 50 nm gold layer. The chromium ensures firm bonding of the gold to the substrate. Contact to the electrodes use microman-ipulated 12 μm titanium probes. For the top electrode deposited on the film, which is slightly larger than the bottom electrode on the substrate, we use a small convex brass pad under the titanium probe. These pads prevent damage to the underlying film that can be caused by the bare probes. Since typically $D/t > 1000$, the fringing of the electric field is negligible.

The resistivity specimens are basically capacitors, and therefore can pick up an electrical charge, which polarizes the film. This quasi-static polarization in high-resistivity materials decays very slowly. Thus, when a voltage is applied to these specimens, the observed current does not initially behave in a simple manner due to the electrical history of the film. Typically, a decaying current dependent on the applied voltage plus a current that is independent of the applied voltage are seen. The latter is called a "background" current. The voltage-dependent current must either be nearly constant (with constant applied voltage) or decay exponentially, converging to a stable value in order to obtain a viable resistance measurement. The behavior of these currents depends very much on the particular polymer being measured and the specimen temperature. Following the recommendation of the Keithley White Paper (Daire 2004), we use the alternating polarity technique to obtain a measurement, where the voltage-dependent current converges to within a few percent of constancy, the "background" current is minimized, and the temperature is constant.

Basically, the applied voltage source in Figure 7.21 is a square wave of period T with equal amplitudes in the positive and negative half-cycles. The presence of background currents introduce systematic errors and noise when computing the resistance by measuring the current i_T with a voltage V applied to the specimen. Averaging a running measurement of the current over a period of time cannot remove a systematic shift. However, the alternating polarity technique for resistance measurements removes the contributions of the background current and provides a convenient method for obtaining reproducible results. By an averaging procedure, any uniformly decaying background current can be canceled out to second order. An equivalent circuit for the measurement setup is shown in Figure 7.22a. Assuming that the decay of the background current can be reasonably approximated by an exponential decay, the total current, i_T, can be written

$$i_T = \frac{V}{R} + i_{C0}e^{-t/\tau_C} + i_{b0}e^{-t/\tau_b} \tag{7.8}$$

where V/R is the dc leakage current due to the specimen volume resistance R, i_{C0} is the peak transient current due to the charging of the capacitance C of the specimen, which decays with time constant $\tau_C = R_C C$ due to the effective resistance R_C "seen" by C in the measurement circuit, i_{b0} is the initial background current with effective background decay time constant τ_b, and t is measured from the beginning of each half-cycle. Taking the Taylor series expansion of the background decay term in Equation 7.8 to second order and averaging appropriately the end currents for $t = T/2$ at four successive half-cycles starting with a positive half-cycle, these decay terms can be canceled out to second order. It is observed that this should start after the first complete cycle, because the first current cycle is typically highly asymmetrical due to the response of the specimen to the initial application of the applied voltage and any initial charge. Starting with a positive half-cycle and referring to Figure 7.22b, this simple averaging process over four successive half-cycles is given by

$$I_{14} = \frac{1}{2}\left\{\frac{1}{2}\left[\frac{(I_1 - I_2)}{2} + \frac{(-I_2 + I_3)}{2}\right] + \frac{1}{2}\left[\frac{(-I_2 + I_3)}{2} + \frac{(I_3 - I_4)}{2}\right]\right\} = (I_1 - 3I_2 + 3I_3 - I_4)/8 \tag{7.9}$$

where I_1 and I_3 are the final currents of two consecutive positive half-cycles, and I_2 and I_4 are the final (algebraic) currents of the two consecutive negative half-cycles that follow the above positive half-cycles as illustrated in Figure 7.22b. This calculation can be done for any four consecutive half-cycles that start with a positive half-cycle; however, for the cancellation to be complete, the time durations of the half-cycles must be equal to within 0.1% (Daire 2004). For example, from Figure 7.22b, which

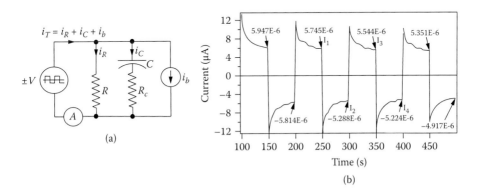

(a)

(b)

FIGURE 7.22 (a) Equivalent circuit for resistivity measurement, where V is the applied alternating potential, i_R is the dc leakage current due to the volume resistance R of the polymer specimen, i_C is the charging/discharging current through C with time constant $\tau_C = R_C C$ of the electrode-polymer-electrode sandwich connected to the measuring circuit, and i_b is the polymer "background" current. (b) Typical alternating polarity (± 5 V) data taken on a 3.52 μm thick, -linked, water-insoluble DNA polymer film at 100°C.

does not show the first cycle, three values of resistivity are obtained, namely, 2.64E9 $\Omega \times cm$, 2.80E9 $\Omega \times cm$, and 2.89E9 $\Omega \times cm$, starting with the first positive value for $I_1 =$ 5.947E−6 A. These data illustrate the typical convergence of the resistivity as later cycles are analyzed. The degree of convergence is determined by the choice of cycle duration and the behavior of the material. The 100 s duration used in Figure 7.22b usually gives better than 5% convergence after five full cycles. The length and the number of cycles used for a resistance measurement are set such that the resultant resistance shows acceptable convergence. At room temperature, we use five 100 s cycles. At temperatures typically over 50°C, when the resistivity decreases by ~100 times due to Arrhenius behavior, it is then possible to reduce to four 80 s or shorter cycles. It should be noted that these are static measurements, where the temperature is held constant during the entire data acquisition time period.

The voltage square wave, the current and voltage measurements, temperature control, resistivity calculations, and data recording are accomplished with a LabView program controlling a Keithley Model 617 Electrometer and a Watlow F4 temperature controller connected to a custom-designed hot plate inside a dry-nitrogen-purged, electrostatically shielded chamber. The hot plate temperature is held constant to within ±0.1°C. This is a special hot plate that has temperature uniformity over its 8" diameter of less than ±1°C. There are five thermocouples that monitor the temperature. All thermocouples are mounted with high-temperature, thermally conductive paste in holes in the bottom of the hot plate that places the junctions within ~1 mm of the hot plate top surface with the center thermocouple connected to the controller. The hot plate and adequate lengths of the thermocouples are mounted inside a tight fitting Teflon "box" around the sides and bottom. Precisely controlled temperature is essential since the resistivity can change over four orders of magnitude between 20°C and 90°C. The specimens to be measured are vacuum dried at ~35°C typically for more than 24 h immediately following curing and then measured in a flowing dry nitrogen gas environment with a measured humidity of 0.0%. If the specimen is exposed to air after curing for a day or more, then the specimen often needs to be dried for several days before measuring. These procedures were found to be absolutely necessary for reproducible measurements, especially with the biopolymers. The electrometer is operated in the *V/I* resistance mode that permits the choice of an applied voltage that is sufficient to produce accurately measureable current levels while avoiding shorting events between the electrodes. Measurements usually start at ~20°C, but in some cases a temperature as low as 15°C was obtained with the cold water flowing through the hot plate cooling coils. The maximum temperature was typically 90°C, but some polymers were taken to higher temperatures with no discernible damage.

7.5.3 Data Analysis

The direct current conduction through insulating polymers takes place by ionic, electronic, or by polarons, or some combination of these mechanisms (Seanor 1982). Ionic conduction typically requires impurities or defects, which create ion vacancies, unattached ions, or weakly attached ions. Electronic transport processes involve electrons or holes, or both, and require some distribution of acceptor and donor states. These processes take place by "hopping" mechanisms in which electrons or holes hop between localized sites and drift toward the appropriate electrode in an applied field. In DNA, it has been proposed that the double helix with a random base pair sequence can be viewed as a 1D disordered system. The disorder leads to electronic localization, and electron/hole hoppings between these localized "impurity" sites along the chain are responsible for the conductivity (Yu and Song 2001).

There have been a great many studies of charge transfer in DNA molecules (Fink and Schönenberger 1999; Porath et al. 2000; Giese et al. 2001; Yoo et al. 2001; Grozema et al. 2002; Cui et al. 2008; Ortmann et al. 2009). The current view of charge transport in DNA is by electron vacancies or holes, which form polarons (Singh 2004). This mechanism is seen as rapid and efficient by virtue of a channel formed by the overlapping π orbitals of the base pairs that provide a means for charge migration of the injected holes in the DNA molecule (Takada et al. 2006). A hole is expected to quickly localize at the nearest guanine radical cation, since the oxidation potential of a guanine site is the lowest in DNA (Voityuk 2008). This localized charge cloud is called a polaron and can extend over a few base pairs. The hole or polaron then becomes mobile by the transport of an electron from a distant G residue to the cation site, which can occur over distances >20 nm. Although there are reported differences in the hole transfer rates between guanine-cytosine (GC) bases, called G hopping, and between adenine-thymine (AT) bases, called A hopping, as well as single-step tunneling, which impact the limiting charge mobility, the transport range is almost unlimited. The main limit to the range is suggested to be the chemical reaction with water, presumably in the grooves of the DNA molecule, which limits the range to distances less than 5 μm (Lakhno et al. 2003). Since the as-received salmon Na-DNA used in this work has a molecular weight of >8000 kDa for which the length is >4.1 μm, then charge transport can likely take place over the entire length of the molecule. By comparison, 1000 kDa sonicated DNA has a length of 0.51 μm. Going one step further, 200 kDa DNA has a length of only 103 nm, which shows the importance of molecular weight in characterizing DNA-based materials.

In the work presented here, however, the DNA materials studied were in the form of micron-size films deposited on glass or on an oxidized silicon wafer by spin casting or drop casting from either organic solvent or aqueous solutions, cured, and thoroughly dried. There was no effort to control the orientation of the DNA molecules and, as such, there will be some random distribution of orientations and "coiling" or nonstraight conformations in a given film. Thus, it is evident that the observed dc conduction mechanism in these films includes both the conduction in or through the DNA molecules and between the DNA molecules. The statistical distribution of molecule orientation in micron-size films can be expected to depend on the molecule lengths. For example, the >4.1 μm as-received Na-DNA formed into a 1.56 μm film can be expected to have the helix axes of a large fraction of the DNA molecules predominately parallel, or nearly so, to the plane of the substrate, and hence, to the electrodes.

Theoretical and experimental studies of the polaron conduction mechanism in the DNA molecule have shown the polaron mobility to be high (for a polymer), ranging from 0.04 to 4.5 cm^2/V-s (Grozema et al. 2002; Lakhno et al. 2003; Cui et al. 2008). We have carried out preliminary studies of the charge transport mobility in our DNA–CTMA and as-received DNA films using the photoconductive time-of-flight technique (Seanor 1982). We found only hole carriers with mobilities $\mu_H \sim$ 1E-6 cm²/V-s. In addition, our resistivity studies show a clear dependence on molecular weight from the as-received value of >8000 kDa to 200 kDa. Since conductivity is proportional to mobility, this leads us to hypothesize that the charge transport through the micron-size DNA films is dependent primarily on the transport *between* the molecules. If this holds true, then it is reasonable to expect that the charge transport along the helix axis is different from that perpendicular to the axis. This expectation appears to be supported by computations of mobility through and across a GC sequence that show the mobility perpendicular to the helix axis is 30 times smaller than along the axis (Ortmann et al. 2009). Since the Raman spectra of DNA(sonicated)-CTMA showed no dependence on molecular weight (Yaney et al. 2008), it is assumed that the observed conductivities of micron-size films of sonicated DNA depend on the statistical distributions of the molecule orientations. These distributions in the thin films used in this work can be expected to be constrained to some degree by the length of the molecules, which is directly proportional to the molecular weight. Although the attached CTMA surfactants enlarge the physical extent of the DNA molecules, this hypothesis should still apply to DNA–CTMA films, which has been at the center of our resistivity measurements. More experiments and detailed modeling are needed to confirm this hypothesis. Implicit in this discussion is that the mechanism of the observed conductivity in these micron-size films is not ionic. It is not evident that conductivity would depend on molecular weight if ionic transport was the operative mechanism.

It has been found that above some corner or cross-over temperature T_c the conductivity of DNA (Eley and Spivey 1962; Kutnjak et al. 2005; Yaney et al. 2006) is described by the Arrhenius ansatz given by

$$\sigma = \sigma_0 e^{-(E_a/kT)} \tag{7.10}$$

where the exponential prefactor σ_0 depends on the number density of the sites involved in the transport (Böttger and Bryksin 1985), E_a is the activation energy, k is the Boltzmann constant, and T is the Kelvin temperature. This region is due to nearest neighbor hopping. At temperatures below T_c, charge transfer is described as due to variable range hopping and leads to a weaker dependence on temperature than given in Equation 7.10. This variable-range hopping model is given by

$$\sigma(T) = \sigma_0 e^{-(T_0/T)^\beta} \tag{7.11}$$

where T_0 is a constant and, typically, $\beta = \frac{1}{2}$ or $\frac{1}{4}$ (Böttger and Bryksin 1985; Yu and Song 2001; Duttal and Mandal 2004), which is often too strong to describe the observed temperature dependence below T_c. Measurements of Li-DNA conductivity down to temperatures $\sim -53°C$ have shown that a useful empirical model is given by (Kutnjak et al. 2005)

$$\sigma(T) = \sigma_B + \sigma_0 e^{-E_a/kT} \tag{7.12}$$

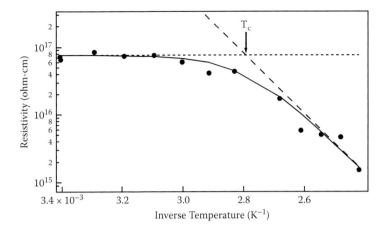

FIGURE 7.23 Illustration of the corner temperature, T_c, using the APC data from Figure 7.24. The dotted line is σ_B, and the dashed line is the Arrhenius term. (Reprinted with permission from Yaney, P.P. et al., *Proc. SPIE* 7403: 74030M-1–9, 2009.)

where σ_B is a constant resistivity at temperatures below the empirical corner or crossover temperature T_c defined as (Yaney et al. 2007)

$$T_c = \frac{E_a}{-k \ln(\sigma_B / \sigma_0)} \tag{7.13}$$

which is the temperature given by Equation 7.12 at which the generalized Arrhenius function equals σ_B. A graphical representation of this "corner" or "crossover" temperature is given in Figure 7.23. Kutnjak et al. were able to fit a portion of their data from measurements on Li-DNA to Equation 7.11 for $\beta = \frac{1}{2}$.

The thrust of these studies was to characterize the temperature behaviors of the resistivity of both as-received DNA and DNA-based films, including as a function of the mean molecular weight, and compare to the behaviors of selected oil-derived insulating polymers. The alternating polarity technique was used to measure the resistivity ρ and characterize the behaviors by fitting Equation 7.12 to the data using the conductivity $\sigma = 1/\rho$. The resistivity is computed from the measured resistance R by $\rho = \pi R D^2/4t$, where D is the bottom electrode diameter and t is the mean thickness measured from the tab across D (see Figure 7.21).

The analysis and graphing program Igor Pro was used to fit Equation 7.12 to the data, which often required some experimentation to be successful for a given data set. The σ_B term was needed for all the data sets consistent with Kutnjak et al. (2005). It was found that only the conductivity form of the data could be consistently fitted. Efforts to fit Equation 7.11 with $\beta < 1$ to data having high T_c were not successful. The long measurement time needed to reach a reasonable convergence for each data point using the alternating polarity technique made it difficult to automate the measurements or to acquire a large number of points for every specimen. Nevertheless, the resulting parameter values give evidence of adequate characterization of the studied materials.

7.5.4   DATA

7.5.4.1   DNA Compared to Nonbiopolymers

The as-received DNA has molecular weight >8000 kDa. Solutions using this DNA can have high viscosity, which are difficult to spin-coat and can give uneven films when drop-cast. Lower viscosity can be obtained by reducing the molecular weight using a high-power sonication probe (see Section 7.2.1), which gives mean molecular weights down to ~200 kDa. (*Note:* The term *mean* molecular weight is the maximum of the assumed symmetrical molecular weight distributions obtained from agarose gel electrophoresis measurements. A more precise characterization might be the *peak* molecular weight. See Figure 7.1.) This is followed by exchanging the surfactant CTMA+ for the Na+ counter ion on the DNA, resulting in hydrophobic DNA–CTMA. The sonication, however, produces broad distributions of molecular weight, typically ±50% or more, which means the distinctiveness of the measurements on films with different mean molecular weights could be somewhat compromised. Measurements were made on many DNA(sonicated)-CTMA films with mean molecular weights ranging from 200 to 1000 kDa. In addition, unsonicated, as-received DNA and as-received-DNA with CTMA were studied. The former is water soluble, which is more of a challenge to produce as a uniform film by spinning or drop casting. A summary of these studies on DNA materials are given in Figure 7.24. The fits

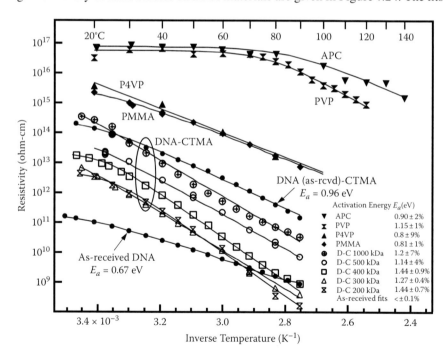

FIGURE 7.24   Resistivities of DNA-based films and selected non-DNA polymer films. Solid curves are fits of Equation 7.10 to the data. Fitting parameters are in Table 7.3. Note the Celsius scale at the top of the graph. (Reprinted with permission from Yaney, P.P. et al., *Proc. SPIE* 7765: 776509-1–8, 2010.)

TABLE 7.3

Parameters Obtained from Fitting Equation 7.12 to the Conductivity Data of the Measurements Given in Figure 7.24 Plus Silk in Figure 7.26[a]

Polymer	ρ @ 20°C (Ω-cm)	E_a (eV)	σ_0 (Ω-cm)$^{-1}$	σ_B (Ω-cm)$^{-1}$	T_c (°C)
DNA (1000 kDa)-CTMA	2.04e14	1.21 ± 7%	2.64e6	3.0e−16	~4
DNA (500 kDa)-CTMA	3.04e13	1.14 ± 4%	9.07e5	1.0e−14	~14
DNA (400 kDa)-CTMA	1.32e13	1.44 ± 0.9%	2.08e11	4.5e−14	22
DNA (300 kDa)-CTMA	4.17e12	1.27 ± 0.4%	1.16e9	4.0e−14	~11
DNA (200 kDa)-CTMA	3.32e12	1.44 ± 0.7%	5.38e11	2.0e−13	24
DNA (>8000 kDa)-CTMA	1.41e14	0.96 ± 0.1%	1.03e2	3.3e−15	19
DNA (>8000 kDa)	1.04e11	0.67 ± 0.1%	2.23	3.0e−12	~12
Tufts' Silk	1.92e12	0.62 ± 5%	3.5e−4	5.1e−13	79
APC	8.0e16	0.90 ± 2%	5.88e−5	7.6e16	86
PVP	5.4e16	1.15 ± 1%	0.738	5.4e16	76
P4VP	4.7e15	0.8 ± 9%	4.64e−2	~2e16	~10
PMMA	1.9e15	0.81 ± 1%	1.88e−2	3.9e15	20

Source: Reprinted with permission from Yaney, P.P. et al., *Proc. SPIE* 7765: 776509-1–8, 2010.

[a] Ω-m = 0.01 Ω-cm; siemens/m = 100 (Ω-cm)$^{-1}$.

of the Arrhenius function plus a constant, Equation 7.12, to the conductivity data, from which the fitting parameters given in Table 7.3 were obtained, are given as resistivity and shown as solid curves in the figure. These data are given in units of ohms-cm rather than siemens/m, because resistance is more useful for device design when the conductivity is very low. The data are plotted with the inverse temperature increasing to the left in order to show the Celsius temperature behavior increasing to the right. The Celsius scale is shown at the top of each graph. This method of plotting with a logarithmic ordinate clearly shows the Arrhenius temperature dependencies of the resistivities.

The data show that processing as-received DNA with CTMA increases the activation energy from 0.67 to 0.96 eV, and sonication furthers increases this parameter to 1.3 eV on average. In addition, there is a clear dependence of resistivity on molecular weight with little change above 1000 kDa. Likewise, there is little difference between 200 kDa and 300 kDa data. Although the corner temperature, T_c, was computed for all the DNA plots in Figure 7.24, these plots did not extend to sufficiently low temperatures to allow for accurate determinations of T_c in all cases; however, for completeness, these values are included in the data of Table 7.3.

The sonication procedure typically involves a periodic series of high-energy ultrasonic pulses separated by cooling periods. This is done typically for 30 min with the total sonic energies up to 200 kJ (see Section 7.2.1). In an effort to determine the impact of continuous sonication on resistivity, a cooled batch of DNA was sonicated

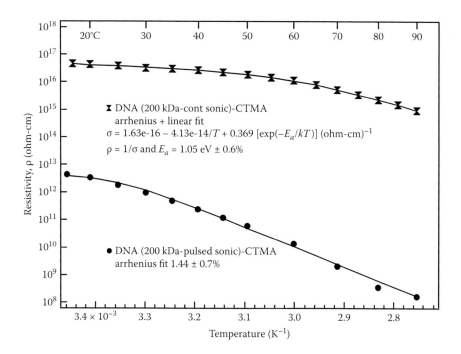

FIGURE 7.25 Resistivity of DNA(200 kDa)-CTMA from continuously sonicated DNA in comparison to normal pulsed sonication. (Reprinted with permission from Yaney, P.P. et al., *Proc. SPIE* 7403: 74030M-1–9, 2009.)

without pulsing for 30 min with occasional interruptions for cooling to give a mean molecular weight of ~200 kDa. Although there was much debris in the solution, most was removed by filtering, and a resistivity device was fabricated and measured. The data are given in Figure 7.25 with pulsed-sonicated DNA(200kDa)-CTMA shown for comparison. This is the first data set for which it was necessary to include a linear term in Equation 7.12 to achieve a quality fit. As can be seen in Figure 7.25, the fit is perfect to the eye with E_a = 1.06 eV ±0.6%. Moreover, the resistivity increased by nearly four orders of magnitude compared to the pulse-sonicated DNA. Micro-Raman spectroscopy studies were carried out on DNA(sonicated)-CTMA and on as-received Na-DNA (Yaney et al. 2008) that showed that the former is B-form DNA independent of the sonicated molecular weight. The Raman spectrum of the continuously sonicated DNA (Yaney 2009, unpublished) showed considerable reduction in the strengths of the vibrational modes of the bases relative to the phosphate mode, a broadening of the phosphate mode, and a broad, irregular Raman band well below the base breathing modes (<650 cm^{-1}) relative to the pulsed sonicated DNA. These changes suggest that the continuous sonication caused significant damage to the DNA molecules and that the DNA material was probably made partially amorphous by the sonication. The similarity of the resistivity behaviors of this heavily sonicated material to that of APC (amorphous polycarbonate; see later text) cannot be ignored.

The measurements on nonbiopolymers poly[methyl methacrylate] (PMMA), amorphous polycarbonate (APC), poly 4-vinylphenol (P4VP), and poly(vinylpyrrolidone)

(PVP) are also included in Figure 7.4 for comparison. There are three sets of data in the APC plot taken on different days. As a result, these data show more scatter, which is not unreasonable considering the extraordinary high resistivity of this polymer. Nevertheless, the three sets still follow the basic trend, confirming reasonable repeatability. The PVP, which is water soluble, also showed a very high resistivity close to APC. Both polymers are distinctive in that the onset of Arrhenius behavior occurs around $T_c = 80°C$. These four high-resistivity polymers were measured with the 12-mm-diameter electrodes to lower the measured resistance. Fitting parameters for these polymers are included also in Table 7.1.

7.5.4.2 Silk

Silk films from Tufts University (Jin et al. 2005) and from AFRL/RXBN (Phillips et al. 2004) processed from bombyx mori silk fibroin were studied. The Tufts' films were provided on our 12-mm-diameter electrode substrates. The silk was prepared from high concentrations of regenerated silk fibroin in aqueous solution (~8 wt.%). After processing, the silk is insoluble in water. The AFRL/RXBN material was dissolved in an ionic liquid, processed, and provided in solution form, from which we spin-coated films. The resistivity data for these two materials are given in Figure 7.26, and the fitting parameters for the Tufts' material are given in Table 7.3. Although the Tufts' silk resistivity on the scale used in the figure appears almost independent of

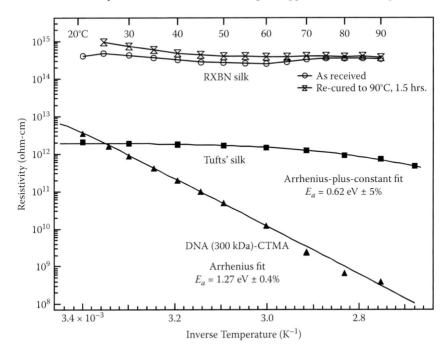

FIGURE 7.26 Resistivities of silk films compared to DNA(300 kDa)-CTMA. Solid curve for Tufts' silk is the fit of Equation 7.10 to the data. The solid lines for the AFRL/RXBN data connect the points for clarity. (Reprinted with permission from Yaney, P.P. et al., *Proc. SPIE* 7403: 74030M-1–9, 2009.)

temperature, the data does show the onset of Arrhenius behavior at $T_c = 79°C$ with an activation energy of 0.62 eV, which is very close to the as-received DNA value of 0.67 eV. The silk film from AFRL/RXBN was initially cured at ~60°C, which produced a somewhat erratic and time-dependent resistance behavior versus temperature, some of which is visible in Figure 7.26. Curing the specimen again, but at 90°C, gave a more smooth and stable temperature dependence. However, the material showed no Arrhenius behavior and, in fact gave a "concave up" curvature in contrast to the Tufts' silk curve and the curves in Figure 7.24. The AFRL/RXBN silk is nearly 1000 times more resistive than the Tufts' silk, which, if nothing else, highlights the complexity of these biopolymers and the importance of the processing techniques. The observed behavior of the AFRL/RXBN silk suggests weak ionic conductivity.

7.5.4.3 Effect of Humidity and Measurement Accuracy

Enhanced conductivity, lower activation energies, and bending of the curve in the 20°C to 40°C range occur when a small amount of excess water is present, that is, when the films are not thoroughly dried. Two sets of data for each of DNA(as-rcvd) (hydrophilic) and DNA(as-rcvd)-CTMA (hydrophobic) given in Figure 7.27 are examples of these effects. These resistivity changes that occur when excess water is present are attributed to H^+ ions from dissociation of the water (Ha et al. 2002;

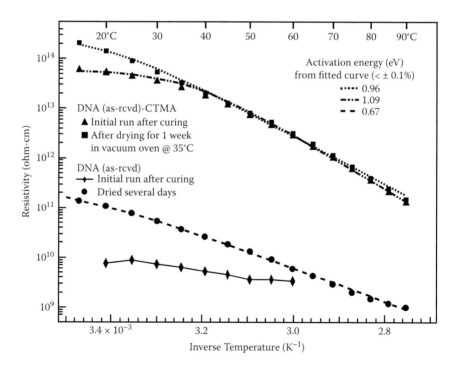

FIGURE 7.27   As-received DNA with Na^+ counter ion and as-received DNA with $CTMA^+$ replacing the Na^+. The effect of inadequate drying after curing the water-based as-received DNA film is shown in the lower plot. (Reprinted with permission from Yaney, P.P. et al., *Proc. SPIE* 7765: 776509-1-8, 2010.)

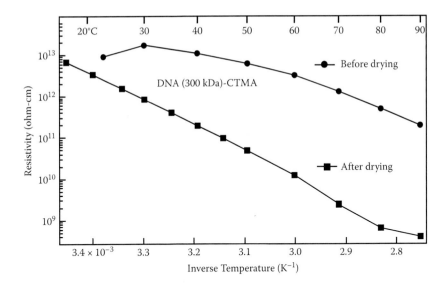

FIGURE 7.28 Effect of long-term exposure to humidity and inadequate drying on the resistivity of a DNA–CTMA film. The lines connect the points for clarity. (Reprinted with permission from Yaney, P.P. et al., *Proc. SPIE* 7403: 74030M-1-9, 2009.)

Briman et al. 2004). The contrast between two data sets taken on the same resistivity device given in Figure 7.28 is an extreme example of the long-term effect of humidity when no effort was made to thoroughly dry the film. The unusual increase in resistivity due to long exposure (many weeks) to room air might be explained by water being *absorbed* into the volume of the film, which interfered with charge transport and charge injection, thus increasing the resistivity. The behaviors of the two materials in Figure 7.27 are more typical when either hydrophobic or hydrophilic films are exposed to humid air for about a day or are not adequately dried, wherein water is probably *adsorbed* onto the polymer molecules in the film-electrode interfacial regions. As reported by Eley and Spivey (1962), the bending of the resistivity curve in the initial plot for the DNA(as-rcvd)-CTMA film near room temperature has been seen often in our DNA–CTMA studies when a film was not adequately dried.

The precision of these measurements is best shown by examples of reproducibility. For example, two different PMMA specimens were measured with the resulting data, and the fits to the data equal to within ±10%. Another example is DNA(500kDa)-CTMA shown in Figure 7.24 as D-C 500kDa. This plot has data from two separate specimens taken at different times. Some plots shown in Figure 7.24 show small deviations from well-defined behavior at the higher temperatures. This is probably due to changes in the polymer or changes in the contact (i.e., the interface) between the gold electrodes and the polymer. The accuracy of the resistivity measurements is difficult to estimate, because of material and processing issues and the lack of precise knowledge of the complete three-dimensional thickness profiles of the films; however, we estimate that the accuracies for the well-behaved data from spin-cast films are equal to or better than ±20% based on our ability to reproduce the measurements, and the accuracies of the Keithley electrometer and the Dektak profiler. The surface temperature of the hot plate was within ±0.3°C to the reading of a calibrated electronic surface thermometer.

7.5.4.4 DNA with Conductive Dopants

There are at least two ways to make materials using DNA that have significant dc conductivity, namely, (1) somehow organize the DNA molecules in a lattice such that the material exhibits true semiconducting behavior, or (2) add conducting chemicals to the DNA with the hope that they will enhance the charge transport sufficiently or increase the charge density to make the material useful for fabricating electronic devices such as BioFETs (see Chapter 10). The following gives a summary of our current efforts to implement method 2.

Three dopants or additives that were investigated are phenyl-C61-butyric acid methyl ester (labeled PCBM in Figure 7.29); polyethylenedioxythiophene-poly-styrenesulfonate, PEDOT-PSS, known as Baytron P or CLEVIOS P (Baytron and CLEVIOS are registered trademarks of H. C. Starck-CLEVIOS GmbH, Goslar, Germany) (labeled BP in Figure 7.30); and ammonium tetrachloroplatinate $(NH_4)_2PtCl_4$) (labeled Pt in Figure 7.31). The BP and Pt aqueous solutions were each mixed with sonicated DNA followed by ion exchange processing with the surfactant CTMA and then dissolved in butanol. PCBM was dissolved in chloroform prior to mixing it with DNA–CTMA in butanol. These solutions were spin-coated on half glass slides with eight gold electrodes (see Figure 7.21), and then coated with the top gold electrodes for resistivity measurements. Figures 7.29–7.31 give the results of the

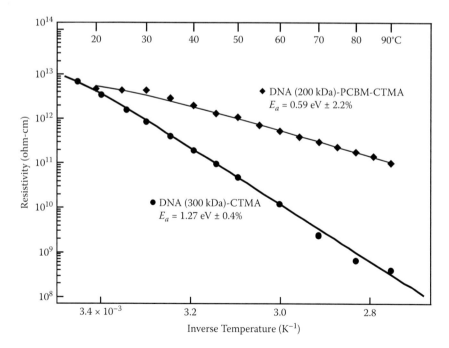

FIGURE 7.29 Behavior of DNA(200 kDa) complexed with dopant PCBM before adding CTMA in comparison to DNA(200 kDa)-CTMA from Figure 7.24.

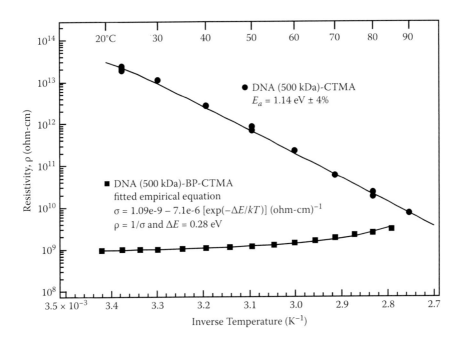

FIGURE 7.30 Behavior of DNA(500 kDa) complexed with dopant BAYTRON P (from H. C. Starck-CLEVIOS) before adding CTMA in comparison to DNA(500 kDa)-CTMA from Figure 7.24.

measurements plus plots of nondoped DNA films for reference. The graphs are plotted with identical scales to facilitate the comparisons.

Although the PCBM-doped data in Figure 7.29 showed clear Arrhenius behavior, the room temperature resistivity was only slightly lower than the nondoped material and had activation energy approximately half of the nondoped material.

In Figure 7.30, the BP-doped film gave over four orders of magnitude lower resistivity at room temperature. Although it shows a very small temperature dependence, it did have an "inverted" Arrhenius (concave upward) behavior, which permitted fitting an empirical "inverted" Arrhenius-type function to the data as shown in the graph.

The Pt data, given in Figure 7.31, had Arrhenius behavior up to 60°C (negative slope), but then the resistivity plot changes to a positive slope. Initially, the specimen had been cured at 60°C, which gave resistivities lower than the nondoped material by about a factor of 10. Because of the deviation from Arrhenius behavior, the specimen was recured at 90°C in the hope that the plot would "straighten." As shown in Figure 7.31, this curing increased the resistivity, probably due to water or ammonia loss, but did not remove the break in the curve at 60°C. It was not evident what property of the material was responsible for this effect, although the behavior appears to be some sort of threshold response.

Instead of adding BP to sonicated DNA before performing the ion exchange with CTMA, a much more conductive version of BP identified as BFE (Baytron™ FE and CLEVIOS™ FE are products of H. C. Starck-CLEVIOS GmbH, Goslar, Germany) was added in increasing dopant levels to as-received DNA in aqueous

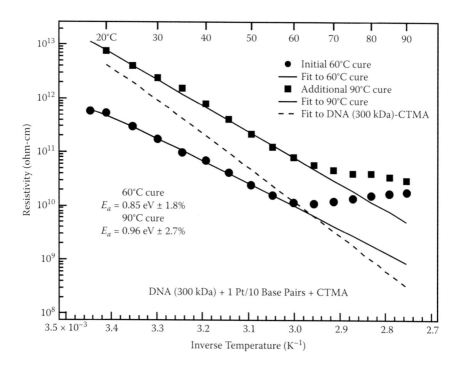

FIGURE 7.31   Behavior of DNA(300 kDa) complexed with ammonium tetrachloroplatinate dopant before adding CTMA after curing at two temperatures in comparison to the fitted curve for the DNA(300 kDa)-CTMA data in Figure 7.24.

solutions without complexing with CTMA and drop-cast into thin films. The main objectives were to compare to DNA–BP–CTMA and to determine how the resistivity behaved with increasing amounts of BFE. It was not possible to perform this kind of study using BP, presumably due to the presence of CTMA. The results are given in Figure 7.32, which show two to five orders of magnitude decrease in resistivity over the neat as-received DNA plotted in Figure 7.27. This behavior is similar to the DNA–BP–CTMA result, including little or no dependence on temperature, but the 11.9 wt% DNA–BFE resistivity was a 1000 times more conductive than DNA–BP–CTMA. This conductivity is the largest observed so far using dopants in DNA. The 5.8 wt% device was measured to 105°C to check for Arrhenius behavior.

7.5.4.5   Discussion

Our studies show that, with the exception of RXBN silk and continuously sonicated DNA–CTMA, the room temperature resistivities of the nondoped bioderived polymers are a factor of 10 or more lower than the oil-derived polymers measured in this work. Moreover, except for PVP, the activation energies of the oil-based polymers average 0.84 eV, as compared to the range of 0.96 to 1.44 eV for the as-received DNA–CTMA and the five molecular weights of DNA–CTMA given in Figure 7.24 and Table 7.3. A second difference is the high corner temperatures of $T_c \sim 80°C$ of APC and PVP. The only other studied vacuum-dried polymers that showed high T_c

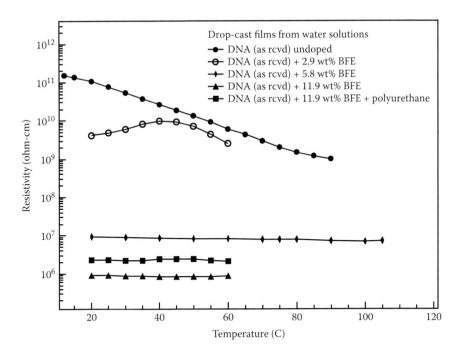

FIGURE 7.32 Behavior of as-received DNA with dopant BAYTRON FE (from H. C. Starck-CLEVIOS). Note the linear temperature scale on the abscissa.

were Tuft's silk and continuously sonicated DNA. One possible connection between these four materials could be that they are all highly amorphous.

Referring to Table 7.3, this leads to a third difference between Group A that has the five molecular weights of DNA (sonicated)-CTMA and Group B that includes DNA(>8000 kDa), DNA(>8000 kDa)-CTMA, Tufts' silk, APC, PVP, P4VP, and PMMA, which is the extraordinary gap between the values for the prefactor σ_0. The A:B ratio of these values range between ~8.8E3 and 9E15. Following the DC hopping model by Miller and Abrahams (1860) reviewed and updated by Böttger and Bryksin (1985), this prefactor for nearest-neighbor hopping is proportional to $\exp[-f(\eta\alpha N^{-1/3})]$ in a disordered or random material, where α is the inverse of the Bohr radius (Bohr radius = 5.29E-2 nm), N is the number density of the sites involved in hopping transport, $N^{-1/3}$ is the mean spacing of pairs of hopping sites, and η is a number on the order of unity. Assuming the proportionality constants are approximately equal for the two groups, then the natural logarithm of the A:B ratio of the prefactors ranges from ~9 to 37 Bohr radii (using the minimum and maximum A:B ratios). This indicates the difference between the mean hopping distances of Group B minus Group A ranges between ~0.5 and 2 nm. These are significant differences on the atomic scale and demonstrate that Group B polymers are better electrical insulators than Group A, and the sonicated DNA materials have higher densities of hopping sites, which presumably could lead to enhanced conductivity if some kind of ordering can be achieved in the material. Based on this approximate analysis, APC and Tufts' silk have the lowest density of hopping sites relative to Group A.

The DNA studies by Kutnjak et al. (2005) included measurements at 78 V/cm of the conductivity of 3–4 mm thick specimens of vacuum-dried, oriented 10,000 kDa Li-DNA prepared from calf thymus, but most of their studies were at 75% humidity. Their vacuum-dried data give a resistivity parallel to the orientational (helix) axis at ~20°C of 5E12 ohm-cm, which is 50 times the value we obtained at ~15 kV/cm for a vacuum-dried 1.56 μm thick film of as-received Na-DNA of molecular weight >8000 kDa. Their data, however, shows for an increase in the field from ~31 to ~78 V/cm, an increase in the conductivity by seven, which suggests that our lower resistivity (higher conductivity) is due partly to our much higher field and to the statistical orientation of Na-DNA in our devices that sets the field largely perpendicular to the helix axes, as discussed later.

An activation energy of 0.98 eV for oriented, vacuum-dried Li-DNA from calf thymus was reported by Kutnjak et al. as compared to 0.67 eV for vacuum-dried, randomly oriented Na-DNA from salmon roe and milt sacs used in our work (see Section 7.2). The small polaron is usually considered the hopping charge carrier in DNA (Yoo et al. 2001; Cui et al. 2008). In this case, the activation energy of a disordered material is the sum of the energy to move the polaron over its site barrier plus the energy needed to hop the distance to another site (Böttger and Bryksin 1985). The mean site barrier energy seen by the polarons, which must hop between molecules, could depend on either the base sequence in the DNA molecules or the counter ion, while the mean hopping distance would depend on the DNA number density (i.e., packing density) in the material. The change in the Raman spectra for B-form DNA with six different guanine-cytosine (GC) to adenine-thymine (AT) ratios reported by Deng et al. (1999) indicates that the salmon DNA has a GC:AT ratio of ~27% (Yaney et al. 2008), whereas calf thymus has a ratio of 42% (Deng et al. 1999). This difference in the GC:AT ratios could account for the different activation energies; however, the counter ion and mean hopping distance cannot be neglected. More carefully designed experimental studies are needed to resolve this issue.

Measurements by Eley and Spivey (1962) on five specimens of vacuum-dried thymus DNA gave an average activation energy $E_a = 1.22$ eV, which compares well to the Group A DNA–CTMA average value of 1.30 eV, but is almost twice the 0.67 eV value for our Na-DNA. They dried their specimens in high vacuum at 100°C, and their lowest temperature was ~80°C. Since their activation energy is also higher than that reported by Kutnjak et al., it would appear that neither base sequence nor counter ion is a factor. The closeness of the DNA–CTMA average activation energy to the Eley and Spivey value would seem to support different number densities, since the DNA molecules with CTMA counterions would be expected to have a smaller DNA number density with the corresponding larger mean spacing of the hopping site pairs. More study is needed to understand these behaviors.

An important issue in designing electrical or electro-optical specimens using sonicated DNA material is the dependence of bulk (or volume) electrical and electro-optical properties on the molecular weight of the sonicated DNA (see Section 7.2.1). The results for dc resistivity given in Table 7.3 show this dependence. As discussed in Section 7.5.3, the observed conductivity in the films studied in this work is due primarily to the molecule-to-molecule charge transport, which can be expected to be constrained by the mean length of the molecules, which is proportional to the molecular weight. This means the statistical distribution of orientations of the molecules in the thin films relative to the electrodes depend on the molecular weight. Okahata et al. (1998) measured the conductivity of dried, aligned DNA strands in a film and reported the conductivity parallel to the strands to be four orders of magnitude higher than that perpendicular to the strands. This behavior can be attributed, at least in part, to anisotropic mobility such as computed for a GC sequence by Ortmann et al. (2009), which will be dependent on a base sequence in the particular DNA molecule under study (Yoo et al. 2001; Cui et al. 2008). Thus, it is reasonable to conclude that changes in resistivity of our films are due to the changes in the orientation distribution of the molecules with molecular weight. As described in Section 7.5.3, since high-molecular-weight DNA would be expected to have a large fraction of the helix axes perpendicular to the applied field, the high molecular weight would have the lowest conductivity (highest resistivity) as reported here. As the molecular weight is decreased, the molecule length decreases, and the orientation distribution will include more molecules aligned parallel to the applied field, thereby increasing the conductivity (decreasing resistivity). Thus, the conductivity is dependent on the inverse of the mean molecule length and therefore on the inverse of the molecular weight. The conductivities at 20°C of DNA–CTMA plotted in Figure 7.33 versus the inverse of the mean molecular weight show increasing conductivity consistent with this model. This behavior follows also at higher temperatures. The trend curve shown in the figure is the likely form of the functional dependence, which is given

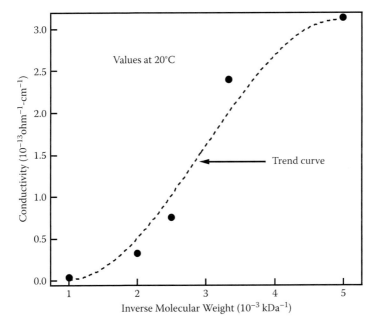

FIGURE 7.33 Dependence of the dc conductivities of DNA(sonicated)-CTMA at 20°C on the inverse of the mean molecular weights. The trend curve is a conjecture showing the expected behavior. (Reprinted with permission from Yaney, P.P. et al., *Proc. SPIE* 7765: 776509-1–8, 2010.)

to illustrate the anticipated flattening out of the dependence for molecular weights above 1000 kDa and below 200 kDa.

This flattening out at the two ends of the Figure 7.33 plot happens for similar reasons. As the length of the molecule increases beyond 0.51 μm (1000 kDa), the orientation distribution of the DNA molecules in the films will approach the point where an increase in molecular weight does not change the orientation distribution significantly from parallel to the electrodes, because of the increasing lengths of the molecules. This description appears supported by the >8000 kDa DNA(as-rcvd)-CTMA curve in Figure 7.24 at the lower temperatures, but shows a change in the flattening behavior with increasing temperature, because of its lower activation energy. At the other extreme, the difference in resistivity between 200 kDa (103 nm length) and 300 kDa (154 nm length) is small, indicating that as the molecular weight decreases, the DNA molecules are randomly oriented over all orientations independent of the length and the resistivity approaches a constant value independent of molecular weight. This analysis applies to the typical spin-cast, micron-size films that are fabricated for DNA resistivity measurements. Of course, there are additional factors that might impact the conformation of the molecules in the film, such as interfacial effects. Additional experiments and modeling are needed.

Fabricating films of DNA–CTMA can be done with some degree of confidence that a given film will be smooth and without pinholes. This depends on the solution viscosity, the spin parameters, temperature of the spin solution, and the temperature of the substrate as well as the effort taken to minimize bubbles in the wet film during and following spinning. The alternating polarity technique typically shows that the measured current is independent of the polarity, which was the reason for choosing non-chemically-reacting gold electrodes.

7.5.5 SUMMARY

One of the key results of these measurements on DNA, Tuft's silk, and oil-derived polymers is that the dc volume resistivity versus temperature behavior of these polymers can be described by the Arrhenius function plus a constant from room temperature up to at least 90°C. This implies that the charge transport mechanism in all of these materials above the corner temperature is dominated by nearest-neighbor hopping. Many of these measurements were repeated a number of times to ensure we have high confidence in this description. These studies have also confirmed that this description of the resistivity behavior can be seen only when the film is thoroughly dried and measured in a totally dry atmosphere. Humidity is the primary factor that can cause the measurements to deviate from the foregoing description. There was some concern early on that the dry nitrogen gas we used could be a factor, particularly in the DNA measurements; however, measurements in a dry argon atmosphere were the same as in the dry nitrogen. Among the materials studied, the DNA materials are the most sensitive to humidity.

From the earliest days of resistivity studies on DNA–CTMA, it has been qualitatively known that high-molecular-weight DNA has a resistivity higher than that of low-molecular-weight DNA. This is confirmed quantitatively for mean molecular weights between 200 and 1000 kDa. An explanation based on the dependence of the

DNA molecule length on the molecular weight and anisotropic mobility relative to the helix axis of the DNA molecules is described.

There is a general increase in the activation energies for the six DNA–CTMA films studied from 0.96 to 1.44 eV, which roughly increases with decreasing molecular weight. This behavior might be due to an increase in the mean spacing of hopping site pairs with decreasing molecular weight, which could point to a decrease in the DNA–CTMA number density, possibly due to CTMA. Although there are significant discrepancies between our activation energy values and those previously reported, our values appear reasonably self-consistent. More measurements and modeling are needed.

Using a simple model for conductivity, large differences in the number density of sites involved in hopping between the DNA(sonicated)-CTMA materials and the oil-derived polymers plus Tufts' silk were revealed in the preexponential parameters obtained from fits of Equation 7.12 to the data. The fact that the as-received DNA did not show these large differences suggests that sonication contributed to these larger hopping site number densities, which is supported by the 200, 300, and 400 kDa sonicated materials having the highest number density values relative to the foregoing materials as determined from the model.

The stark differences in the resistivity behaviors of the two silk films underlines the dependence on the origin and processing history of these complicated biomaterials. Using the DNA results as a reference, it can be suggested that the Tufts' material shows hopping-type conduction, because it has Arrhenius behavior for $T_c > 60°C$ with 0.62 eV activation energy, while the RXBN material shows no Arrhenius-type behavior, and therefore, no evidence of a nearest-neighbor hopping transport. The very high resistivity of this material and the temperature behavior suggests that this material is more amorphous than the Tufts' material and might be showing weak ionic conductivity.

The PCBM and Pt dopants or additives that were complexed with the DNA show well-defined activation energies. Both dopants lowered the activation energy, but PCBM lowered it by more than a factor of two, which is a desirable feature, because it suggests that PCBM reduces either the hopping site barrier or the mean spacing between the hopping sites in the doped DNA material. The Pt dopant did not provide any significant improvement in the conduction behavior over that of DNA(300 kDa)-CTMA. More study is needed to clarify these issues.

The BP and BFE dopants, however, reduced the room temperature resistivity by a factor of 10^4 or more. These are the lowest resistivities with doped DNA we have observed at room temperature. Interestingly, the BP film had a well-defined weak "inverted" Arrhenius-type behavior wherein the resistivity increased with increasing temperature. The increase was only about a factor of two over 20°C to 90°C, meaning that this polymer has nearly constant resistivity similar to the BFE dopant. This temperature independence or near independence of the resistivity in the BP and BFE dopants does not fit any of the usual hopping models for the charge transfer mechanism. This suggests that either nearly free ions or electrons are responsible for charge transport. Although it appears that the DNA functions as a host for controlling the concentration of the highly conductive BP or BFE dopants, more information on how BP and BFE molecules interact with the DNA molecule is needed before this behavior can be understood. AC impedance studies of these dopant–host systems would be useful toward this end.

ACKNOWLEDGMENTS

The authors thank Matthew Dickerson and Rajesh Naik, AFRL/RXBN, for a sample of their silk solution and Fiorenzo Omenetto of Tufts University for sample films of their silk material. We also thank Kristi Singh, AFRL/RXBN, for carrying out the continuous sonication experiment. We are always grateful to Naoya Ogata, Chitose Institute of Science and Technology (Japan), for providing the purified DNA and many fruitful ideas. We appreciate the many efforts of Gerry Landis, UDRI, in providing photolithography and electrode coating for our specimens and Dan Sallenberger, UDRI, for helping us with our electronics. We thank the student and professional members of the Biotronics group of AFRL for their lab work, useful discussions, and suggestions.

REFERENCES

Ahamed, F. and G. Subramanyam. 2004. Design of a Si MMIC compatible ferroelectric varactor shunt switch for microwave applications. *Proc. IEEE Int. Symp. on Applications of Ferroelectrics* 285–288.

Ausubel, F. et al., eds, 1999. Agarose gel electrophoresis. In *Short Protocols in Molecular Biology*, 4th Ed., 2.14–2.16. New York: John Wiley & Sons.

Babcock, K. and C. Prater. 2004. *Phase Imaging: Beyond Topography*. Veeco Instruments. Application Notes.

Bartsch, C. M. 2007. Development of a Field Effect Transistor Using a DNA-Biopolymer as the semiconducting layer. Ph.D. Dissertation, University of Dayton, Dayton, OH.

Bartsch, C. M., G. Subramanyam, J. G. Grote et al. 2007. Bio-organic field effect transistors. *Proc. SPIE* 6646: 66460K.

Bartsch, C. M., G. Subramanyam, J. G. Grote, F. K. Hopkins, L. L. Brott, and R. R. Naik. 2007a. Dielectric and electrical transport properties of biopolymers. *Proc. SPIE* 6470: 64700C.

Bartsch, C. M., G. Subramanyam, J. G. Grote, F. K. Hopkins, L. L. Brott, and R. R. Naik 2007b. A new capacitive test structure for microwave characterization of biopolymers. *Microwave Opt. Technol. Lett.* 49: 1261.

Bartsch, C. M., G. Subramanyam, J. G. Grote, F. K. Hopkins, L. L. Brott, and R. R. Naik. 2006. Microwave dielectric properties of biopolymers. *Proc. SPIE* 6401: 640107.

Böttger, H. and V. V. Bryksin 1985. *Hopping Conduction in Solids*. Berlin: Akademie-Verlag.

Briman, M., N. P. Armitage, E. Helgren et al. 2004. Dipole relaxation losses in DNA. *Nanoletters* 4: 733–736.

Cui, P., J. Wu, G.-Q. Zhang et al. 2008. Hole polarons in poly(G)-poly(C) and poly(A)-poly(T) DNA molecules. *Sci. China Ser. B-Chem.* 51: 1182–1186.

Daire, A. 2004. Improving the Repeatability of Ultra-High Resistance and Resistivity Measurements. Keithley White Paper, Keithley Instruments, Inc.: http://www.keithley.com/support.

Deng, H., V. A. Bloomfield, J. M. Benevides et al. 1999. Dependence of the Raman signature of genomic *B*-DNA on nucleotide base sequence. *Biopolymers* 50: 656–666.

Dutta1, P. and S. K. Mandal. 2004. Charge transport in chemically synthesized, DNA-doped polypyrrole. *J. Phys. D: Appl. Phys.* 37: 2908–2913.

Eley, D. D. and D. I. Spivey. 1962. Semiconductivity of organic substances. *Trans. Faraday Soc.* 58: 411–415.

Fink, H.-W. and C. Schönenberger. 1999. Electrical conduction through DNA molecules. *Nature* 398: 407–410.

Giese, B., J. Amaudrut, A.-K. Köhler et al. 2001. Direct observation of hole transfer through DNA by hopping between adenine bases and by tunneling. *Nature* 412: 318–320.

Grote, J. G., E. M. Heckman, J. A. Hagen et al. 2006. DNA: new class of polymer. *Proc. SPIE* 6117: 61170J.

Grote, J. G., E. M. Heckman, D. E. Diggs et al. 2005. DNA-based materials for electro-optic applications: current status. *Proc. SPIE* 5934: 593406.

Grote, J. G., E. M. Heckman, J. A. Hagen et al. 2004a. Deoxyribonucleic acid (DNA)-based optical materials. *Proc. SPIE* 5621: 16–22.

Grote, J. G., J. Hagen, J. Zetts et al. 2004b. Investigation of polymers and marine-derived DNA in optoelectronics. *J. Phys. Chem. B* 108: 8584–8591.

Grote, J. G., N. Ogata, J. A. Hagen et al. 2003. Deoxyribonucleic acid (DNA)-based nonlinear optics. *Proc. SPIE* 5211: 53–62.

Grozema, F. C., L. D. A., Siebbeles, Y. A. Berlin et al. 2002. Hole mobility in DNA: effects of static and dynamic structural fluctuations. *ChemPhysChem.* 6: 536–539.

Ha, D. H., H. Nham, K.-H. Yoo et al. 2002. Humidity effects on the conductance of the assembly of DNA molecules. *Chem. Phys. Lett.* 355: 405–409.

Hagen, J. A., J. G. Grote, K. M. Singh, R. R. Naik, T. B. Singh, and N. S. Sariciftci. 2007. Deoxyribonucleic acid biotronics. *Proc. SPIE* 6470: 64700B-4.

Hagen, J., W. Li, A. Steckl, and J. Grote. 2006. Enhanced emission efficiency in organic light-emitting diodes using deoxyribonucleic acid complex as an electron blocking layer. *Appl. Phys. Lett* 88: 171109-1–3.

Heckman, E. M. 2006. The Development of an All-DNA-Based Electro-Optic Waveguide Modulator. Ph.D. Dissertation, University of Dayton, Dayton, OH.

Heckman, E. M., P. P. Yaney, J. G. Grote, and F. K. Hopkins. 2006a. Development and performance of an all-DNA-based electro-optic waveguide modulator. *Proc. SPIE* 6401: 640108.

Heckman, E., P. Yaney, J. Grote, F. Hopkins, and M. Tomczak. 2006b. Development of an all-DNA-surfactant electro-optic modulator. *Proc. SPIE* 6117: 61170K.

Heckman, E. M., J. G. Grote, F. K. Hopkins, and P. P. Yaney. 2006c. Performance of an electro-optic waveguide modulator fabricated using a deoxyribonucleic-acid-based biopolymer. *Appl. Phys. Lett.* 89: 181116.

Heckman, E., P. Yaney, J. Grote, and F. Hopkins. 2005a. Poling and optical studies of DNA NLO waveguides. *Proc. SPIE* 5934: 593408.

Heckman, E. M., J. A. Hagen, P. P. Yaney, J. G. Grote, and F. K. Hopkins. 2005b. Processing techniques for deoxyribonucleic acid: Biopolymer for photonics applications. *Appl. Phys. Lett.* 87: 211115.

Heckman, E. M., J. G. Grote, P. P. Yaney, and F. K. Hopkins. 2004. DNA-based nonlinear photonic materials. *Proc. SPIE* 5516: 47–51.

Jin, H.-J., J. Park, V. Karageorgiou et al. 2005. Water-stable silk films with reduced-sheet content. *Adv. Funct. Mater.* 15: 1241–1247; the silk-coated resistivity slides were provided by Fiorenzo Omenetto of Tufts University.

Kuchel, P. and G. Ralston. 1998. *Schaum's Outline of Theory and Problems of Biochemistry.* New York: McGraw-Hill.

Kutnjak, Z., G. Lahajnar, C. Filipic et al. 2005. Electrical conduction in macroscopically oriented deoxyribonucleic and hyaluronic acid samples. *Phys. Rev. E* 71: 041901-1–8.

Lakhno, V. D. and N. S. Fialko. 2003. Hole mobility in a homogeneous nucleotide chain. *JTEP Lett.* 78: 336–338.

Miller, A. and B. Abrahams. 1960. Impurity conduction at low concentrations. *Phys. Rev.* 120: 745.

Okahata, Y., T. Kobayashi, K. Tanaka et al. 1998. Anisotropic electric conductivity in an aligned DNA cast film. *J. Am. Chem. Soc.* 120: 6156–6166.

Ortmann, F., K. Hannewald, and F. Bechstedt. 2009. Charge transport in guanine-based materials. *J. Phys. Chem.* 113: 7367–7371.

Ouchen, F., S. N. Kim, M. Hay et al. 2008. DNA-Conductive polymer blends for applications in biopolymer based field effect transistors (FETs). *Proc. of SPIE* 7040: 704009-1.

Ouchen, F., P. P. Yaney, and J. G. Grote. 2009. DNA thin films as semiconductors for BioFET. *Proc. SPIE* 7403: 74030F.

Phillips, D. M., L. F. Drummy, D. G. Conrady et al. 2004. Dissolution and regeneration of Bombyx mori silk fibroin using ionic liquids. *J. Am. Chem. Soc*. 126: 14350–14351; solution supplied by Matthew Dickerson and Rajesh Naik of AFRL/RXBN, WPAFB, OH.

Porath, D., A. Bezryadin, S. de Vries et al. 2000. Direct measurement of electrical transport through DNA molecules. *Nature* 403: 635–637.

Pozar, D. M. 1998. *Microwave Engineering.* New York: Wiley.

Prater, C., P. Maivald, K. Kjoller, and M. Heaton. 2004. *Tappingmode Imaging Applications and Technology.* Veeco Instruments, Application Notes.

Rodger, A. and B. Norden. 1997. DNA-ligand interactions. In *Circular Dichroism and Linear Dichroism*, 30–31. New York: Oxford University Press.

Russell, P., D. Batchelor, and J. T. Thornton. 2004. *SEM and AFM: Complementary Techniques for High Resolution Surface Investigations.* Veeco Instruments, Application Notes.

Samoc, A., M. Samoc, J. G. Grote et al. 2006. Optical properties of deoxyribonucleic acid (DNA) polymer host. *Proc. SPIE* 6401: 640106-1–10.

Seanor, D. A. 1982. *Electrical Properties of Polymers.* Orlando, FL: Academic Press, chap. 1.

Serry, F., K. Kjoller, J. Thornton, R. Tench, and D. Cook. 2004. *Electric Force Microscopy, Surface Potential Imaging, and Surface Electric Modification with the Atomic Force Microscope.* Veeco Instruments, Application Notes.

Singh. 2004. Polaron transport mechanism in DNA. *J. Biomater. Sci. Polymer Edn.* 15: 1533–1544.

Subramanyam, G., C. M. Bartsch, J. G. Grote et al. 2009. Effect of external electrical stimuli on DNA-based biopolymers. *NANO: Brief Rep. Rev.* 4: 69–76.

Subramanyam, G., E. Heckman, J. Grote et al. 2005. Microwave dielectric properties of marine DNA based polymers. *Microwave Opt. Technol. Lett.* 46: 278–82.

Takada, T., K. Kawai, M. Fujitsuka et al. 2006. Rapid long-distance hole transfer through consecutive adenine sequence. *J. Am. Chem. Soc.* 128: 11012–11013.

Tinoco, I., K. Sauer, J. Wang, and J. Puglisi. 1995. Circular dichroism of nucleic acids and proteins. In *Physical Chemistry: Principles and Applications in Biological Sciences*, 585–588. Saddle River, NJ: Prentice Hall.

Voityuk, A. 2008. Conformations of poly{G}-poly{C} stacks with high hole mobility. *J. Chem. Phys.* 128: 045104-1–6.

Yaney, P. P., F. Ouchen, and J. G. Grote. 2009. Characterization of polymer, DNA-based, and silk thin film resistivities and of DNA-based films prepared for enhanced electrical conductivity. *Proc. SPIE* 7403: 74030M.

Yaney, P. P., F. Ahmad, and J. G. Grote. 2008. Raman microprobe spectroscopic studies of solid DNA-CTMA films. *Proc. SPIE* 7040: 70400N.

Yaney, P. P., E. M. Heckman, and J. G. Grote. 2007. Resistivity and electric-field poling behaviors of DNA-based polymers compared to selected non-DNA polymers. *Proc. SPIE* 6646: 664605.

Yaney, P. P., E. M. Heckman, A. A. Davis et al. 2006. Characterization of NLO polymer materials for optical waveguide structures. *Proc. SPIE* 6117: 61170W.

Yoo, K.-H., D. H. Ha, J.-O. Lee et al. 2001. Electrical conduction through poly(dA)-poly(dT) and poly(dG)-poly(dC) DNA molecules. *Phys. Rev Lett.* 87: 198102-1–4.

Yu, Z. G. and X. Song 2001. Variable range hopping and electrical conductivity along the DNA double helix. *Phys. Rev. Lett.* 86: 6018–6021.

Zamenhof, S. 1957. Preparation and assay of deoxyribonucleic acid from animal tissue. In *Methods in Enzymology*, ed. S. Colowick and N. Kaplan, 696–703. New York: Academic Press.

Zhang G., L. Wang, J. Yoshida, and N. Ogata. 2001. Optical and optoelectronic materials derived from biopolymer, deoxyribonucleic acid (DNA). *Proc. SPIE* 4580: 337–346.

Zhang, G. et al. 2002. Nonlinear optical materials derived from biopolymer (DNA)-surfactant-azo dye complex. *Proc. SPIE* 4905: 375–380.

Zhou, Y., Y. Zhou, D. J. Klotzkin et al. 2007. Stimulated emission of sulforhodamine 640 doped DNA distributed feedback (DFB) laser specimens. *Proc. SPIE* 6470: 64700V.

8 Applications of DNA to Photonics and Biomedicals

Naoya Ogata

CONTENTS

8.1 Introduction 231
8.2 Photonic Applications of DNA 232
    8.2.1 Stability Improvements of DNA Photonic Devices by Blending
        with Synthetic Polymers 233
        8.2.1.1 Experimentals 233
        8.2.1.2 Results and Discussion 235
    8.2.2 Chelation of DNA with Novel Metals or Rare Earth Metal
        Compounds 237
    8.2.3 Conclusion 242
8.3 Biomedical Application of DNA Films 245
    8.3.1 UV Cross-Linking of DNA Films 245
    8.3.2 Cell Culture on DNA Films 246
    8.3.3 Wound-Healing Effect of DNA and Cross-Linked DNA Films 247
8.3 Conclusion 250
References 253

8.1 INTRODUCTION

Deoxyribonucleic acid (DNA) carries the genetic information of all living things and is well known to form a double-helical structure in which layers of four nucleic acids, namely, adenine, thymine, guanine, and cytosine, are stacked. DNA has a huge molecular weight of over billions and can form a clear film, while DNA is water soluble with sodium counterions that are not appropriate for applying DNA to material sciences such as electronic devices. However, DNA molecules become insoluble in water, yet become soluble in polar organic solvents such as ethanol, when sodium cations are replaced with quaternary ammonium salts, lipids that contain long alkyl chains to form DNA–lipid complexes, and clear and tough films are easily obtained by solvent casting of ethanol solutions (Wang et al. 2001).

Pure DNA was available from salmon roe in amounts of over 1,000 ton/year in Hokkaido, Japan, and industrial processes for the isolation and purification of DNA from salmon roe were already established in 1998. A semicommercial plant was built in 2000 at Hakodate in Hokkaido, Japan, so that materials are now available for applications of DNA in such areas as photonics, separation process, or biomedical materials.

The most characteristic features of DNA for photonic applications are light amplification of optical dyes that are intercalated (inserted) into the double helix of DNA molecules. The light amplification of optical dyes by DNA is so large that the fluorescent light of dyes is amplified to over 150 times in comparison with the case when optical dyes are dissolved in normal polymers such as poly(methyl methacrylate) (PMMA). Thus, recent research results on DNA–lipid complexes have shown various attractive applications for such photonic devices as optical memories, switches, and sensors (Wang et al. 2001; Yoshida et al. 2004, 2005; Heckman et al. 2004; Yaney et al. 2005; Watanuki et al. 2005).

DNA molecules are essentially biopolymers derived from salmon roe, and are biocompatible and nontoxic for all living things, including human beings. This chapter describes further applications of DNA for material sciences in the areas of not only photonics and electronics but biomedicals, including cell culture and wound-healing effects.

8.2 PHOTONIC APPLICATIONS OF DNA

Although DNA photonics and electronics are very attractive for device applications, there are problems with DNA optical devices related to moisture absorption of DNA molecules, which are very much hydrophilic. Adsorbed water influences the dye-intercalated structures of DNA molecules. Therefore, it is necessary to protect the dye-intercalated state of DNA molecules by sealing off water penetration. It was previously reported (Yoshida et al. 2005) that a novel hybridization method of the dye-intercalated DNA molecules by means of the so-called sol–gel process was effective to increase stabilities and durability of DNA photonic devices under environmental changes. The concept of the sol–gel process is applied to DNA devices as follows: encapsulation of the dye-intercalated DNA–lipid complex by the sol–gel process was carried out by dissolving the dye-intercalated DNA–lipid complex into tetraethoxy silane $(EtO)_4Si$ (TEMOS) containing UV-curable acrylate groups with stirring at room temperature, followed by cross-linking reactions by irradiating UV lamp to encapsulate DNA photonic devices, as schematically shown here (Scheme 8.1).

Although the sol–gel process for the encapsulation of DNA devices is effective, there are problems for limited dissolution of DNA complexes in sol derivatives, and special types of chiral lipids such as l-alanine-derived lipid as counterions of DNA were used to obtain homogeneous sol-containing DNA complexes to prepare gel films.

On the other hand, hybridizations or blending methods of DNA–lipid–dye complexes with hydrophobic synthetic polymers were much easier to prepare devices by a simple casting process. This paper describes novel blending methods for the encapsulation of DNA devices by synthetic polymers such as hydrophobic fluorinated poly(methyl methacrylate) in order to attain stability improvements of DNA devices.

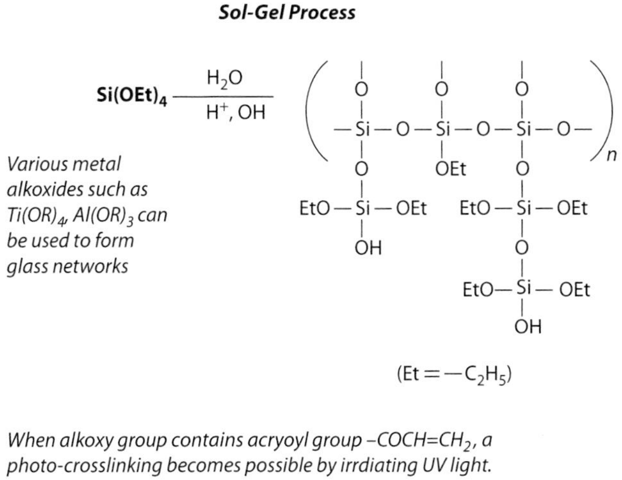

Sol-Gel Process

$$\text{Si(OEt)}_4 \xrightarrow[\text{H}^+,\,\text{OH}]{\text{H}_2\text{O}}$$

Various metal
alkoxides such as
$Ti(OR)_4$, $Al(OR)_3$ can
be used to form
glass networks

$(Et = -C_2H_5)$

When alkoxy group contains acryoyl group –COCH=CH$_2$, a
photo-crosslinking becomes possible by irrdiating UV light.

SCHEME 8.1   Preparation of crosslinkable organo polysilicates.

8.2.1   Stability Improvements of DNA Photonic Devices by Blending with Synthetic Polymers

8.2.1.1   Experimentals

8.2.1.1.1   Preparation of DNA–Lipid Complex Films (Yamaoka and Ogata 2004; Wu et al. 2004)

Figure 8.1 shows the preparative method of DNA–lipid complex films. A single-chain trimethyl ammonium-type lipid (CTMA hereafter) was used to form DNA–lipid complexes. First, refined DNA was dissolved in distilled water. The lipid solution dissolved in distilled water was mixed with the DNA aqueous solution, and then the DNA–lipid complex was washed in distilled water, followed by a drying process in a vacuum oven for 24 h at 40°C. After the drying process, the DNA–lipid complex was dissolved in a mixed solution of EtOH:CHCl$_3$ = 1:4, together with optical dye compounds. Finally, the solution was poured onto a Teflon-coated dish, followed by evaporation of the solvent to obtain films, which is described schematically in Figure 8.1.

8.2.1.1.2   Chelation of DNA with Novel Metals or Rare Earth Metal Compounds

Chelation of DNA with novel metal compounds was carried out by dissolving pure DNA in water in an amount of 1wt%, followed by the addition of novel metal compounds such as HAuCl$_4$, NaAuCl$_4$, and PdCl$_2$, which were dissolved in water in amounts of equal moles of metal ions to 1 mol DNA (calculated as unit of 4 nucleic acid bases [adenine thymine, cytosine, and guanine]). Also, rare earth metal compounds such as EuCl$_3$, TbCl$_3$, and NdCl$_3$ were used for the chelation with DNA in a similar way. No precipitation occurred for DNA–Au ion or rare earth metal ion complexes, while a pinky and bulk precipitate was formed for the DNA–Pd complex. However, DNA remained in these aqueous solutions as UV-Vis spectra

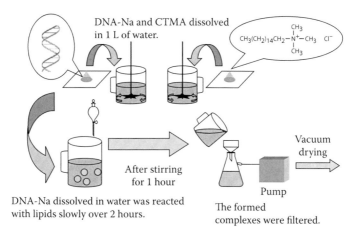

FIGURE 8.1   Preparative method of DNA–lipid complex films.

indicated absorptions owing to DNA molecules, which is indicated in Figure 8.2. However, absorption peaks for DNA–Eu or DNA–Pd were shifted toward lower wavelengths in comparison with DNA–Na, indicating the chelating effect of these metal ions with DNA.

DNA–CTMA was dissolved in ethanol in an amount of 1 wt%, followed by the addition of novel metal or rare earth metal ions as described earlier. Addition of $PdCl_2$ to the ethanol solution of DNA–CTMA caused a pink, bulky precipitate that was eliminated to measure UV-Vis spectra, as shown in Figure 8.3.

It is seen in Figure 8.3 that the DNA–CTMA–Pd complex showed a much lower absorption peak in comparison with the DNA–CTMA–Eu complex, indicating a strong chelate effect of Pd to DNA–CTMA molecules as in the case of DNA in an aqueous solution.

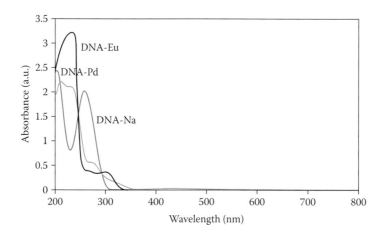

FIGURE 8.2   UV-Vis spectra of aqueous solutions of DNA–metal complexes.

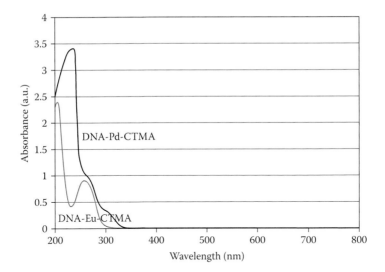

FIGURE 8.3   UV-Vis spectra of DNA–CTMA–metal complexes in ethanol solutions.

8.2.1.2   Results and Discussion

8.2.1.2.1   Blending of DNA–Lipid–Dye Complexes with Synthetic Polymers

Table 8.1 summarizes the results of testing for dissolution in various solvents in a concentration of 1 g of DNA–CTMA complex in 100 mL solvent at room temperature under stirring with a magnetic stirrer. When a solvent could dissolve the DNA–CTMA complex to give a clear solution, an evaporation process was carried out under slightly reduced pressure to prepare a film of the DNA–CTMA complex. When a phase separation of a film occurs, no clear and transparent films were obtained. Table 8.1 indicates hexafluoroisopropanol is the best solvent to obtain a clear and homogeneous film of the DNA–CTMA complex.

It was reported earlier that the DNA–CTMA–dye complexes could uniformly blend with poly(methyl methacrylate) (PMMA) by a solution-blending method to form a clear film. However, PMMA still absorbs a little water under 100% relative humidity. When the methyl ester group is replaced with the benzyl group, it is expected that poly(benzyl methacrylate) (PBMA) is more hydrophobic than PMMA. Based on this expectation, an in situ polymerization of benzyl methacrylate (BMA) was carried out by dissolving the DNA–CTMA–dye complex into monomeric benzyl methacrylate in the presence of a photoinitiator and by irradiating UV light for the in situ polymerization. The DNA–CTMA–dye complex could dissolve in BMA in an amount of 20 wt%, and a clear and transparent film was obtained after the in situ polymerization by UV irradiation.

Water absorption behaviors of the DNA–CTMA–dye–PBMA film were measured by weight increases of the film under various relative humidities, and the results are shown in Figure 8.4.

TABLE 8.1

Blending of Dye-Intercalated DNA–CTMA with Synthetic Polymers

Synthetic Polymers	Ethanol	Chloroform	Dichloroethane	Hexa F-2-Isopropoanol	Film's Appearance
DNA–CTMA	O	X	X	O	Clear film
Polycarbonate SP	X	O	O	X	Half-transparent film
Nylon CM8000	X	O	O	O	White film
PMMA	X	O	O	O	Clear film
Parmax2000	X	X	O	X	White film

Note: O: soluble, X: insoluble, Parmax 2000: poly(*p*-phenylene) attached with phenoxy benzophenone

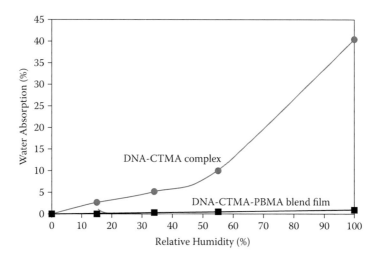

FIGURE 8.4 Water absorption of DNA–CTMA–PBMA blend film.

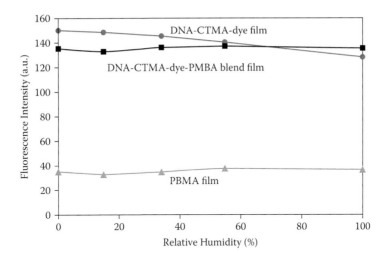

FIGURE 8.5 Fluorescence intensities (a.u) of DNA–CTMA–dye–PMBA films under various relative humidities.

Fluorescence intensities (a. u.) as functions of relative humidity are shown in Figure 8.5, which showed no changes under various humidity conditions. Thus, DNA–CTMA–dye sealed in PBMA becomes very stable even under 100% relative humidity.

It is expected that more hydrophobic fluorinated PMMA may be more effective for environmental changes. Fluorinated PMMA (3FMA) kindly provided by Prof. H. Koike of the Keio University was used for the solution blending of the DNA–CTMA–dye complexes by using hexafluoroisopropanol (HFIP) as a solvent. A clear and transparent film was obtained after evaporating HFIP.

Water absorption of the DNA–CTMA–dye–3FMA film containing 20 wt% of the DNA–CTMA–dye complex was measured under various relative humidities as shown in Figure 8.6, which was determined by weight increases of the completely dried film. It can be seen in Figure 8.6 that no water was absorbed in the blend film.

Figure 8.7 indicates fluorescence intensities as functions of relative humidity that show no change when the DNA–CTMA–dye complex was blended with 3FMA, which does not absorb water. Thus, remarkable stability improvements of the DNA devices were attained by blending the DNA devices with hydrophobic and miscible synthetic polymers.

8.2.2 CHELATION OF DNA WITH NOVEL METALS OR RARE EARTH METAL COMPOUNDS

Doping of various Eu compounds with DNA–CTMA was performed in an ethanol solution, followed by evaporating ethanol to form films, and fluorescence quantum yields were measured by excitation with YAG laser in comparison with PMMA as a base polymer. The results are summarized in Table 8.2.

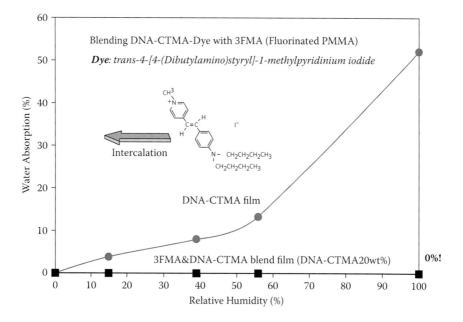

FIGURE 8.6 Water absorption of the blend film under various relative humidities.

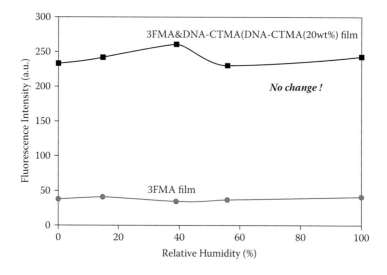

FIGURE 8.7 Fluorescence intensities as functions of relative humidity.

TABLE 8.2
Fluorescence Quantum Yields of Eu-Doped DNA–CTMA and PMMA

Doping of DNA with Rare Earth Metal Compound
(Comparison between DNA and PMMA
for fluorescence quantum yields)

	Eu-TTA			Eu-TFC	
	PMMA	DNA-CTMA		PMMA	DNA-CTMA
	12.2%	17.5%		0.75%	6.8%
	Eu-FOD			Eu-HFC	
	PMMA	DNA-CTMA		PMMA	DNA-CTMA
	4.3%	14.3%		0.57%	3.6%
	Eu-DPM			Eu-PTA	
	PMMA	DNA-CTMA		PMMA	DNA-CTMA
	0.33%	0.5%		5.7%	9.1%

It is seen in Table 8.2 that fluorescence quantum yields in DNA–CTMA films slightly depended on structures of Eu chelate compounds, while they are much higher than those compositions in which PMMA was used as a substrate. Thus, the chelate effect of Eu metal to DNA–CTMA enhanced fluorescence emission much more than PMMA.

When EU–FOD was doped into DNA–CTMA, a very strong and bright emission of fluorescence light was observed as shown in Figure 8.8.

It was reported by Prof. J.-I. Jin (Kwon et al. 2009) of Korea University that a chelation occurred when DNA was doped by $HAuCl_4$ (see Scheme 8.2).

The chelation of DNA with Au^{+++} induced formation of a single-chain DNA, which was verified by measuring the CD spectra as a function of Au^{+++} concentration, as shown in Figure 8.9.

He described that the chelation of DNA with Au^{+++} cation induced a magnetic property that is very interesting in terms of applications of DNA to electronics such as field-effect transistors. When Pd^{++} or Eu^{+++} cations were added to DNA aqueous solutions, CD spectra became completely flat as shown in Figure 8.10, indicating that the DNA double-helical chain lost a single chain.

A single chain formation of DNA could be confirmed by intercalating ethidium bromide (EtBr) to DNA since the fluorescence intensity of EtBr is greatly enhanced by the intercalation of EtBr into the double-helical structure of DNA. Figure 8.11 indicates the fluorescence intensity of EtBr in the presence of Eu^{+++} or Pd^{++}, and it is seen in Figure 8.11 that less fluorescence intensity was observed, indicating a single chain formation of DNA.

FIGURE 8.8 (**See color insert.**) Fluorescence emission of DNA–CTMA film doped with Eu–FOD.

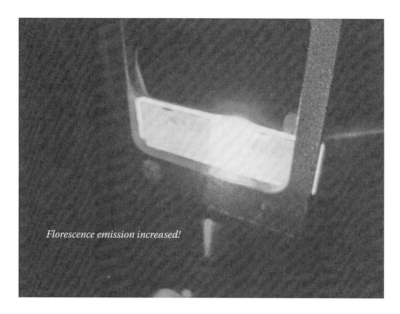

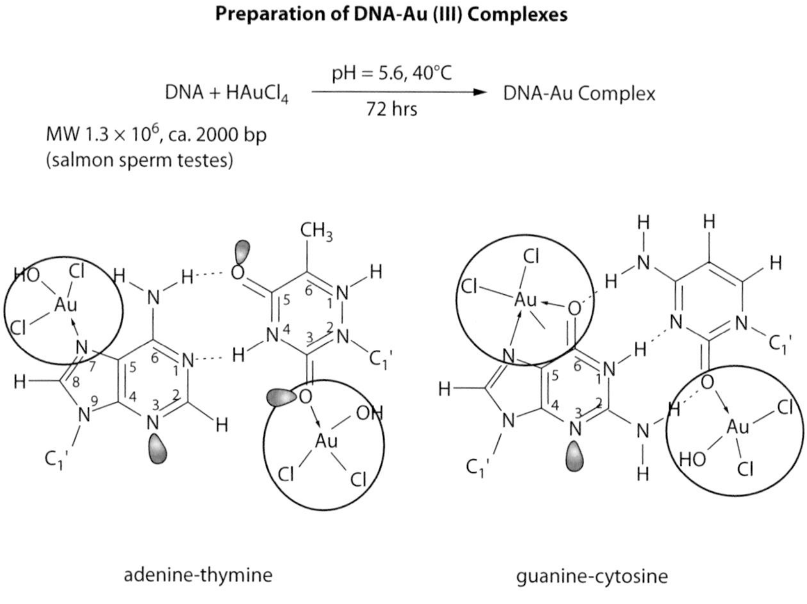

SCHEME 8.2 Chelating cites of DNA bases.

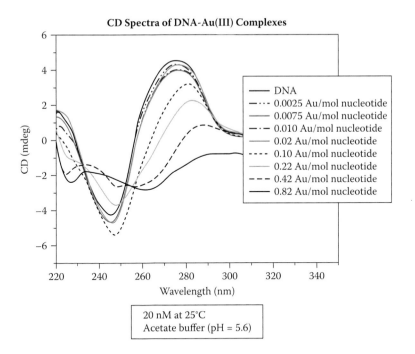

FIGURE 8.9   CD spectra of DNA–Au^{+++} complexes. (Reprinted with permission from Y.-W. Kwon and J.-I. Jin, unpublished result.)

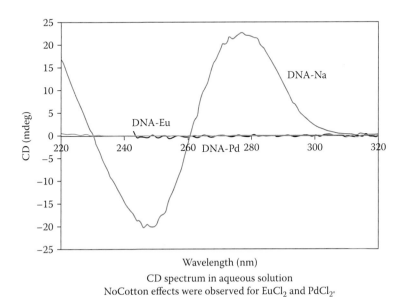

FIGURE 8.10   CD spectra of DNA in the presence of Eu^{+++} or Pd^{++}.

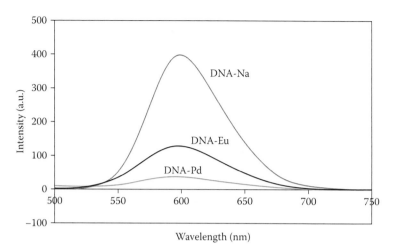

Each concentration: 0.01 wt%, EtBr: 0.01 wt%
Large decreases in fluorescence intensity. No intercaltion of EtBr in DNA

FIGURE 8.11   Fluorescence intensity of DNA in the presence of Eu^{+++} or Pd^{++}.

Fluorescence intensities of an aqueous solution of DNA in the presence of Pd^{++}, Eu^{+++}, Tb^{+++}, and Nd^{+++} were measured, and the results are summarized in Figures 8.12–8.15, respectively. These figures indicate large amplifications of fluorescence light in the presence of novel metal or rare earth metal cations; especially, Eu^{+++} or Tb^{+++} showed a very large fluorescence emission with increasing amounts of DNA. Figure 8.16 shows the effects of DNA to rare earth metal cation ratios, which indicate that a large fluorescence emission occurred with increasing ratios of DNA to rare earth metal cations, especially in the range of more than two of DNA to Eu^{+++} or Tb^{+++}. These results strongly suggest that electron shells would be twisted, and the electron state, including electron spin, would be changed by interactions with nucleic acid bases of DNA. These results suggest that the DNA magnet would be feasible, as Prof. Jin suggested. The way will be paved for novel electronic applications of DNA by these results, and further research on DNA electronics is required.

8.2.3 CONCLUSION

Blending methods of more hydrophobic fluorinated PMMA and DNA–CTMA–dye complexes were effective for stability improvements of DNA devices by environmental changes. Fluorinated PMMA (3FMA) and DNA–CTMA–dye–3FMA were miscible to form a clear and transparent film. The film containing 20 wt% of DNA–CTMA–dye complex did not absorb water under 100% relative humidity, and fluorescence intensities as functions of relative humidity did not change at all. Thus, stability improvements of the DNA devices were attained by blending the DNA devices with hydrophobic and miscible synthetic polymers.

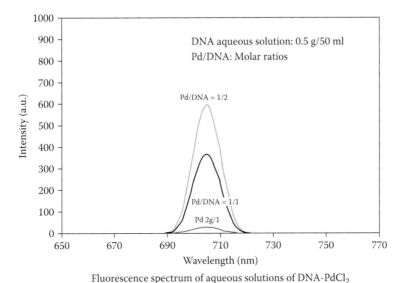

Fluorescence spectrum of aqueous solutions of DNA-PdCl$_2$

FIGURE 8.12　Fluorescence intensity of DNA/Pd^{++} solution with increasing ratios of Pd^{++}.

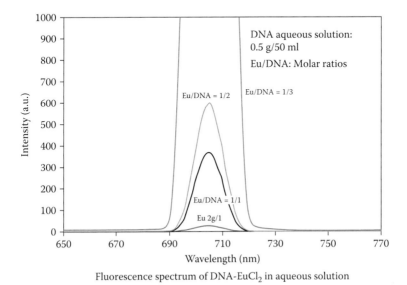

Fluorescence spectrum of DNA-EuCl$_2$ in aqueous solution

FIGURE 8.13　Fluorescence intensity of DNA/Eu^{+++} solution with increasing ratios of Eu^{+++}.

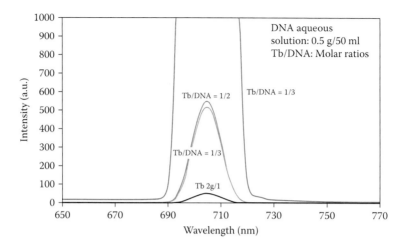

FIGURE 8.14 Fluorescence intensity of DNA/Tb^{+++} solution with increasing ratios of Tb^{+++}.

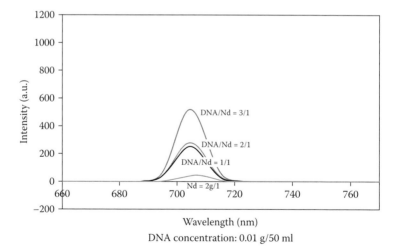

FIGURE 8.15 Fluorescence intensity of DNA/Nd^{+++} solution with increasing ratios of Nd^{+++}.

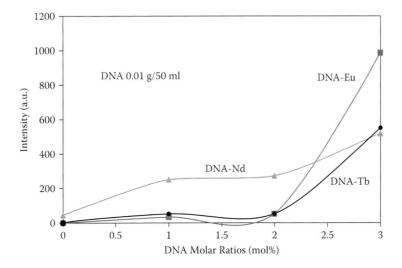

FIGURE 8.16 Fluorescence intensity as functions of DNA ratios.

Large amplifications of fluorescence light were attained in the presence of novel metal or rare earth metal cations; especially Eu^{+++} or Tb^{+++} showed a very large fluorescence emission with increasing amounts of DNA. These results strongly suggest that electron shells would be twisted and the electron state including electron spin would be changed by interactions with nucleic acid bases of DNA. Novel electronic applications of DNA will be expanded.

8.3 BIOMEDICAL APPLICATION OF DNA FILMS

Salmon-derived DNA is soluble in water, so it is possible to prepare clear and transparent films of the DNA by evaporating an aqueous solution of the DNA under a reduced pressure at 50°C. Since the DNA is biocompatible and nontoxic, it is expected that DNA films can be applied for biomedical applications such as cell culture. However, since the DNA is soluble in water, it is necessary to insolubilize the DNA films in water because an aqueous cell culture medium is used for cell growth. Several methods of water-soluble DNA for insolubilization in water such as replacements of sodium cation of DNA by quaternary ammonium cation or metal polycations such as calcium cation. However, these insolubilization methods of the DNA may be accompanied by toxic problems for cell growth. The best insolubilization method would be UV irradiation of the DNA films for cross-linking reactions, so UV irradiation conditions of the DNA films were investigated in terms of structural analyses.

8.3.1 UV CROSS-LINKING OF DNA FILMS

High-molecular-weight DNA of over 6.6 million can be prepared as a transparent film by casting an aqueous solution (5 wt%) of the DNA and by slowly evaporating water at 40°C on a Teflon-coated glass plate. It has been known that UV light

TABLE 8.3
UV-Curing Conditions of DNA Films

Initiators (1 wt%)	Time of Irradiation	Solubility in Water
Darocure	30 s, 1, 3, 6, 10, 20 min.	All insoluble
Ilgacure	30 s.,1, 3, 6, 10, 20 min.	All insoluble
None	1, 2, 3, 5, 10, 15, 30 min.	All insoluble

Note: UV power: 160 mW/cm^2; irradiation distance between lamp and DNA films: 10 cm.

irradiation of DNA causes cross-linking reactions among DNA molecules to insolu-bilize DNA molecules in water. It has been assumed that the cross-linking reactions of the DNA molecules might be related with a dimerization reaction of thymine within the DNA molecules. However, since the DNA molecules are well known to form a stable double-helical structure, the dimerization reaction of two different thymine molecules may not occur as the thymine molecules are fixed in the double helix of the DNA molecules through hydrogen bonding with adenine. Precise reaction conditions for the cross-linking reactions have not been ever reported, so the UV-curing conditions were investigated in terms of applications of water-insoluble DNA films for biomedical applications such as cell culture membranes.

A Molitex UV lamp was used for the cross-linking reaction molecules. This UV lamp is so powerful that no UV-curing agents are necessary to start the cross-link-ing reactions. UV cross-linking reactions of the DNA containing UV-curing agents (1 wt%) and the pure DNA films were compared in terms of insolubilization in water, and the results are shown in Table 8.3, which indicates that the cross-linking reaction of DNA occurred by irradiating for only 1 min without curing agents when the UV powerful lamp was used.

Differential infrared spectra of nonirradiated and irradiated DNA films were mea-sured as shown in Figure 8.17, which indicated strong enhanced peaks at 1226 and 1062 cm^{-1}. These peaks correspond to the –O-P-O- bond. Therefore, the UV-cross-linking reaction of DNA is assumed to be owing to an excitation of the P=O bond to form O-P-O bonds, resulting in cross-linking among DNA molecules.

Surface pictures by a scanning electron microscope of these UV-cured DNA films are shown in Figure 8.18, which shows the surface of DNA film of 5 and 20 in thicknesses, respectively.

It is clearly seen in Figure 8.18 that the surface of the thick DNA film of 20 μm showed much rough surface structure in comparison with the surface of the DNA film of 5 μm. It is presumed that a thick DNA film would cause much shrinkage with increasing thickness of the DNA films.

8.3.2 CELL CULTURE ON DNA FILMS

Cell culture on the UV-irradiated DNA films was carried out by using rat carti-lage cells (ATDC5) at 37°C in an incubator under CO_2 atmosphere. As shown in

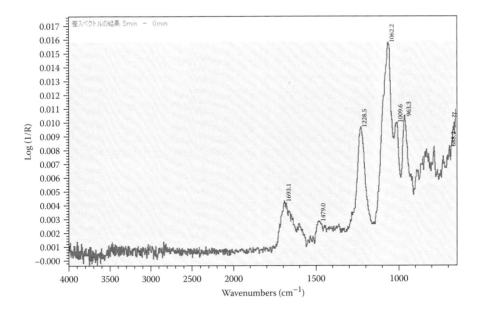

FIGURE 8.17 Differential infrared spectrum between nonirradiated and irradiated DNA films.

Figure 8.19, cells grew on the surface of the DNA film in comparison with a control, that is, a normal glass plate. The cell growth rates were measured by counting number of cells that grew on the DNA film. And results are summarized in Figure 8.20. It is seen in Figure 8.20 that the growth rates were dependent on the thickness of the DNA films, indicating that thinner DNA films were better in terms of the growth rate of the cells.

The thickness dependence of the cell growth would be related to the surface structure of the DNA films as described in the previous section. More precise research is needed in terms of cell adhesion and surface structures of DNA films

8.3.3 WOUND-HEALING EFFECT OF DNA AND CROSS-LINKED DNA FILMS

It was confirmed before that gas permeation of DNA and cross-linked DNA films was very high in terms of oxygen, carbon dioxide, and water vapor gases, and also, no DNA toxicity was found, which was expected since the DNA was isolated from salmon roe. Animal testing for wound-healing effects of DNA films was carried out by using the wounded skin of rats, and the results are summarized in Figure 8.21 (Ogata 2009). The wounded rat skin without attached DNA film did not recover after 14 days, while the wounded rat skin with attached DNA almost recovered as shown in Figure 8.21.

A UV-cured DNA film was attached to one half of the wounded skin, and the other half of the skin was left as it was. The recovery states of the wounded skin were compared as shown in Figure 8.22. The DNA-film-attached part of the wounded skin recovered much faster than the part without the DNA attachment. A cross-section of

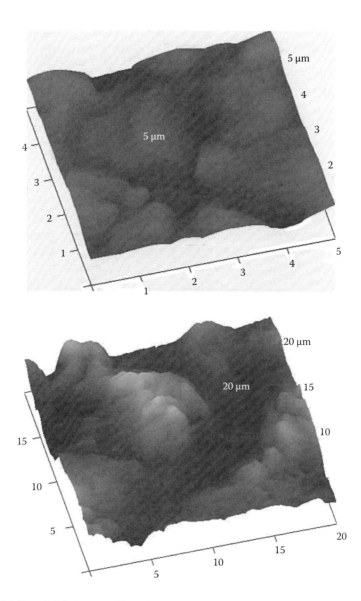

FIGURE 8.18   AFM pictures of irradiated DNA.

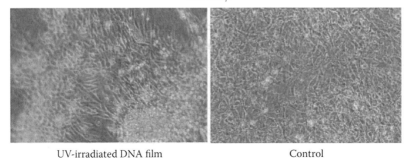

ATDC5 (Rat Cartilage) Cell Culture
on UV Irradiated DNA Films
After 15 Days

UV-irradiated DNA film   Control

FIGURE 8.19   Cell culture of ATDC5 cells on UV-cured DNA film.

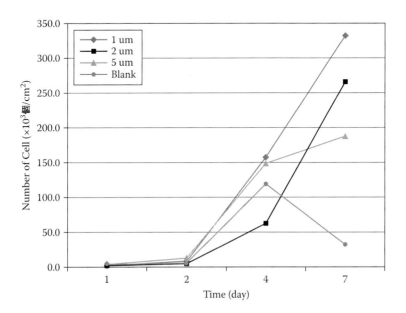

FIGURE 8.20   Cell (ATDC5) growth number as functions of incubation days.

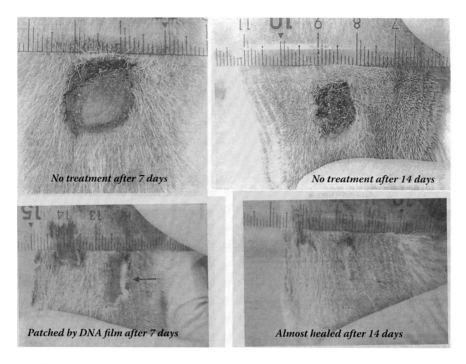

No treatment after 7 days

No treatment after 14 days

Patched by DNA film after 7 days

Almost healed after 14 days

FIGURE 8.21   Wounded skin recovery of a rat skin.

the wounded skin tissue of a rat is also shown in Figure 8.22, which indicates that the DNA-film-attached part of the wounded skin almost recovered. These animal testing results strongly suggest that DNA films are very effective for wounded skin recovery. The rapid recovery of the wounded skin could be related to good permeation of oxygen, carbon dioxide, and water through the DNA films.

Gas permeation data of DNA films are summarized in Figure 8.23 (Sada et al. 2006), which indicates that oxygen and carbon dioxide gas can permeate DNA films comparable to water vapor; the film derived from DNA–CTMA has an especially high oxygen permeation. The high oxygen gas permeation may assist in rapid healing of the skin wound. Further medical testing may be necessary for long-term uses, especially on humans.

8.3   CONCLUSION

DNA films are applicable for various biomedical uses such as cell culture and wound healing because of excellent biocompatibility. The applications of DNA will be expanded to not only tissue engineering for regeneration of organs but also drug delivery systems (DDS).

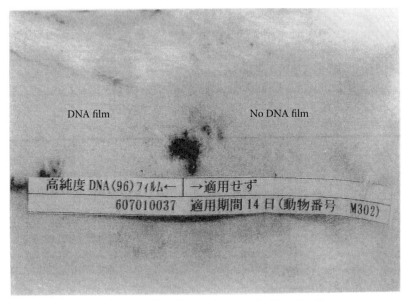

DNA film   No DNA film

高純度 DNA（96）フィルム←　│→適用せず
607010037　適用期間 14 日（動物番号　M302）

Wound of a rat skin after 14 days, patched by DNA film

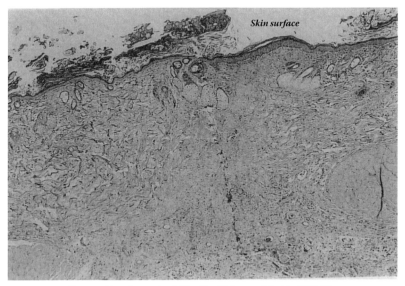

Skin surface

Cross section of wound rat skin after 14 days, patched by DNA film

FIGURE 8.22 (**See color insert.**) Comparison of DNA-film-attached wounded skin and non-DNA-film-attached skin.

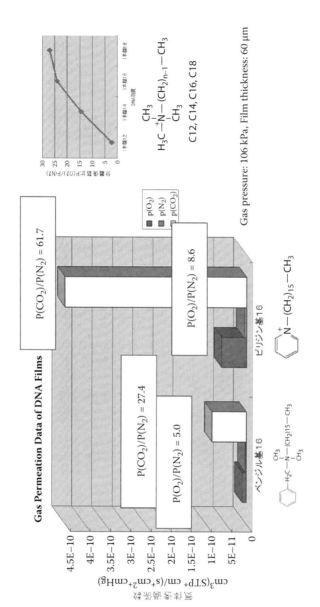

FIGURE 8.23 Gas permeation data of various DNA films.

REFERENCES

Heckman, W. M., J. Grote, P. P. Yaney, and F. K. Hopkins. 2004. DNA-based nonlinear photonic materials. *Proc. SPIE* 5516: 47–51.

Kwon, Y.-W., C. H. Lee, D. H. Choi, and J.-I. Jin. 2009. Material chemistry of DNA. *J. Mat. Chem.* 19: 1353–1380.

Ogata, N. 2009. Novel applications of DNA materials. *Proc. SPIE* 7403: 740305-1.

Sada, T. T., M. Yoshikawa, and N. Ogata. 2006. Oxygen permselective membranes from DNA-lipid complexes. *Membrane* 31: 281–283.

Wang, L., J. Yoshida, N. Ogata, S. Sasaki, and T. Kamiyama. 2001. Self-assembled supramolecular films derived from marine deoxyribonucleic acid (DNA)—cationic lipid complexes: large-scale preparation and optical and thermal properties. *Chem. Mater.* 13: 1273–1281.

Watanuki, A., J. Yoshida, S. Kobayashi, H. Ikeda, and N. Ogata. 2005. Optical and photochromic properties of spiropyran-doped marine-biopolymer DNA-lipid complex films for switching applications, *Proc. SPIE* 5724: 234–241.

Wu, C. M., W. Kiou, H. L. Chen, T. L. Lin, and U. S. Jeng. 2004. Self-assembled structure of the binary complex of DNA and cationic lipid. *Macromolecules* 37: 4974–4980.

Yamaoka, K. and N. Ogata. 2004. Effect of lipids on physical properties of DNA-lipid complexes. *Kobunshi Ronbunsh.,* 61: 384–390.

Yaney, P. P., E. M. Heckman, D. E. Diggs, F. K. Hopkins, and J. Grote. 2005. Development of chemical sensors using polymer optical waveguides fabricated with DNA. *Proc. SPIE.* 5724: 224–233.

Yoshida, J., L. Wang, S. Kobayashi, G. Zhang, H. Ikeda and N. Ogata. 2004. Optical properties of photochromic-compound derived from dye-doped marine-biopolymer DNA-lipid complex films for switching applications. *Proc. SPIE* 5351: 260–268.

Yoshida, J., A. Watanuki, S. Kobayashi, H. Ikeda, and N. Ogata. 2005. Potential switching application based on the photochromism of spiropyran-doped marine-biopolymer DNA-lipid complex films. *Tech. Digest, 10th Optoelectronics and Communication Conference (OECC2005).* 342–343, Seoul, Korea.

9 DNA-Based Thin-Film Devices

Carrie M. Bartsch, Joshua A. Hagen,
Emily M. Heckman, Fahima Ouchen,
and James G. Grote

CONTENTS

9.1 All-DNA-Based Electro-Optic Waveguide Modulator...................256
    9.1.1 Fabrication 256
    9.1.2 Electro-Optic Coefficient..257
    9.1.3 Device Testing and Performance 259
    9.1.4 Conclusions 262
9.2 Field-Effect Transistors 263
    9.2.1 Principles of Operation...263
    9.2.2 Polymer FETs 264
        9.2.2.1 All-Polymer FETs 265
        9.2.2.2 Sensor Applications 265
    9.2.3 DNA Biopolymer as the Semiconducting Layer 266
        9.2.3.1 Measurement Setup..266
        9.2.3.2 Initial Bottom-Gate BioFET.................................267
        9.2.3.3 Improvements to the BioFET.................................271
9.3 Development of a BioLED: DNA as an Electron-Blocking Layer in Organic Light-Emitting Diodes 273
    9.3.1 Materials Used for the Fabrication of BioLEDs...................273
        9.3.1.1 Emitting Molecules Used in BioLEDs 273
        9.3.1.2 Hole Transport Layers Used in BioLEDs 273
        9.3.1.3 Hole Blocking Layer Used in BioLEDs...................274
        9.3.1.2 ETL Used in BioLEDs..274
        9.3.1.3 EBL Used in BioLEDs..274
    9.3.2 Fabrication of BioLEDs...276
        9.3.2.1 Anode Patterning and Deposition..........................276
        9.3.2.2 Solvent-Based Deposition of HTL and EBL............277
        9.3.2.3 Molecular Beam Deposition 277
    9.3.3 Green (Alq_3)-Emitting BioLED Results 278
    9.3.4 Comparison of DNA–CTMA to Other Optoelectronic Polymers 280
    9.3.5 Lifetime of BioLED and Baseline Devices 283

9.4   Conclusions..286
References...286

Thin-film organic-based devices (electrical, optical, and electro-optic) have many advantages over their inorganic counterparts, such as low cost and low-temperature fabrication, leading to flexible, transparent, and biocompatible devices with high tunability. With these new organic structures comes a significant need for new materials to drive the development of high-performance devices. One such material is genomic deoxyribonucleic acid (DNA). The structure of the DNA molecule, as well as its high optical transparency and tunable electrical conductivity, as discussed in Chapter 8, make it a very useful material for a number of electro-optic (EO) devices. Three of these devices are detailed in this chapter: all-DNA-based electro-optic waveguide modulators (Heckman 2006, 2006), field-effect transistors (Bartsch et al. 2007), and bio-organic light-emitting diodes (Hagen 2006; Hagen et al. 2007; Hagen et al. 2006).

9.1   ALL-DNA-BASED ELECTRO-OPTIC WAVEGUIDE MODULATOR

This research focuses on integrating a DNA-based biopolymer, for the first time, as both the cladding and core layers of an electro-optic polymer waveguide modulator (Heckman 2006). The central thesis of the work on this device is that it will seed important advances in the field of photonics research. It introduces a new bio-derived material with many potential performance enhancements over current polymer materials (Grote et al. 2003, 2004, 2005; Heckman et al. 2004). As a core EO polymer layer, the DNA was doped with the EO chromophore disperse red 1 (DR1) and electrically poled to induce an EO coefficient. To date, this result was the first and only known example of a poled DNA-based polymer film (Heckman et al. 2005; Heckman 2006; Yaney et al. 2006a; Heckman, Yaney et al. 2006b; Heckman, Grote et al. 2006). The first all-DNA-based three-layer optical waveguide was also demonstrated. The waveguide used a DNA-based film for both cladding layers and an unpoled DNA–DR1-based film for the core layer.

9.1.1   Fabrication

Both the waveguide core and cladding layers were prepared from 200 kilo-Dalton (kDa) DNA–CTMA dissolved in butanol at a concentration of 110 mM (Heckman et al. 2005; Heckman 2006). As previously mentioned in Chapter 8, a chemical cross-linker, poly(phenylisocyanate)-co-formaldehyde (PPIF), was used to cross-link the DNA–CTMA films to allow successive spin-coating of multiple layers without solvent damage to the existing layer. The core layer had 20 wt% (81.2 mol%) PPIF cross-linker and 3 wt% (12.8 mol%) DR1, and the cladding layer had 10 wt% (36.1 mol%) PPIF cross-linker. A summary of the refractive indices for films with varying concentrations of PPIF and DR1 are shown in Table 9.1 (Heckman 2006). All percentages are given with respect to the initial amount of DNA–CTMA in each solution. The DR1 was limited to 3 wt% to limit absorption losses at the 690 nm measuring wavelength. Each layer was successively spin-coated onto a gold-coated silicon (Si) substrate at 800 rpm for 10 s

TABLE 9.1
Refractive Indices of Various DNA–CTMA (DC) Core and Cladding Materials

Material	Use	n (690 nm)	n (1152 nm)	n (1300 nm)	n (1523 nm)
DC	Clad	1.5135	1.5054	1.5042	1.5029
DC–PPIF$_{10}$	Clad	1.5292	1.5201	1.5191	1.5181
DC–PPIF$_{15}$	Clad	1.5325	1.5225	1.5212	1.5199
DC–PPIF$_{10}$–DR1$_4$	Core	1.5526	1.5350	1.5335	1.5321
DC–PPIF$_{10}$–DR1$_5$	Core	1.5476	1.5292	1.5272	1.5252
DC–PPIF$_{20}$–DR1$_5$	Core	1.5637	1.5452	1.5433	1.5415
DC–PPIF$_{20}$–DR1$_3$	Core	1.5544	1.5390	1.5370	1.5350
PMMA	Buffer	1.4881	1.4840	1.4836	1.4832
APC	Buffer	1.5662	1.5548	1.5534	1.5521

Source: Reproduced with permission from E. Heckman, Ph.D Dissertation 2006.

Note: Cross-linker is PPIF, and DR1 is the chromophore dye. The numbers in subscripts are the amounts of each substance in wt% with respect to DNA–CTMA. The indices listed at 1152 and 1523 nm are measured values, while the indices listed at 690 and 1300 nm are found from a Cauchy fit to the measured values. Also listed are APC and PMMA, materials that could be used as a buffer layer.

with a 5 s ramp. Initial curing to remove the solvent took place for 5 min at 80°C, and the samples were then transferred to a 175°C vacuum oven for 15 min. Before spin-coating the top cladding layer, a low-tack tape was used to mask off both ends of the substrate so that only a 2 cm middle section of the core layer was exposed. This facilitates prism coupling into the core layer to launch the waveguide mode. Finally, a 7.5 wt% polymethylmethacrylate (PMMA) solution in cyclopentanone was spin-coated at 1000 rpm for 10 s with no ramp over the top cladding layer (again using low-tack tape to protect the exposed core layer) and cured at 80°C for 2 h to form a ~0.5 μm buffer or passivation layer. The top electrodes were deposited through sputtering.

The waveguide has a bottom cladding layer thickness of 6.0 μm, a core layer thickness of 3.8 μm, and a top cladding layer thickness of 11.7 μm plus the PMMA buffer layer. This yields a total thickness of 22 μm. All thicknesses were measured with a Veeco Dektak profilometer. This waveguide was not designed to be single mode and supports four TE and four TM modes. Although ultimately single-mode operation is preferred, the multimode design was chosen for ease of fabrication and testing. A schematic of the modulator design is shown in Figure 9.1.

9.1.2 ELECTRO-OPTIC COEFFICIENT

Electrode poling was used to pole both the cross-linked and non-cross-linked DNA–CTMA-chromophore films (Heckman et al. 2005). Preliminary poling studies were made on single-layer films. The films are spin-coated onto indium tin oxide (ITO)-coated glass slides with spinning parameters optimized to achieve 2–3 μm thick

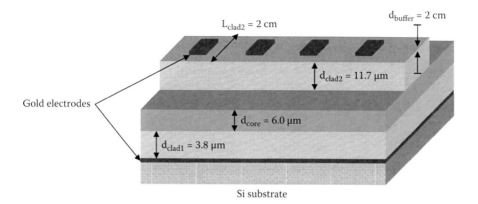

FIGURE 9.1 Schematic of 3-layer DNA-based waveguide modulator. Actual Si substrate is round with a 2″ diameter. (Reprinted with permission from E. Heckman, Ph.D Dissertation 2006.)

films. After the films have cured, gold electrodes are deposited on top of the film with a 100 Å layer of chromium applied first to promote adhesion. Poling conditions ranged from 60–100 V/μm applied fields at a poling temperature of 60°C–80°C in a flowing nitrogen environment. The EO coefficients were measured using a common ellipsometeric technique, modified Teng and Man (Teng and Man 1990).

A summary of the poling results achieved with single-layer films is presented in Table 9.2. Because DNA is unique among polymer materials, it presents poling challenges not seen with other polymers. A clear T_g could not be measured for either cross-linked or non-cross-linked DNA–CTMA films using differential scanning calorimetry (DSC). The temperatures used for poling the DNA-based films were determined through systematic study, and 65°C was found to be optimum. The molecular weight (MW) of the starting DNA plays a significant role in the poling efficiency for both cross-linked and non-cross-linked DNA–CTMA films. An EO coefficient could not be obtained for DNA with a MW >1000 kDa; a MW <500 kDa was necessary to induce an EO coefficient. Two different poling techniques were tried, and it was found that applying the poling temperature first and then incrementally applying the field achieved better results than applying the field first, then raising the temperature. These are referred to as the *T then V* and *V then T* techniques, respectively.

EO coefficients comparable to poled PMMA–DR1 films were achieved with the DNA biopolymer–DR1 films for similar concentrations of DR1. The results in Table 9.2 are for a single-layer film only. When poling a three-layer waveguide, the resistivities of each layer must be accounted for to determine the actual poling field across the core region (Grote et al. 2001). The resistivities of the core and cladding layers were previously measured to be the same; the small differences in cross-linker and DR1 concentrations do not produce a measurable effect (Yaney et al. 2006). Neglecting the buffer layer, the voltage across each layer can be approximated as the total applied voltage divided by the layer thickness. Because the waveguide is relatively thick (22 μm), the poling voltage across the core region was limited due to the available voltage supply (1 kV). Additionally, the full available voltage range was

TABLE 9.2
Summary of Poling Results for Single-Layer DNA-Based Films

Sample ID	Mean MW (kDa)	Cross-Linker (wt%)	DR1 (wt%)	Poling Field (V/μm)	Temp. (°C)	r_{33} (pm/V)	Poling Technique
NXL1	200	0	5	56	65	0.72 ± 0.10	T then V
NXL2	200	0	5	56	65	0.70 ± 0.01	T then V
NXL3	200	0	5	64	65	0.48 ± 0.02	T then V
NXL4	500	0	5	41	65	0.65 ± 0.01	T then V
XL1	200	10	5	70	65	2.02 ± 0.03	T then V
XL2	200	10	5	64	65	1.51 ± 0.02	T then V
XL3	200	20	5	91	65	1.04 ± 0.07	T then V
XL4	200	10	5	75	65	0.83 ± 0.03	V then T
XL5	200	10	5	50	65	0.60 ± 0.02	V then T
XL6	200	10	5	50	100	0.71 ± 0.01	V then T
XL7	200	20	5	75	65	0.74 ± 0.02	V then T
XL8	500	20	5	75	65	1.00 ± 0.04	V then T
XL9	500	20	5	100	80	1.33 ± 0.03	V then T
XL10	500	20	5	100	90	1.12 ± 0.01	V then T

Source: Reproduced with permission from E. Heckman, Ph.D Dissertation 2006.

Note: Under Sample ID, "NXL" designates non-cross-linked DNA–CTMA films, and "XL" designates cross-linked DNA–CTMA films. All r_{33} results were measured at 690 nm using the ellipsometric technique. Under a poling technique, "*T* then *V*" designates the temperature was applied before the voltage, and "*V* then *T*" designates the voltage was applied before the temperature.

not used to prevent the possibility of shorting the sample. The sample was heated to a poling temperature of 65°C, and a total of 700 V was incrementally applied across the three-layer stack. This corresponds to a poling field of 32 V/μm across the core region. Based on previous poling data, it is estimated that under these poling conditions for 3 wt% DR1, an EO coefficient r_{33} of 0.5 pm/V was to be expected (Heckman et al. 2005).

9.1.3 DEVICE TESTING AND PERFORMANCE

The common crossed-polarizer technique for a transverse EO phase modulator was used to evaluate the performance of the modulator (Heckman 2006; Heckman et al. 2006; Yariv and Yeh 1994; Wang et al. 1994). The modulator is placed between a polarizer–analyzer pair oriented at 45° and −45° so that with no voltage applied, there is no transmitted signal. When a voltage is applied, a phase change between the TE and TM components is induced. Using the isotropic material model (and neglecting overlap issues), the index of refraction change for the TM mode is three times larger than that of the TE mode ($r_{33} = 3r_{13}$) (Singer et al. 1987). In this configuration, the resulting equation for the half-wave voltage V_π is (Yariv and Yeh 1984; Wang et al. 1994)

$$V_\pi = \frac{\lambda d}{n^3(r_{33} - r_{13})L} = \frac{\lambda d}{n^3\left(\frac{2}{3}r_{33}\right)L} = \frac{\lambda d}{n^3 r_{eff} L} \qquad (9.1)$$

where λ is the wavelength, d is the core thickness, n is the index of the core, L is the length of the electrode, and r_{eff} is the effective EO coefficient for this modulator configuration. For a polymer device, which has different indices in the planes perpendicular and parallel to the poling axis, the addition of a compensator before the analyzer is needed to adjust the polarization of the out-coming light from elliptical back to linear. Because the laser is linearly polarized, a polarization rotator is used in place of the polarizer to orient the incident beam to 45°. A schematic of this configuration is shown in Figure 9.2. Prism coupling was used to couple the beam into and out of the modulator (Yaney et al. 2006). A picture of the prism-injected waveguide mode propagating through the modulator both under and next to the top electrode is shown in Figure 9.3.

After launching 45° linearly polarized light into the modulator, it was found that the amplitude of the TE mode was larger by a factor of three than that of the TM mode. This prevents cancellation of the two polarization components at the output, regardless of the compensator setting. To ensure equal TE and TM amplitudes at the output, the polarization rotator was adjusted toward TM until the output TE and TM amplitudes were equal, and good cancellation was observed. With no voltage applied across the sample, the compensator was adjusted to find minimum transmission. A dc voltage was then applied across the electrodes to induce a phase change between the TE and TM components. A lock-in amplifier was used to measure the resulting signal change.

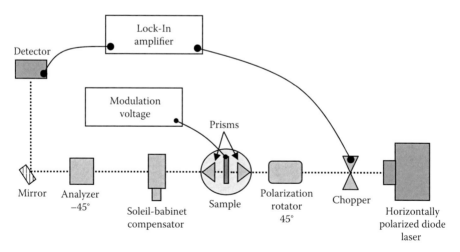

FIGURE 9.2 Schematic of testing configuration for transverse EO polymer-phase modulator. The horizontally polarized laser beam is rotated by 45°, and an analyzer is set to −45°. The compensator is used to adjust the phase from elliptical back to linearly polarized light. A chopper is used to reference the lock-in amplifier.

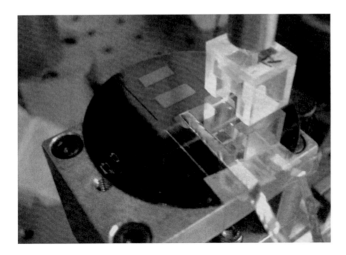

FIGURE 9.3 (**See color insert.**) Two beams propagating through three-layer poled DNA-based waveguide. One beam is propagating under the electrode, and the other is next to the electrode. A second beam between the electrodes is also shown. (Reprinted with permission from E. Heckman, Ph.D Dissertation 2006.)

Because of the 22 μm total modulator thickness, it was not possible to apply a dc voltage large enough to observe a π phase change. Based on our estimate of r_{33} = 0.5 pm/V, and given the modulator parameters of L = 0.8 cm, d = 3.8 μm, n = 1.5544, and λ = 690 nm, a V_π of 262 V was expected. To observe the π phase change with this half-wave voltage would require a total voltage of 1517 V across the 22 μm three-stack. To avoid risking electric field breakdown of the materials making up the modulator, the voltage was limited to 400 V. A sine-squared function was fit to the data; however, the data spanned an insufficient range to conclusively determine the function fitting parameters. This precludes the possibility of determining the values of V_π and r_{33}, as there are an infinite number of sine-squared fits to these data. However, by using our predicted value of V_π = 262 V and the square root of the modulation signal as a weighting factor (assumed proportional to the standard deviation in the curve fitting), a good fit was found to the data. This fit yielded a V_π = 263 V with an uncertainty of ±10% (Heckman 2006). Figure 9.4 shows the change in the lock-in signal as a function of applied dc voltage with the sine-squared fit. Although this analysis does not give an experimentally determined value of V_π, and thus r_{33}, it does confirm the expected sine-squared behavior within a reasonable uncertainty.

An unpoled electrode on the same specimen was also tested to confirm that the observed effect was an EO effect and not another voltage-dependent effect such as piezoelectric or electroabsorption. No change in signal was observed as a result of the applied voltage, confirming that the observed effect is EO. The poled electrode on the modulator was tested several times over a 5-day time span. Qualitative ac modulation was also observed. For ac modulation, the chopper was removed and the lock-in was driven by the applied ac signal. A dc bias of 200 V was applied in addition to the ac signal. A clear ac modulation was detected with both the lock-in and a

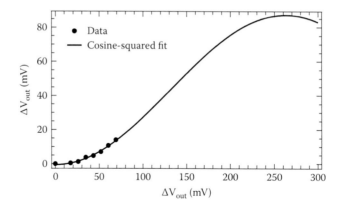

FIGURE 9.4 DC modulation of an all-DNA-based EO modulator, extended view. A sine-squared fit is used to find the DC input across the core layer required for a π phase change. (Reprinted with permission from E. Heckman, Ph.D Dissertation 2006.)

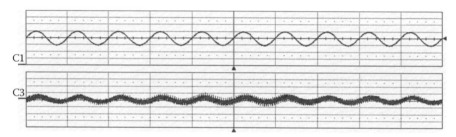

FIGURE 9.5 Oscilloscope screen showing ac modulation of all-DNA-based EO modulator. The top trace is the applied voltage of 200 V dc plus an ac component of 35 V rms. The modulated signal from the photodetector is shown in the bottom trace with a value of 0.6 mV rms. The channel labels at the left (C1 and C3) indicate the zero voltage positions. (Reprinted with permission from E. Heckman, Ph.D Dissertation 2006.)

digital oscilloscope. Figure 9.5 shows the oscilloscope screen with both the applied ac signal and the observed ac modulation of the laser beam. No ac modulation was observed with the unpoled modulator.

9.1.4 CONCLUSIONS

An all-DNA-based EO waveguide modulator has been successfully fabricated and tested, and found to exhibit EO modulating behavior. Using an estimated value of r_{33} = 0.5 pm/V and V_π = 262 V, a sine-squared fit to the modulating data was obtained that yielded a V_π = 263 V with an uncertainty of ±10%. While these performance parameters are far below the desired industry standard of subvolt operation already demonstrated by other polymer modulators (Wang et al. 1994), it is predicted that future DNA-based biopolymer Mach–Zehnder modulators can meet or exceed current polymer device performance (and would utilize the full value of r_{33}). This can potentially be accomplished by using an increased concentration of a higher-quality chromophore to achieve a greater EO coefficient. Reducing the thickness of the core and cladding layers and using lower resistivity cladding layers, relative to the core, will also allow for higher poling fields to further increase the EO coefficient, in addition to allowing single-mode operation. Additionally, an all-DNA-based Mach–Zehnder device offers several promising advantages over conventional polymer devices including reduced optical losses, refractive index tunability for enhanced waveguide design, and increased EO activity through chromophore alignment. It is projected that this device demonstration will lead the way to attaining all these potential advantages of the DNA biopolymer.

9.2 FIELD-EFFECT TRANSISTORS

9.2.1 PRINCIPLES OF OPERATION

Thin-film field-effect transistors (FETs) are three-terminal devices that use a small input signal to control a large output (Singh 2001). The three terminals of the FET are typically called the *gate*, *source*, and *drain*. Each FET is designed to have a particular spacing between the source and drain. The gate is separated from the source and drain by a layer of electrically insulating material. The first organic thin-film transistor (TFT), reported by Weimer in 1962, is based on polycrystalline semiconductor films deposited onto an insulating substrate (Weimer 1962).

The FET structure used in this work is that of a metal–insulator–semiconductor field-effect transistor (MISFET). Current flows from the source to the drain of the MISFET when voltage is applied across the source and drain (V_{DS}). The resulting current (I_{DS}) can be influenced by voltage applied to the gate (V_{GS}). V_{GS} causes a layer of charges that is just a few nanometers thick to form at the interface between the insulator and the semiconductor thin film. The layer of accumulated charges at the interface of the semiconductor and insulator is the channel. This channel connects the source to the drain and permits the flow of charges when the transistor is active. The maximum amount of charge that flows through the channel at any particular time is controlled by the voltage applied to the gate. FETs can operate in two different modes called *enhancement mode* and *depletion mode*. A depletion-mode, or normally on, FET has a large channel where current is free to flow when there is no gate bias applied ($V_{GS} = 0$ V). Therefore, to stop the flow of current through a depletion-mode FET, a negative or positive bias (depending if the semiconductor is an n- or p-type) is used to deplete the channel of charge carriers (Sedra and Smith 1998). The pinch-off voltage in a depletion-mode FET is the voltage that causes the channel to be completely shut off, and current no longer flows. An enhancement-mode, or normally off, FET has no current flowing when there is no gate bias (no accumulation layer); therefore, for current to flow, a channel first needs to be created (Sedra and Smith 1998). The threshold voltage in an enhancement-mode FET is the voltage where the transistor changes from having no current flowing in the channel to having current flow. In an enhancement-mode FET, applying a bias with a greater magnitude than the threshold voltage ($V_{GS} > V_t$) to the gate causes the channel to expand such that more current flows. The ratio between I_{DS} in the off state (no accumulation layer is present) and I_{DS} in the on state is called the on/off ratio, and this provides a measure of the performance of the FET as an electronic switch. The drain current equation for an FET in the linear region is given by Equation 9.2.

$$I_{DS} = \mu_{lin}C_i \frac{W}{L}\left[(V_{GS} - V_t)V_{DS} - \frac{1}{2}V_{DS}^2\right] \tag{9.2}$$

and in the saturation region (at higher V_{DS}), the drain current equation is given by Equation 9.3:

$$I_{DS} = \mu_{sat}C_i \frac{W}{2L}[(V_{GS} - V_t)^2] \tag{9.3}$$

where I_{DS} is the drain current, W/L is the ratio of channel width to channel length, μ_{sat} is the charge carrier mobility in the saturation region, μ_{lin} is the charge carrier mobility in the linear region, V_{GS} is the gate-to-source voltage, V_{DS} is the drain-to-source voltage, C_i is the capacitance per unit area of the electrical insulator, and V_t is the threshold voltage. The transconductance, g_m, in the linear region, is given by Equation 9.4

$$g_m = \frac{\delta I_D}{\delta V_G} = \frac{W}{L}C_i\mu_{lin}V_D \tag{9.4}$$

where the partial derivative of I_{DS} with respect to V_{GS} is taken for a constant value of V_{DS}, and the variables are as defined earlier.

9.2.2 POLYMER FETS

The FETs of interest in this chapter are those that have one or more polymer layers. Polymers can be used for any part of a FET. When a polymer makes up any layer of an FET, from the electrodes to the semiconducting layer, the FET is often referred to as a polymer FET, an organic FET (OFET), or a TFT. The advantages of organic-based FETs include low cost, abundance of raw materials, and most importantly the possibility of designing large area systems by using simple techniques such as spin-coating and vacuum evaporation. The development of OFETs is presently hindered by poor device performance, as compared to their inorganic counterparts such as crystalline or amorphous Si transistors. This is mainly due to the very low charge carrier mobilities in organic semiconductors. Most of the organic FETs described are constructed as TFTs with silicon dioxide, SiO_2, as the insulator.

The first solid-state OFET using an organic semiconductor on an inorganic dielectric is reported in a paper published two decades ago (Ebisawa et al. 1983). A. Tsumura et al. demonstrate a transistor that is fabricated utilizing a film of an organic macromolecule, polythiophene, as a semiconductor (~1400 Å thick) and thermally grown SiO_2 (~ 3000 Å thick) as a gate insulator. They report an on/off ratio around 10^2 with a field-effect mobility of $\sim 10^{-5}$ cm^2/(Vs) and a transconductance of 3 nS (Tsumura et al. 1986).

Yamamoto et al. examine a spin-coated polymer semiconducting layer made of copolymers in which standard transistor behavior is observed. Copolymers are polymers that are capable of operating in both n-type and p-type FETs (Yamamoto et al. 2005). TFTs using semiconducting polymers show that performance is dependent upon polymer film thickness (Deen et al. 2004). Polymers can be used as gate dielectrics, but this has caused a large hysteresis in the transfer characteristics. Uemura et al. report that using a layer of mineral clay to protect the semiconductor layer eliminates the hysteresis (Uemura et al. 2003).

9.2.2.1 All-Polymer FETs

All-polymer FETs are typically fabricated by spin-coating, vapor deposition, and other common lithography techniques with all the layers made of polymer materials. They are shown to be stable in air without protective coatings, which is important for low-cost, high volume applications (Rost et al. 2004). In comparing all-polymer FETs to fully inorganic FETs by changing one layer at a time of the inorganic FET to a polymer material, it is reported that very slight degradation of performance is seen when the polymer layer replaces the silicon substrate and oxide layer (Backlund et al. 2005). Organic FETs need to be lightweight, thin, and pliable to obtain flexible electronic devices. Pliable materials need to be used for the active and insulating layers in order to achieve this (Yoshida et al. 2003). Thin-film semiconducting polymers can be used as the active material in FETs. However, undoped semiconducting materials have high electrical resistivities. Edman et al. report that doped semiconducting polymers have lower electrical resistivities, and higher doping levels yield lower resistances. Therefore, polymers with high electron and hole mobilities that can support doping are useful for transistors (Edman et al. 2004).

The mobility and device performance of FETs with organic gate insulators are affected by the specific polymer used (Kang et al. 2005). X. Peng et al. (Peng et al. 1990) reports the first all-organic TFTs made of alpha-sexithienyl (α6T) as the semiconducting layer with various polymeric insulating layers. They compare the field-effect currents from five different polymers as gate insulators. These five polymers are polystyrene (PS), polymethymethacrylate (PMMA), polyvinyl alcohol (PVA), polyvinyl chloride (PVC), and cyanoethylpullulan (CYEPL). Their results ranged from no field-effect current to an amplified current that surpassed the SiO_2-based devices.

Ashizawa et al. report on successfully fabricating and testing FETs where the source, drain, and gate electrodes and also the channel are made of the same doped conducting polymer. The FET is normally on, with a positive gate voltage causing it to act in depletion mode (Ashizawa et al. 2005). Thin-film polymer FETs have also been used for optical emission applications. The gate bias controls the location of the light emission (Swensen et al. 2005; Sakanoue et al. 2004). In these light-emitting FETs, high performance is obtained from the simultaneous injection of holes from the drain and electrons from the source (Sakanoue et al. 2004).

9.2.2.2 Sensor Applications

Many polymer FETs described in the literature are used for sensing applications. Sensors are devices that measure a particular signal, such as voltage, temperature, gas, and light. Krishna et al. describe a polymer FET with a conducting polymer gate that is fabricated as a chemical sensor using standard complimentary metal–oxide–semiconductor (CMOS) processing (Krishna et al. 2003). Kazanskaya et al. report on FET-based sensors with a polymer membrane that works as a potentiometric sensor detecting ammonium ions or as a biosensor detecting urea. The polymer membrane used with this sensor is light sensitive, and its elements are easily changed so that it can detect a variety of biological substances (Kazanskaya et al. 1996). Dutta and Narayan report on a polymer FET that detects light and operates as a memory device. Information is optically introduced and electrically removed from this FET (Dutta and Narayan 2004). Deen and Kazemeini describe the use of photosensitive semiconducting polymers in photodetectors. These photodetectors are highly sensitive with no gate bias (Deen and Kazemeini 2005). The low cost and comparable semiconducting characteristics of polymer semiconducting materials to silicon make their use appealing for sensor applications. However, polymer semiconducting materials tend to have low charge carrier mobility and strong temperature dependence (Gao et al. 2003), which may not be ideal for some sensor applications.

Conducting polymers are typically the active gate material in gas or vapor sensors. They operate at room temperature, consume lower power than non-polymer-based FET sensors, and are dependent upon the humidity present (Covington et al. 1999). They are more sensitive than resistive gas sensors, and they respond quickly (Lee et al. 2005). Using composite conducting polymers prevents humidity from affecting the magnitude of the response but shows significant unwanted temperature dependence. In addition, it is easy to combine the polymer deposition with current CMOS technology (Covington et al. 2001).

9.2.3   DNA Biopolymer as the Semiconducting Layer

FETs are used to demonstrate the use of DNA biopolymers as the semiconducting layer. Biologically-based field-effect transistors (BioFETs) utilizing a DNA-biopolymer as the semiconducting layer are studied. First, a brief description of the measurement system used is given. Next, a bottom gate BioFET structure is described, and the results from the first DNA-based semiconducting layer in a transistor are analyzed and discussed (Bartsch 2007; Bartsch et al. 2007). Finally, improvements to this initial device and the results obtained from this improved device are described and analyzed.

9.2.3.1   Measurement Setup

A Keithley 4200 Semiconductor Characterization System is connected to the transistor under study using an on-wafer micromanipulator needle probe station, shown in Figure 9.6.

The semiconductor characterization system controls and measures the currents and voltages in all the transistor structures described in this chapter. This semiconductor characterization system has a current compliance level that can be set for each terminal. The current compliance level is a limit set on the measured current level that limits the power that can be delivered to the device. Using this characterization system, contact is made to the source, drain, and gate by direct contact with needle probes, unless the gate is located on the bottom of the substrate. The sample holder

FIGURE 9.6 **(See color insert.)** The probe station setup used for measuring the DNA-based transistors. (Reprinted with permission from C. Bartsch, Ph.D Dissertation 2007.)

on the needle probe station is gold-coated and is, therefore, conductive, so when the gate is on the bottom of the substrate, the needle probe to the gate is placed directly on the sample holder rather than contacting the gate directly. The body on the top-gate transistor structure can be contacted in the same way as the gate on the bottom-gate structure. The semiconductor characterization system is setup to function as a curve tracer, measuring, recording, and plotting current–voltage (I–V) curves for the BioFETs while also producing a spreadsheet that contains all the voltages and currents being controlled and measured during the test scan.

9.2.3.2 Initial Bottom-Gate BioFET

One of the standard polymer transistor gate insulators seen in the literature is SiO_2 used in a bottom-gate FET structure. In the literature, this structure is built from low electrical resistivity silicon wafers with a SiO_2 layer on top of the silicon. The gate contact is on the unpolished side of the silicon wafer, with the SiO_2, serving as the gate insulator. The source and drain are patterned on top of the SiO_2 and on top of the source, and drain the semiconducting polymer is deposited (Deen et al. 2004). The cross section of the fabricated BioFET structure is shown in Figure 9.7.

A bottom-gate BioFET structure similar to that found in the literature, containing DNA: poly(3,4-ethylenedioxythiophene) (PEDOT):CTMA as the semiconducting layer, is fabricated using silicon with a very thin naturally occurring layer of SiO_2 on a silicon wafer instead of a thicker deposited layer of SiO_2 on low electrical resistivity silicon. As previously described in Chapter 8, DNA:PEDOT:CTMA is formed from the raw DNA obtained from CIST, poly(3,4-ethylenedioxythiophene) poly(styrenesulfonate) (PEDOT:PSS), and CTMA. This processing renders an organically soluble semiconducting complex (Hagen et al. 2007). This structure consists of the gate contact on the unpolished side of a 304.8 μm thick silicon wafer, while the

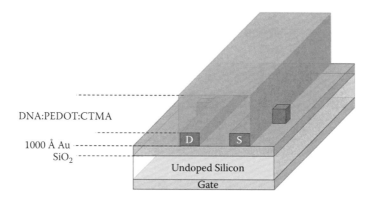

DNA:PEDOT:CTMA

1000 Å Au

SiO$_2$

D   S

Undoped Silicon

Gate

FIGURE 9.7 A cross section of the bottom-gate BioFET structure. (Reprinted with permission from C. Bartsch, Ph.D Dissertation 2007.)

native SiO$_2$ on the polished side of the silicon wafer acts as the insulating layer. The source and drain are patterned on top of the SiO$_2$ by sputtering 100 Å of Cr, for adhesion, followed by 1000 Å of gold through the same shadow mask, forming a 4.0 mm long and 10.16 mm wide channel. On top of the source and drain, a 1 μm thick layer of DNA:PEDOT:CTMA is applied by spin-coating to act as the semiconductor.

Figure 9.8 plots the I–V characteristics obtained from a bottom-gate BioFET device on a silicon wafer as described earlier (Bartsch 2007).

The current compliance limit is set to 0.5 mA on each terminal. This figure shows that, for drain voltages from 0.0 to 0.1 V, the drain currents have a steep increase that goes through the origin for all the gate voltages 0.0 to −20.0 V. As the gate voltage increases toward zero, the drain current decreases at positive drain voltages, as one would expect for a depletion-mode transistor. These measurements are of particular interest because the gate currents are substantially lower than the source and drain currents for all drain voltages under all gate bias conditions. The gate, source, and drain currents are plotted as a function of drain voltage with −5.0 V on the gate in Figure 9.8. This figure shows that the gate current is an order of magnitude smaller than the source or drain current for all drain voltages. It also shows that the source and drain currents are always nearly equal in magnitude and opposite in sign, as expected for a transistor when positive current is defined as current leaving the terminal. Both parts of Figure 9.8 are interesting because together they show that if the gate can be successfully isolated from the source and drain, the gate modulates the drain to the source current and transistor behavior is observed.

This data obtained from the bottom-gate BioFET described earlier containing DNA:PEDOT:CTMA shows promise. However, this structure often produces higher than expected gate currents. These gate currents are potentially due to the presence of mobile charges in the DNA-biopolymers. Drs. Singh and Sariciftci, collaborators on this research at Johannes Kepler University of Linz, are investigating the use of blocking layers to control these mobile charges in DNA-based devices.

Analyzing the transistor curves presented earlier, the linear regions show that these BioFETs that are fabricated on a very thin layer of SiO$_2$ have a calculated electrical resistance, that is, between 5.4 and 22.9 kΩ. Using a multimeter to measure

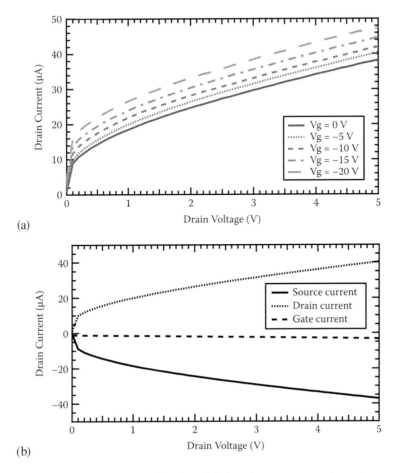

FIGURE 9.8 (a) Drain current as a function of drain voltage as measured on a bottom-gate BioFET on a silicon wafer with a very thin silicon dioxide layer as the gate insulator. (b) Gate, drain, and source currents are plotted as a function of drain voltage for this device, verifying that the gate current is substantially lower than either the source or drain currents. Note that positive current is leaving the terminal. (Reprinted with permission from C. Bartsch, Ph.D Dissertation 2007.)

the sheet resistance of the device yields values that vary dramatically. This is likely due to poor adhesion of the gold source and drain to the very thin layer of SiO_2 and, therefore, one can safely assume that the lower resistance values are more accurate. Making this assumption, the measured resistance of these devices is between 16 and 36 kΩ.

The device mobility for this biopolymer containing FET is determined theoretically from the experimentally obtained current and voltages. Specifically, the drain current equations in the linear and saturation regions, given earlier in this chapter, can be used to determine the linear and saturation mobilities. However, a complicating factor for these calculations is the fact that the mobilities in these regions are not always the same, and often the linear mobility is lower (Newman et al. 2004). This means that calculating the mobility in the linear region will only produce a lower boundary for the mobility.

For the data reported here, the threshold voltage cannot be accurately determined. Therefore, the drain current equation in the linear region cannot be used to determine the linear region mobility. Instead, using the transconductance, g_m, in the linear region, as defined earlier in this chapter to determine the mobility of the device, the linear mobility is calculated to be 0.018 cm²/(Vs). The mobility of PEDOT:PSS, Baytron P, is known to be 10 cm²/(Vs) (Park et al. 2003). The value obtained for the linear mobility of the FET device is a reasonable value for an insulator doped with a Baytron P conducting polymer as it is reasonable to expect the mobility to be lower than the highest mobility material in the complex. Unfortunately, this value is an overestimate of the mobility of DNA:PEDOT:CTMA, because the width-to-length (W/L) ratio for the bottom-gate device described here is 2.63, and to minimize the effects of fringe currents, the W/L ratio should be at least 10.0 (Dimitrakopoulos and Malenfant 2002).

Even without knowing the threshold voltage for the data, the drain current equation in the saturation region, given in Equation 9.3, can be used to calculate the saturation region mobility. Since there are two unknowns, μ_{sat} and V_t, two equations are needed to solve for the saturation mobility. Therefore, the drain currents corresponding to two different gate voltages at the same drain voltage will be used. Figure 9.9 shows how the effective saturation mobility, calculated from the drain currents at 0 and −10 V applied to the gate, varies with drain voltage (Bartsch et al. 2007).

This figure shows that the saturation mobility decreases with increasing drain voltage in the saturation region of operation for the transistor. Therefore, the maximum saturation mobility occurs at the lowest drain voltage in the saturation region. For the transistor reported earlier to have low gate currents and produce transistor curves, the effective saturation mobility occurs when the drain voltage is 0.2 V. For

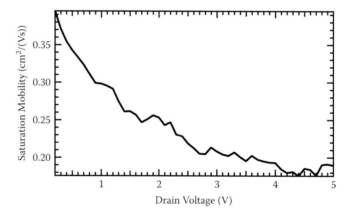

FIGURE 9.9 Saturation mobility as a function of drain voltage as calculated from the drain current versus drain voltage data on a bottom-gate BioFET with a very thin layer of native silicon dioxide as the gate insulator. The data for this plot is calculated using the drain currents associated with gate voltages of 0 and −10 V, when the drain voltage is 0.2 V, obtained from Figure 9.8. (Reprinted with permission from C. Bartsch, Ph.D Dissertation 2007.)

gate voltages of 0.0 and −10.0 V, the maximum saturation mobility is calculated to be 0.45 cm²/(Vs).

The mobility values obtained using these methods are overestimates due to the low W/L ratio of the device. Therefore, the true mobility for this material is still unknown. However, the preceding calculations suggest that the mobility of DNA:PEDOT:CTMA is two orders of magnitude less than the mobility of PEDOT:PSS, the conducting polymer that is combined with DNA to make this semi-conductive biopolymer (Bartsch 2007).

9.2.3.3 Improvements to the BioFET

The same basic structure described earlier continues to be used. Now, a heavily p⁺ doped silicon wafer is used as the gate contact, and a 300 nm thick layer of SiO_2 is thermally grown on Si for use as the gate insulator. Cr/Au electrodes that are approximately 100 nm thick are deposited by thermal evaporation on the SiO_2. These transistors also have a larger W/L ratio, for the source and drain electrodes, of about 5. All of the electrical characterization is carried out under ambient conditions. Additionally, all I–V measurements are taken with V_{DS} varying from 0 to 20 V, as the thicker gate insulator allows higher voltages to be applied before breakdown occurs.

For devices made using DNA:PEDOT:CTMA, the I–V curves continue to exhibit current amplifier-like behavior as shown in Figure 9.10.

The measured device parameter (V_{DS}, V_{GS}, and I_{DS}) values for DNA:PEDOT:CTMA containing transistors are comparable to the values obtained for state-of-the-art pentacene- and poly(3-hexylthiophene) (P3HT)-based field-effect organic transistors (Dimitrakopoulos and Malenfant 2002).

Field-effect mobility and threshold voltages are calculated using the saturation region current equation for standard MOSFETs, using the steepest slope of the $I_{DS}^{1/2}$ versus V_{GS} plot with V_{DS} equal to 20 V. The I_{DS}–V_{DS} characteristics are determined for applied V_{DS} ranging from 0 to 20 V with gate biases varying from −10 to 10 V. These results show a slight deviation in the I_{DS}–V_{DS} relationship from the ideal case.

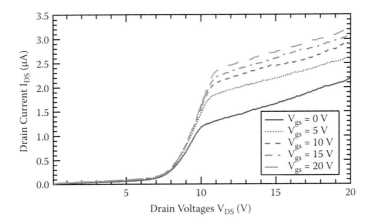

FIGURE 9.10 Output characteristics of DNA:PEDOT:CTMA-based blends. (Reprinted with permission from Ouchen, F., P.P. Yaney, and J.G. Grote. *Proc. SPIE* 7403: 74030F-1–8, 2009.)

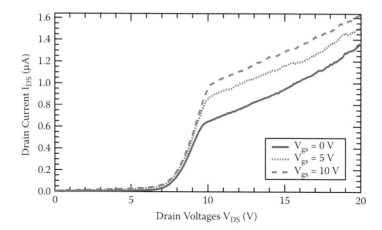

FIGURE 9.11 Output characteristics of FETs using P3HT as a semiconducting film. (Reprinted with permission from Ouchen, F., P.P. Yaney, and J.G. Grote. *Proc. SPIE* 7403: 74030F-1-8, 2009.)

The DNA:PEDOT:CTMA containing OFET exhibits high I_{DS} when no gate voltage is applied. Xu and Berger attribute this behavior to the presence of charged impurities in the semiconducting material (Xu and Berger 2004). DNA:PEDOT:CTMA has a significantly higher overall conductivity than DNA–CTMA due to the blending of DNA with the conductive polymer, PEDOT:PSS.

The increase in drain current, I_{DS}, with an increase in positive gate voltages, V_{GS}, shows enhancement-mode behavior with p-type conductivity in DNA:PEDOT:CTMA. A significant offset voltage is measured in the output characteristics of these devices. Though well-behaved saturation characteristics are measured at the output of P3HT-based FETs, these devices also produce a slightly lower but significant offset voltage as shown in Figure 9.11.

S. P Tiwari et al. report an offset voltage on pentacene-based thin-film FETs and attributes such behavior to high contact resistance due primarily to the polymer–electrode interface effects and channel dimensions W/L (Tiwari et al. 2007). The BioFET designs described here have W/L<10 that makes the contribution of the contact resistance considerable compared to the channel resistance. This low W/L ratio is most likely responsible for the offset in the drain voltages.

Additionally, the effective free charge carrier mobility is around 8×10^{-4} cm²/(Vs) as extracted from the transfer characteristics when V_{DS} is fixed to 20 V. There are several reasons why the approximated mobility for these improved devices is lower than the approximated mobility for the initial device that contained DNA:PEDOT:CTMA as the semiconducting layer. These include the different V_{DS} values at which the mobilities were calculated (recall that it was shown earlier that mobility decreases with increasing drain voltage), a factor of two difference in the W/L ratio for the transistors, and the higher voltage range over which the device data was measured for the improved devices. Each of these factors reduces the transistor mobility and, therefore, the combination of them will further reduce the mobility.

9.3 DEVELOPMENT OF A BioLED: DNA AS AN ELECTRON-BLOCKING LAYER IN ORGANIC LIGHT-EMITTING DIODES

The field of organic light-emitting diodes (OLED) has been growing significantly in recent years due to its applications in high-efficiency solid-state lighting and high-quality flat-panel displays (Service 2005). However, in order for OLEDs to achieve performance levels to rival and exceed their inorganic counterparts, new materials must be developed. There are a number of different material needs in developing OLEDs, such as charge-blocking layers (electron and hole), charge transport layers, and emitters. The charge transport and blocking layers must have very specific electrical properties as well as high optical transparency for efficient photon transfer through the device. As described in this section, genomic DNA derived from salmon has the ideal properties to be present in an OLED as an electron-blocking layer (EBL). Using DNA as the EBL significantly increases the efficiency of these devices and shows great promise for the future of OLEDs (Hagen 2006; Hagen et al. 2006).

9.3.1 MATERIALS USED FOR THE FABRICATION OF BioLEDs

The following sections review all common materials used in the research of BioLEDs. For all devices, indium tin oxide (ITO) was used for the anode with a work function of 4.7 eV, and aluminum (Al) along with a thin layer of lithium fluoride (LiF) was used as the cathode with a work function of 3.1 eV.

9.3.1.1 Emitting Molecules Used in BioLEDs

The goal of this research was to analyze the performance of DNA–CTMA as a non-emitting material in fluorescent-based OLEDs, so a common, well-known emitter was chosen. The molecule used for this research was the emitter described in the initial work of Tang and VanSlyke 1987; Tang et al. 1989, tris-(8-hydroxyquinoline) aluminum (Alq_3). This material has been widely researched (Gu et al. 1977; Adachi et al. 1991) and emits strongly in the green region. To determine the effectiveness of DNA–CTMA as a nonemitting material in OLEDs, multiple emitting materials were used in separate device structures.

9.3.1.2 Hole Transport Layers Used in BioLEDs

The hole transport layer (HTL) used for all devices was poly(3,4-ethylene dioxy-2,4-thiophene)–polystyrene sulfonate, commonly known as PEDOT–PSS, along with the trade name Baytron P from Bayer Corporation. PEDOT–PSS has also been used as a transparent anode on flexible substrates (Gao et al. 1997). It has a highest occupied molecular orbital (HOMO) of 5.2 eV and a lowest unoccupied molecular orbital (LUMO) of 3.5 eV. The minimal gap between the 4.7 eV work function of ITO and HOMO of 5.2 eV makes it a good HTL.

N,N'-diphenyl-*N,N'*-*bis*(1-naphthylphenyl)-1,1'-biphenyl-4,4'-diamine (NPB) has a HOMO of 5.25 eV and a LUMO of 2.3 eV. The close matching of the 5.25 eV

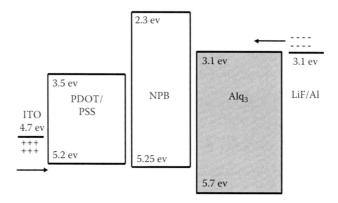

FIGURE 9.12   Energy level diagram of OLED with HTLs. (Reprinted with permission from J. Hagen, Ph.D Dissertation 2006.)

HOMO of NPB and the 5.2 eV HOMO of PEDOT–PSS makes it another excellent HTL. The energy level diagram of a device with PEDOT–PSS and NPB HTLs and Alq$_3$ as the emitting layer is shown in Figure 9.12.

9.3.1.3   Hole Blocking Layer Used in BioLEDs

Transport layers allow holes and electrons to be efficiently injected into the device. However, the balance of electrons and holes is not always equal. This inequality still allows device emission, but at a much higher current than is necessary. Many of these charges can flow through the device without recombining, which results in a low efficiency. A way of balancing these charges is by using a hole blocking layer (HBL). This layer has a high HOMO value, which acts as a barrier between the emitting layer and the anode. For the BioLED devices, 2,9-dimethyl-4,7-diphenyl-1,10-phenanthroline (BCP) was used as the HBL. The HOMO value for BCP is 6.7 eV, which provides at least a ~1 eV barrier for holes to overcome to leak through the device, while it has a LUMO value of 3.2 eV, which still allows for efficient electron injection from the electron transport layer (ETL). The advantages of using BCP as a HBL are seen clearly in the energy diagram in Figure 9.13.

9.3.1.2   ETL Used in BioLEDs

Much like the HTL, the ETL facilitates the injection of charges into the device, electrons in this case. The LUMO value is key to this material and must match the work function of the cathode closely. In this case, the emitting material, Alq$_3$, also happens to have the preferred energy levels of an ETL with a HOMO of 5.7 eV and a LUMO of 3.1 eV. The resulting energy level diagram for an OLED with HTLs, Alq$_3$ as the emitting layer, an HBL, and an ETL is shown in Figure 9.14 in a typical fluorescent OLED structure.

9.3.1.3   EBL Used in BioLEDs

The heart of this research was the incorporation of DNA–CTMA into common OLED structures as a nonemitting layer. Many devices were fabricated with DNA–CTMA as a host molecule doped with emitting small molecule dyes but were unsuccessful.

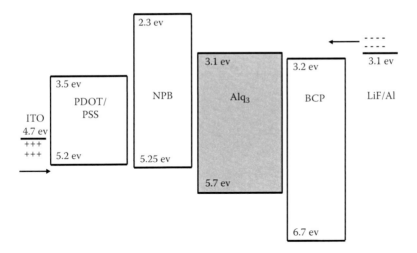

FIGURE 9.13 Energy level diagram of OLED with HTLs and HBL. (Reprinted with permission from J. Hagen, Ph.D Dissertation 2006.)

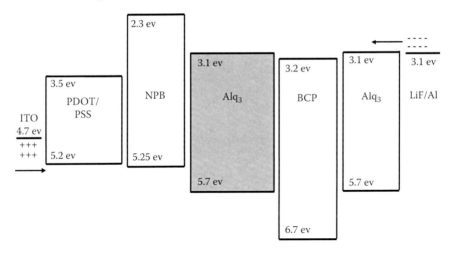

FIGURE 9.14 Energy level diagram of OLED with HTLs, HBL, and ETL. (Reprinted with permission from J. Hagen, Ph.D Dissertation 2006.)

DNA-based BioLEDs previously reported (Kobayashi et al. 2001; Koyama et al. 2002) have incorporated DNA-based thin films as hosts for lumophores indicating the feasibility of the concept, but without significant improvement in device performance over conventional OLEDs. Hirata et al. (2004) have investigated the properties of DNA as a charge transport layer in several device configurations elucidating their properties through their effect on the device current–voltage (I–V) characteristics. None of the previous work in BioLEDs showed any improvement in device performance by including DNA, nor did any device show performance close to well-published OLEDs. By analyzing the unsuccessful BioLEDs fabricated at UC-Nanoelectronics Laboratory using DNA–CTMA as the host layer for the

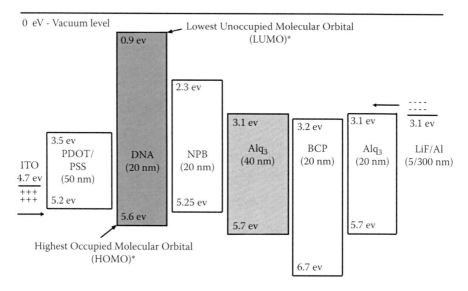

FIGURE 9.15 Energy level diagram of the Alq$_3$-emitting BioLED. (Reprinted with permission from J. Hagen, Ph.D Dissertation 2006.)

emitter, and closely investigating the paper by Hirata et al. (2004), it became clear that DNA–CTMA may have the properties necessary for an EBL. Adamovich et al. (2002, 2003) first noted the use of EBL layers in OLEDs to prevent electron–hole recombination from occurring in adjacent HTL. DNA–CTMA has excellent optical and electrical properties, as described in Chapter 8, and has a HOMO of 5.6 eV and a LUMO of 0.9 eV. The HOMO level matches closely to the HOMO of PEDOT–PSS with only a 0.4 eV gap between the two, and the significantly low LUMO provides a large energy barrier to block the flow of electrons and increase the hole–electron recombination in the emitting layer. The overall structure of the highly efficient green (Alq$_3$)-emitting fluorescent BioLED is shown in Figure 9.15.

9.3.2 FABRICATION OF BioLEDs

9.3.2.1 Anode Patterning and Deposition

Fabrication of BioLEDs was done using a number of techniques in a clean room laboratory, sputtering, spin-coating, and molecular beam deposition (MBD). The devices were built layer by layer, starting with a 2″ round transparent glass substrate. These were bottom-emitting devices, so they were fabricated starting with the transparent anode, ITO. The substrate was patterned using dry lithography into a unique geometry that resulted in an emitting area of 2 × 2 mm and had large areas for contacting with electrodes. This patterned substrate was then placed into a sputtering chamber where ITO was deposited at 100 W power in an Argon plasma at 2 mtorr for 6 min. This resulted in a ~40 nm transparent film of ITO, which was then thermally annealed at 450°C for 2 min to increase transparency and reduce electrical resistivity.

9.3.2.2 Solvent-Based Deposition of HTL and EBL

The anode-patterned device was then cleaned thoroughly in a mixture of solvents (methanol and acetone), followed by rinsing in deionized water and drying with nitrogen. The first layer deposited was the PEDOT–PSS HTL in all device structures. PEDOT–PSS was purchased from the Bayer Corporation as Baytron P, which is an OLED grade material, and was diluted with 5 parts isopropyl alcohol to 1 part PEDOT–PSS to aid in film quality. This mixture was filtered using a 0.2 µm PTFE syringe filter and was then ready for spin-coating. Spin-coating was done using a Laurell spin-coater, and the PEDOT–PSS HTL was deposited by dropping the solution onto the substrate and spinning at 2000 rpm for 20 s. This uniform thin film was then baked in an atmospheric oven at ~130°C, which evaporated all solvents from the film, and further protects the film from solvent attack generated by additional spin-coated layers. For all BioLED devices, the next layer deposited was DNA–CTMA as the EBL. For all comparison "baseline" devices, this step was omitted, and the remainder of device fabrication remained identical. Low molecular weight (~145 kg/mol) DNA–CTMA was dissolved in butanol at a concentration of 1% (by weight of DNA–CTMA). After the solid material was fully dissolved, this solution was filtered with a 0.2 µm PTFE syringe filter, and the solution was ready for spin-coating. The desired film thickness of ~20 nm required spin-coating at 6000 rpm, and this was done for 10 s. At this point, the anode, HTL, and EBL (in the case of the BioLED) have been deposited, and the fabrication shifts to MBD.

9.3.2.3 Molecular Beam Deposition

The remaining layers of the BioLED and baseline devices were deposited using an SVT Associates MBE system operated at the UC-Nanoelectronics Laboratory. The advantages to using this system include high vacuum deposition at 10^{-8} torr and well-controlled growth conditions using Eurotherm temperature controllers and high-quality effusion cells. Another advantage is the large growth chamber that allows for up to three 2″ substrates to be deposited simultaneously. This allowed for deposition of the BioLED and baseline device simultaneously for the most accurate analysis of the benefits of using the DNA–CTMA EBL. Each small molecule material was separately heated up to a set temperature that was calibrated for a set deposition rate. This provided careful control of film thicknesses from run to run. Each of the devices described in the following sections has a slightly different structure based on the desired color output, or emitting layer. All of the film thicknesses reported have been based on the calibrated temperature/deposition rate tables that are constantly updated with the MBE system.

For the green (Alq_3)-emitting devices, a 20 nm layer of NPB was deposited at a temperature of 265°C as an HTL, and a 40 nm layer of Alq_3 was deposited at 230°C as the emitting layer. For the blue (NPB)-emitting devices, a 40 nm layer of NPB was deposited first, and the 40 nm Alq_3 layer in the green-emitting device was omitted. Similarly, for the blue–red (NPB–Eu)-emitting devices, a 40 nm layer of NPB–Eu was obtained by a co-deposition of NPB and europium (deposited at 240°C), and again the 40 nm Alq_3 layer in the green-emitting device was omitted.

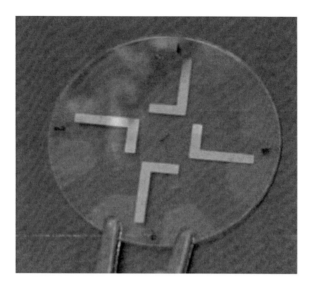

FIGURE 9.16 (**See color insert.**) Photograph of the finished device. (Reprinted with permission from J. Hagen, Ph.D Dissertation 2006.)

All of the remaining steps in the device fabrication were identical after deposition of the emitting layer. The next layer deposited was the HBL material BCP at 450°C with a thickness of 20 nm. All of the layers up to this point were deposited with a circular mask that leaves the edge of the device free from material deposition for connecting electrodes directly to the anode and cathode of the device. After the HBL was deposited, the substrates were removed from the MBE system, and a second mask was placed over the substrates. This mask has four L-shaped patterns that allow the cathode to be deposited resulting in the 2 × 2 mm emitting area. The two final device layers deposited were for the cathode. A very thin layer (~6 nm) of LiF was deposited at 730°C, followed by a ~100 nm deposition of the Al cathode at 1200°C. The device was complete at this point, as shown in Figure 9.16.

9.3.3 Green (Alq₃)-Emitting BioLED Results

The green-emitting BioLED structure described in Figure 9.15 was the first highly successful device. The impact of DNA–CTMA as the EBL was analyzed by fabricating a BioLED device and a baseline device simultaneously in the MBE system. Each device was tested by applying dc voltage while collecting electrical current and brightness. The voltage and current were controlled by LabView, while the brightness was measured through a Minolta colorimeter. Figures 9.17 and 9.18 show luminance and current density versus voltage (LJV diagram) for the baseline and BioLED devices, respectively.

Both devices were cycled from 0 to 25 V, with a maximum luminance of 21,100 cd/m² for the BioLED and a maximum luminance of 12,400 cd/m² for the baseline device. The emission spectra of both devices were identical to the characteristic

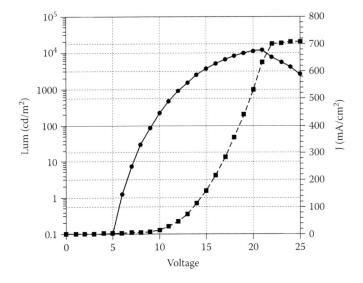

FIGURE 9.17 Luminance current–voltage characteristics for the green-emitting baseline device. (Reprinted with permission from J. Hagen, Ph.D Dissertation 2006.)

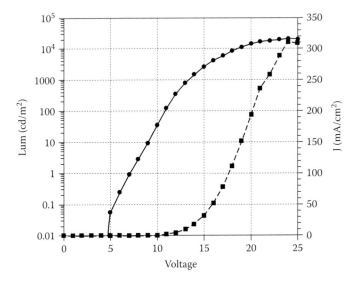

FIGURE 9.18 Luminance current–voltage characteristics for the green-emitting BioLED device. (Reprinted with permission from J. Hagen, Ph.D Dissertation 2006.)

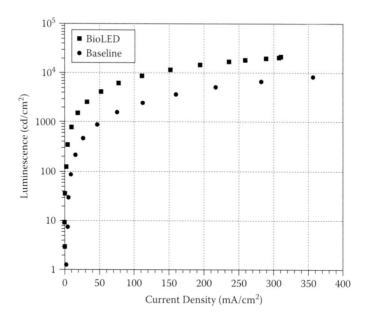

FIGURE 9.19 Luminance versus current density for BioLED and baseline devices. (Reprinted with permission from J. Hagen, Ph.D Dissertation 2006.)

emission of Alq$_3$. The brightness enhancement of the device by adding a DNA–CTMA EBL is evident in Figures 9.19 and 9.20.

Figures 9.18 and 9.19 also show that the BioLED device achieved the enhanced brightness at a much lower current. Efficiency, as defined previously, describes the number of photons emitted per electrons injected, so it is evident that using DNA–CTMA as the EBL also significantly enhanced the device efficiency. The comparison of efficiency versus applied voltage for both the BioLED and baseline devices is plotted in Figure 9.21.

The increase in efficiency for the BioLED device was significant, with ~4× improvement. Operating a device at a set brightness with a low current running through also has an impact on power consumption and possible device lifetime, which is addressed later in this chapter.

9.3.4 COMPARISON OF DNA–CTMA TO OTHER OPTOELECTRONIC POLYMERS

For comparison, OLEDs were fabricated with no EBL ("Baseline" device) and with EBLs using common polymers (PMMA—polymethylmethacrylate or PVK—polyvinyl carbazole).

The energy level diagram previously shown in Figure 9.15 intuitively shows that DNA–CTMA will act as an EBL in the device. With a DNA LUMO level (Hirata et al. 2004) of 0.9 eV, electrons in the NPB layer will experience an energy barrier of 1.4 eV. The DNA–CTMA HOMO level of 5.6 eV should not inhibit hole transport. Long-distance hole transfer has also been reported for DNA in solution (Takada et

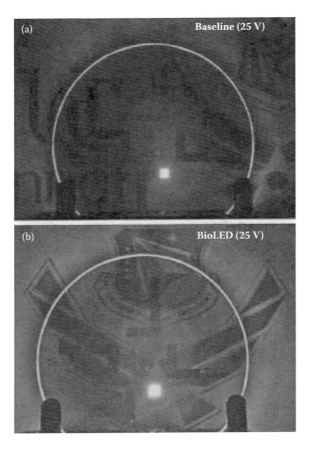

FIGURE 9.20 (**See color insert.**) Photographs of green-emitting BioLED and baseline devices in operation. (Reprinted with permission from J. Hagen, Ph.D Dissertation 2006.)

al. 2004). PVK has a LUMO level of 2.3 eV and a HOMO level of 5.8 eV, which does not suggest that the material will effectively block electrons. The PMMA LUMO and HOMO levels are not reported, but the energy gap is estimated at 5.6 eV from the optical absorption of the material. Such a wide energy gap suggests that PMMA could be a charge-blocking material. PMMA dissolved in cyclopentanone and PVK dissolved in chlorobenzene were spin-deposited to a thickness of ~20 nm. The remaining layers were deposited by molecular beam deposition at pressures ~1 × 10⁻⁷ torr.

The current density versus voltage (J–V) curves in forward bias for the green (Alq₃ emitting) OLEDs as a function of EB material are shown in Figure 9.22. Under reverse bias, a current rectification ratio of >20× at ±20 V was measured, and no light emission was observed. PVK is a commonly used conductive polymer that acted as an HTL in this device structure as shown by the increase in current as compared to the device with no EB material. Conversely, since PMMA is electrically insulating, the current was greatly reduced when it is used as the EBL material. DNA is reported (Fink and Schonenberger 1999; Porath et al. 2000; Cai et al. 2000) to act as a semiconductor

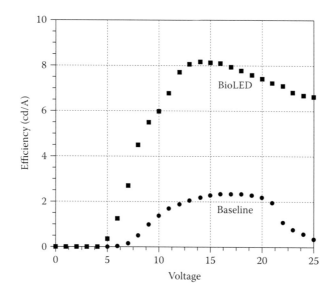

FIGURE 9.21 Efficiency versus voltage for green-emitting BioLED and baseline devices. (Reprinted with permission from J. Hagen, Ph.D Dissertation 2006.)

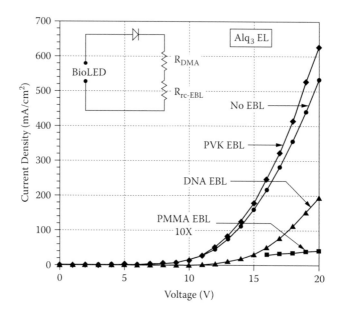

FIGURE 9.22 Current density versus voltage in green Alq$_3$ LEDs: baseline device, DNA EBL, PVK EBL, and PMMA EBL. Inset shows a simple equivalent circuit for the DNA EBL device. (Reprinted with permission from J. Hagen, Ph.D Dissertation 2006.)

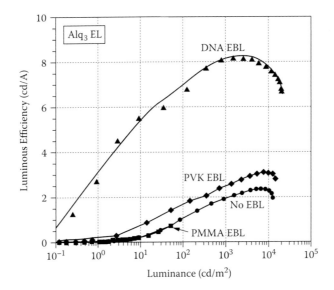

FIGURE 9.23 Luminous efficiency versus luminance in Alq$_3$ EL devices: baseline device, DNA EBL, PVK EBL, and PMMA EBL. (Reprinted with permission from J. Hagen, Ph.D Dissertation 2006.)

material that accounts for the fact that there was a charge transport through the device, but at a lower current than the device without DNA–CTMA. A simple equivalent circuit for the BioLED (DNA EBL) is shown in the insert of Figure 9.22, which isolated the series resistance associated with the DNA layer (R_{DNA}) and the series resistance (R_{no-EBL}) due to all the other layers as measured in the baseline (no-EBL) case. By comparing the J–V slope after turn-on for the two devices, R_{DNA} of ~ 300 Ω is calculated. This corresponds to a DNA–CTMA resistivity of ~ 6 × 10^6 Ω-cm.

The luminous efficiency of the green (Alq$_3$) devices with various EBLs is shown as a function of luminance in Figure 9.23. The DNA–CTMA increased the efficiency by a factor of ~2–10× compared to devices with PVK, PMMA, and no-EB layers.

The hypothesis for the performance of the other optoelectronic polymers PMMA and PVK proved to be true in the experiment. PVK, a conductive polymer, allowed more current to run through the device, but the LUMO level of 2.3 did not effectively block electrons, thus did not improve device efficiency. PMMA is an electrically insulating material that greatly reduced the current flow in the device. However, it did not effectively block electrons to increase efficiency either, which could be due to the low current flow and/or the actual HOMO and LUMO levels of PMMA. DNA effectively blocks electrons and, because of its low electrical resistivity, does not impede current flow, which makes it unique.

9.3.5 Lifetime of BioLED and Baseline Devices

The efficiency increase of OLEDs using DNA–CTMA as an EBL has been well documented earlier. This increase should translate into an increase in the lifetime of the

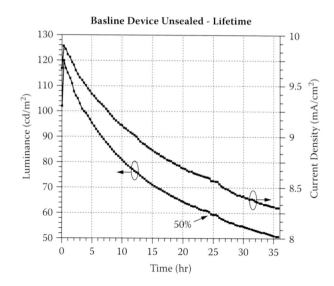

FIGURE 9.24 Lifetime curves for the baseline green device: luminance and current density. (Reprinted with permission from J. Hagen, Ph.D Dissertation 2006.)

devices as well, mainly due to the fact that there is a lower current density running through the materials. This theory was tested by analyzing the lifetimes of a BioLED and baseline device, again grown simultaneously with the common green-emitting structures in Figures 9.14 and 9.15. The devices were set up to operate at an initial brightness of ~100 cd/m² with a constant voltage applied. Current and brightness were recorded in 20 min intervals, and lifetime was defined as the amount of time for the device to decay to 50% of the initial brightness. Since duplicate device testing equipment was not available, the lifetime measurements were done in a series. To avoid the impact of any changing conditions, such as humidity, each device was isolated with a plastic cap sealed with a UV-curable epoxy in atmospheric conditions. As a conservative approach, the baseline device was tested first, meaning the BioLED device was exposed to the humid and oxygen-rich atmospheric conditions for a longer duration. The lifetime results for the baseline device are shown in Figure 9.24.

The baseline device had a maximum brightness of 120 cd/m², which was reached very quickly after an increase from 100 cd/m² at time = 0. Then the device steadily decayed to the 50% brightness point at ~25 h. Figure 9.25 shows the brightness curves for the BioLED device.

The BioLED device had an initial brightness of ~110 cd/m², which was reached after an initial increase from 100 cd/m² at time = 0. The device again steadily decayed to the 50% brightness point, but did not reach this value until ~71 h, which approached a lifetime increase of 3×.

The efficiency of both the baseline and BioLED devices versus time are analyzed in Figure 9.26. The BioLED had an efficiency increase of ~4.5× over the baseline device, which agreed with the ~3× increase in lifetime, in spite of the additional 48 h of atmospheric exposure of the BioLED.

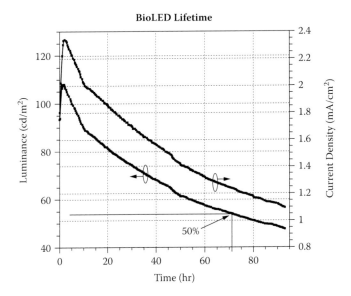

FIGURE 9.25 Lifetime curves for the BioLED green device: luminance and current density. (Reprinted with permission from J. Hagen, Ph.D Dissertation 2006.)

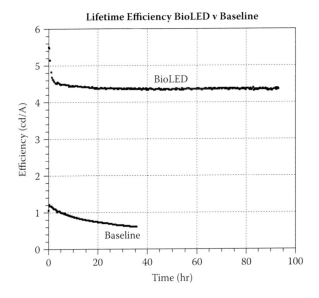

FIGURE 9.26 Lifetime efficiency curves: BioLED and baseline. (Reprinted with permission from J. Hagen, Ph.D Dissertation 2006.)

9.4 CONCLUSIONS

This chapter describes the first DNA-based thin-film modulator, the first FETs using DNA-biopolymers as the semiconducting layer, and the first DNA-containing OLEDs. The results obtained from these devices hold much promise for the future use of DNA-biopolymers in thin-film devices. While not all of the results described earlier are competitive with existing thin-film devices, this work lays the groundwork for some exciting future device development using DNA-based thin films.

REFERENCES

Adachi, C., T. Tsutsui, and S. Saito. 1991. *Optoelectron: Devices Technol.* 6: 25.

Adamovich, V., J. Brooks, A. Tamayo et. al. 2002. High efficiency single dopant white electro-phosphorescent light emitting diodes. *New J. Chem.* 26: 1171–1178.

Adamovich, V., S. Cordero, P. Djurovich et al. 2003. New charge-carrier blocking materials for high efficiency OLEDs. *Org. Electron.* 4: 77–87.

Ashizawa, S., Y. Shinohara, H. Shindo, Y. Watanabe, and H. Okuzaki. 2005. Polymer FET with a conducting channel. *Synth. Metals* 153: 41–44.

Backlund, T.G., H.G.O. Sandberg, R. Osterbacka, H. Stubb, T. Makela, and S. Jussila. 2005. Towards all-polymer field-effect transistors with solution processable materials. *Synth. Metals* 148: 87–91.

Bartsch, C.M. 2007. Development of a field-effect transistor using DNA biopolymer as the semiconductor layer. Ph.D. Dissertation, University of Dayton.

Bartsch, C.M., G. Subramanyam, J.G. Grote, K.M. Singh, R.R. Naik, B. Singh, and N.S. Sariciftci. 2007. Bio-organic field effect transistors. *Proc. SPIE* 6646: 66460K.

Cai, L., H. Tabata, and T. Kawai. 2000. Self-assembled DNA networks and their electrical conductivity. *Appl. Phys. Lett.* 77: 3105–3106.

Covington, J.A., J.W. Gardner, D. Briand, and N.F. de Rooij. 2001. A polymer gate FET sensor array for detecting organic vapours. *Sensor. Actuat. B: Chem.* 77: 155–162.

Covington, J.A., J.W. Gardner, and J.V. Hatfield. 1999. Conducting polymer FET devices for vapor sensing. *Smart Structures and Materials 1999: Smart Electronics and MEMS.* 3673: 296–307.

Deen, M.J. and M.H. Kazemeini. 2005. Photosensitive polymer thin-film FETs based on poly(3-octylthiophene). *Proc. IEEE* 93: 1312–1320.

Deen, M.J., M.H. Kazemeini, Y.M. Haddara et al. 2004. Electrical characterization of polymer-based FETs fabricated by spin-coating poly(3-alkylthiophene)s. *IEEE Trans. Electron Dev.* 51: 1892–1901.

Dimitrakopoulos, C.D. and P.R.L Malenfant. 2002. Organic thin film transistors for large area electronics. *Adv. Mater.* 14: 99–117.

Dutta, S. and K.S. Narayan. 2004. Gate-voltage control of optically-induced charges and memory effects in polymer field-effect transistors. *Adv. Mater.* 16: 2151–2155.

Ebisawa, F., T. Kurokawa, and S. Nara. 1983. Electrical properties of polyacetylene/polysiloxane interface. *J. Appl. Phys.* 56: 3255–3259.

Edman, L., J. Swensen, D. Moses, and A.J. Heeger. 2004. Toward improved and tunable polymer field-effect transistors. *Appl. Phys. Lett.* 84: 3744–3746.

Fink, H. and C. Schonenberger, C. 1999. Electrical conduction through DNA molecules. *Nature* 398: 407–410.

Gao, C., X. Zhu, J.W. Choi, and C.H. Ahn. 2003. A disposable polymer field effect transistor (FET) for pH measurement. *The 12th Int. Conf. on Solid State Sensors Actuators and Microsystems (Transducers '05,* 2: 1172–1175.

Gao, Y., Yu, G., Zhang, C., Menon, R., and A. J. Heeger. 1997. Polymer light-emitting diodes with polyethylene dioxythiophene-polystyrene sulfonate as the transparent anode, *Syn. Met.* 87: 171.

Grote, J., J. Zetts, R. Nelson et al. 2001. Effect of conductivity and dielectric constant on the modulation voltage for optoelectronic devices based on nonlinear optical polymers. *Opt. Eng.* 40: 2464.

Grote, J.G., E.M. Heckman, J.A. Hagen, P.P. Yaney, D.E. Diggs, G. Subramanyam, R.L. Nelson, J.S. Zetts, D.Y. Zang, B. Singh, N.S. Sariciftci, and F.K. Hopkins. 2006. DNA: New class of polymer. *Proc. SPIE* 6117: 61170J.

Grote, J.G., E.M. Heckman, D.E. Diggs, J.A. Hagen, P.P. Yaney, A.J. Steckl, S.J. Clarson, G.S. He, Q. Zheng, P.N. Prasad, J.S. Zetts, and F.K. Hopkins. 2005. DNA-based materials for electro-optic applications: Current status. *Proc. SPIE* 5934: 593406.

Grote, J.G., E.M. Heckman, J.A. Hagen, P.P. Yaney, G. Subramanyam, S.J. Clarson, D.E. Diggs, R.L. Nelson, J.S. Zetts, F.K. Hopkins, and N. Ogata. 2004. Deoxyribonucleic acid (DNA)-based optical materials. *Proc. SPIE* 5621: 16–22.

Grote, J.G., N. Ogata, J.A. Hagen, E. Heckman, M.J. Curley, P.P. Yaney, M.O. Stone, D.E. Diggs, R.L. Nelson, J.S. Zetts, F.K. Hopkins, and L.R. Dalton. 2003. Deoxyribonucleic acid (DNA)-based nonlinear optics. *Proc. SPIE* 5211: 53–62.

Gu, J., M. Kawabe, K. Masuda, and S. Namba. 1977. Electroluminescence of anthracene with powdered graphite electrodes and ambient gas effects on the electrodes. *J. Appl. Phys.* 48: 2493–2495.

Hagen, J.A. 2006. Enhanced luminous efficiency and brightness using DNA electron blocking layers in bio-organic light emitting diodes. Ph.D. Dissertation, University of Cincinnati, Cincinnati OH.

Hagen, J.A., J.G. Grote, K.M. Singh, and R.R. Naik. 2007. Deoxyribonucleic acid biotronics. *Proc. SPIE* 6470: 64700B.

Hagen, J., W. Li, A. Steckl, and J. Grote. 2006. Enhanced emission efficiency in organic light-emitting diodes using deoxyribonucleic acid complex as an electron blocking layer. *Appl. Phys. Lett* 88: 171109.

Heckman, E. 2006. The development of an all-DNA-based electro-optic waveguide modulator. Ph.D. Dissertation, University of Dayton, Dayton, OH.

Heckman, E., P. Yaney, J. Grote, F. Hopkins, and M. Tomczak. 2006c. Development of an all-DNA-surfactant electro-optic modulator. *Proc. SPIE* 6117: 61170K.

Heckman, E., J. Grote, F. Hopkins, and P. Yaney. 2006a. Performance of an electro-optic waveguide modulator fabricated using a deoxyribonucleic-acid-based biopolymer. *Appl. Phys. Lett.* 89: 181116.

Heckman, E.M., P.P. Yaney, J.G. Grote, and F.K. Hopkins. 2006b. Development and performance of an all-DNA-based electro-optic waveguide modulator. *Proc. SPIE* 6401: 640108.

Heckman, E., P. Yaney, J. Grote, and F. Hopkins. 2005a. Poling and optical studies of DNA NLO waveguides. *Proc. SPIE* 5934: 593408.

Heckman, E.M., J.A. Hagen, P.P. Yaney, J.G. Grote, F.K. Hopkins. 2005b. Processing techniques for deoxyribonucleic acid: Biopolymer for photonics applications. *Appl. Phys. Lett.* 87: 211115.

Heckman, E.M., J.G. Grote, P.P. Yaney, and F.K. Hopkins. 2004. DNA-based nonlinear photonic materials. *Proc. SPIE* 5516: 47–51.

Hirata, K., T. Oyamada, T. Imai, H. Sasabe, C. Adachi, and T. Kimura. 2004. Electroluminescence as a probe for elucidating electrical conductivity in a deoxyribonucleic acid-cetyltrimethylammonium lipid complex layer. *Appl. Phys. Lett.* 85: 1627–1630.

Kang, G.W., K.M. Park, J.H. Song et.al. 2005. The electrical characteristics of pentacene-based organic field-effect transistors with polymer gate insulators. *Curr. Appl. Phys.* 5: 297–301.

Kazanskaya, N., A. Kukhtin, M. Manenkova et.al. 1996. FET-based sensors with robust photosensitive polymer membranes for detection of ammonium ions and urea. *Biosens. Bioelectron.* 11: 253–261.

Kobayashi, N., S. Umemura, K. Kusabuka, T. Nakahira, T., and H. Takashi. 2001. An organic red-emitting diode with a water-soluble DNA-polyaniline complex containing Ru(bpy)$_3{}^{2+}$. *J. Mater. Chem.* 11: 1766–1768.

Koyama, T., Y. Kawabe, and N. Ogata. 2002. Electroluminescence as a probe for electrical and optical properties of deoxyribonucleic acid. *Proc. SPIE* 4464: 248–255.

Krishna, T.V., J.R. Jessing, D.D. Russell et al. 2003. Modeling and design of polythiophene gate electrode chemFETs for environmental pollutant sensing. University/Government/Industry Microelectronics Symposium. *Proc. 15th Biennial* 271–274.

Lee, S.M., S.J. Uhm, J.I. Bang et al. 2005. A field effect transistor type gas sensor based on polyaniline. *The 13th Int. Conf. on Solid-State Sensors, Actuators and Microsystems (Transducers '05,* 1935–1938.

Newman, C.R., C.D. Frisbie, D.A. daSilvaFilho, J.L. Bredas, P.C. Ewbank, and K.R. Mann. 2004. Introduction to organic thin film transistors and design of n-channel organic semiconductors. *Chem. Mater.* 16: 4436–4451.

Ouchen, F., P.P. Yaney, and J.G. Grote. 2009. DNA thin films as semiconductors for BioFET. *Proc. SPIE* 7403: 74030F-1–8.

Park, J.H., O. Waldmann, F.C. Hsu et al. 2003. Fabrication and IV characteristics of PEDOT-PSS based field effect devices and their applications to electric circuits. American Physical Society Annual Meeting.

Peng, X., G. Horowitz, D. Fichou, and F. Garnier. 1990. All organic thin film transistors made of alpha-sexithienyl semiconducting and various polymeric insulating layers. *Appl. Phys. Lett.* 57: 2013–2015.

Porath, D., B. Bezryadin, S. de Vries, and C. Dekker. 2000. Direct measurement of electrical transport through DNA molecules. *Nature* 403: 635–638.

Rost, H., J. Ficker, J.S. Alonso, L. Leenders, and I. McCulloch. 2004. Air-stable all-polymer field effect transistors with organic electrodes. *Synth. Metals* 145: 83–85.

Sakanoue, T., E. Fujiwara, R. Yamada, and H. Tada. 2004. Visible light emission from polymer based field-effect transistors. *Appl. Phys. Lett.* 84: 3037–3039.

Service, R. 2005. Organic LEDs look forward to a bright, white future. *Science* 310: 1762–1763.

Sedra, A.S. and K.C. Smith. 1998. *Microelectronic Circuits.* New York: Oxford University Press.

Singer, K.D., M.G. Kuzyk, and J.E. Sohn. 1987. Second-order nonlinear-optical processes in orientationally ordered materials: Relationship between molecular and macroscopic properties. *J. Opt. Soc. Am. B* 4: 968–976.

Singh, J. 2001. *Semiconductor Devices: Basic Principles.* New York: Wiley.

Swensen, J.S., C. Soci, and A.J. Heeger. 2005. Light emission from an ambipolar semiconducting polymer field-effect transistor. *Appl. Phys. Lett.* 87: 253511–3.

Takada, T., K. Kawai, M. Fujitsuka, and T. Majima. 2004. Direct observation of hole transfer through double-helical DNA over 100Å. *PNAS* 101: 14002–14006.

Tang, C. and VanSlyke, S. 1987. Organic electroluminescent diodes. *Appl. Phys. Lett.* 51: 913–915.

Tang, C. S. VanSlyke, and C. Chen. 1989. Electroluminescence of doped organic thin films. *J. Appl. Phys.* 65: 3610–3616.

Teng, C.C. and H.T. Man. 1990. Simple reflection technique for measuring the electro-opto coefficient of poled polymers. *Appl. Phys. Lett.* 56: 1734.

Tiwari, S.P., E.B. Namdas, Ramgopal R.V., D. Fichou, and S.G. Mhaisalkar. 2007. Solution-processed n-type organic field-effect transistors with high ON/OFF current ratios based on fullerene derivative. *IEEE Electron Dev. Lett.* 28: 880–883.

Tsumura, A., H. Koezuka, H., and T. Ando. 1986. Macromolecular electronic device: Field effect transistor with a polythiophene thin film. *Appl. Phys. Lett.* 49: 1210–1212.

Uemura, S., M. Yoshida, S. Hoshino, T. Kodzasa, and T. Kamata. 2003. Investigation for surface modification of polymer as an insulator layer of organic FET. *Thin Solid Films* 438–439: 378–381.

Wang, W., D. Chen, H. Fetterman, Y. Shi, W. Steier, and L. Dalton. 1994. Traveling wave electro-optic phase modulator using cross-linked nonlinear optical polymer. *Appl. Phys. Lett.* 65: 929–931.

Weimer, P.K. 1962. An evaporated thin film triode. *Proc. IRE-AIEE* 50: 1462.

Xu, Y. and P.R. Berger. 2004. High electric-field effects on short-channel polythiophene polymer field-effect transistors. *J. Appl. Phys.* 95: 1497–1502.

Yamamoto, T., T. Yasuda, Y. Sakai, and S. Aramaki. 2005. Ambipolar field-effect transistor (FET) and redox characteristics of a π-conjugated thiophene/1,3,4-thiadiazole CT-type copolymer. *Macromol. Rapid Commun.* 26: 1214–1217.

Yaney, P., E. Heckman, A. Davis et al. 2006. Characterization of NLO polymer materials for optical waveguide structures. *Proc. SPIE* 6117: 61170W.

Yariv, A. and P. Yeh. 1984. *Optical Waves and Crystals.* New York: Wiley.

Yoshida, M., S. Uemura, T. Kodzasa, T. Kamata, M., Matsuzawa, and T. Kawai. 2003. Surface potential control of an insulator layer for the high performance organic FET. *Synth. Metals* 137: 967–968.

10 Nucleic Acids-Based Biosensors

Sara Tombelli, Ilaria Palchetti, and Marco Mascini

CONTENTS

10.1 Introduction 291
10.2 DNA-Based Biosensors for Diagnostics 293
10.3 DNA-Based Biosensor for Environmental Application 297
10.4 New Frontiers in Nucleic Biosensors: Aptamer-Based Biosensors 299
References 305

10.1 INTRODUCTION

The elaboration of biosensors is probably one of the most promising ways to solve some of the problems concerning the increasing demand for sensitive, fast, and cheap measurements in different fields of application.

The definition of a biosensor was recently selected by IUPAC (Thevenot et al. 2001), but a more "modern-time appropriate" definition was chosen by Prof. Turner and Newman in the report "Biosensors: A Clearer View" (Newman et al. 2004). They referred to a biosensor as *a compact analytical device incorporating a biological or biologically derived sensing element either integrated within or intimately associated with a physicochemical transducer.*

The earliest biosensors were catalytic systems that integrated especially enzymes with transducers that convert the biological response into an electronic signal. The next generation of biosensors, affinity biosensors, took advantage of different biological elements, such as antibodies, receptors (natural or synthetic), or nucleic acids. In all of these interactions, the binding between the target analyte and the immobilized biomolecule on the transduction element is governed by an affinity interaction such as the antigen–antibody (Ag-Ab), the DNA–DNA, or the protein–nucleic acid binding.

Research in the field of biosensors was initiated by Clark, whose study on the oxygen electrode was published in 1956 (Clark 1956). Since then, a huge number of biosensors have appeared in the literature, a great number of them with applications in medical diagnostics; more than 80% of the commercial devices based on biosensors are utilized in this domain (Dzyadevych et al. 2008).

Many examples related to the analysis of clinical or environmentally relevant analytes by immunosensors have been reported in the last 20 years, when this approach first began (Lin and Ju 2005). More recently, in the last decade, DNA-based sensing has appeared for real applications to clinical diagnostics and to environmental and food analysis, to detect the presence of pathogenic species, to identify genetic poly-morphisms, and to detect point mutations (Dell'Atti et al. 2006).

A nucleic acid (NA) biosensor is defined as an analytical device incorporating an oligonucleotide, even a modified one, with a known sequence of bases, or a complex structure of NA (like DNA from calf thymus) either integrated within or intimately associated with a signal transducer (Labuda et al., 2010). NA biosensors can be used to detect DNA/RNA fragments or either biological or chemical species. Most nucleic acid biosensors are based on the highly specific hybridization of complementary strands of DNA or RNA molecules; this kind of biosensor is also called a genosen-sor. The probe, immobilized onto the transducer surface, acts as the biorecognition molecule and recognizes the target DNA, while the transducer is the component that converts the biorecognition event into a measurable signal. Assembly of numerous (up to a few thousand) DNA biosensors onto the same detection platform results in DNA microarrays (or DNA chips), devices that are increasingly used for large-scale transcriptional profiling and single-nucleotide polymorphism (SNP) discovery.

In NA biosensors, the detection of the hybridization event has been carried out through different detection technologies, from label-free methods, such as piezo-electric and SPR transduction, to other methods often requiring labels, such as elec-trochemical techniques. Several reviews have recently appeared in the literature (Lucarelli et al. 2008; Wang 2006) elucidating all the critical aspects related to the transduction step.

As the specificity of the hybridization reaction is essentially dependent on the biorecognition properties of the capture oligonucleotide, design of the capture probe is undoubtedly the most important preanalytical step. The probes can be linear oligonu-cleotides or structured (hairpin) oligonucleotides, which are being used with increas-ing frequency (Zhang, Peng et al. 2010; Lin et al. 2009; Zhang, Zhou et al. 2010).

Design of linear probes takes great advantage of many commercially available software products that can design capture oligonucleotides within the hypervariable or highly conserved regions of different genomes after their assembly and alignment. Candidate sequences, usually 18–25 nucleotides in length, are finally tested for theo-retical melting temperature (T_m), hairpins and dimer formation, and for homologies using a Basic Local Alignment Search Tool (BLAST) search (Lucarelli et al. 2008).

The experimental variables affecting the hybridization event at the transducer–solution interface are referred to as stringency, and they generally include hybridiza-tion and post-hybridization-washing buffer composition and reaction temperature. When dealing with complex sets of probes, the basic requirement for a functional system is the ability of all the different probes to hybridize their target sequences with high affinity and specificity under the same stringency conditions.

In addition, a number of probes, with varied chemical composition and confor-mational arrangement, have been used to assemble DNA biosensors. PNAs are DNA mimics in which the nucleobases are attached to a neutral N-(2-aminoethyl)-glycine pseudopeptide backbone. If compared to the conventional oligonucleotide probes, PNAs appear to be particularly interesting for the development of electrochemical genosensing, the main reason being the drastically different electrical characteris-tics of their molecular backbone. Moreover, some reports described the synthesis and hybridization of a novel nucleotide termed LNA (Obika et al. 1998; Chen et al. 2008). LNA is a nucleic acid analogue of RNA in which the furanose ring of the ribose sugar is chemically locked by the introduction of a methylene linkage between 2′-oxygen and 4′-carbon. The covalent bridge effectively "locks" the ribose in the N-type (3-endo) conformation that is dominant in A-form DNA and RNA. This conformation enhances base stacking and phosphate backbone preorganization and results in improved affinity for complementary DNA or RNA sequences, with each LNA substitution increasing the melting temperatures (T_m) by as much as 3.0°C–9.6°C (Yang et al. 2007, Wengel 1999). LNA bases can be interspersed with DNA bases, allowing binding affinity to be tailored for individual applications. Due to the very high affinity of the LNA molecules, it demonstrates that LNA probes hybridize with very high affinity to perfectly complementary targets, and at the same time show an extraordinary specificity to discriminate the targets that differ by a single base (Lin et al. 2010).

As well as specificity, sensitivity is also a key factor in the performances of a biosensor: sensitive detection of a specific gene sequence on the basis of the hybridization reaction can be achieved by increasing the immobilization amount and controlling over the molecular orientation of the probes. For this purpose, nanomaterials have been increasingly introduced in the fabrication of biosensors in order to increase the immobilization amount of the DNA probe and to magnify the detection signal and lower the detection limit. A summary of the different kinds of nanomaterials that have been used in DNA biosensors is reported in Table 10.1, for the very latest published biosensors.

Particularly interesting are those methods that combine different kinds of nanomaterials (nanoparticles and nanowires or nanoparticles and nanotubes), creating innovative architectures that can have high immobilization efficiency and improved sensitivity (Ryu et al. 2010) (Figure 10.1).

In this chapter, we will overview the most recent published nucleic acids-based biosensors, limiting the discussion to genosensors, with different applications, from clinical diagnostics to environmental and food analysis. Moreover, a report on the most innovative nucleic acids-based biosensors, aptamer-based biosensors, also will be included.

10.2   DNA-BASED BIOSENSORS FOR DIAGNOSTICS

The genetic characteristics of patients, such as genotypes or presence of clinically significant polymorphisms, can be used for the identification of a risk of disease (Frank and Hargreaves 2003).

In recent years, the monitoring methods of clinical disease diagnosis and prognosis for genotypes or polymorphisms include those of chromosome analysis, fluorescence in situ hybridization (FISH), flow cytometry (FCM), real-time quantitative reverse transcription PCR(RT-PCR), or RFLP (restriction fragment length polymorphism) (Wang et al. 2009b).

But these methods have some limitations, such as longtime consumption, poor precision, and high expense. Compared to other methods, genosensors have the prominent advantages of being simple, portable, rapid, and inexpensive and they can be considered as a promising solution for the rapid and inexpensive diagnosis of genetic diseases.

TABLE 10.1

Examples of Different Kinds of Nanomaterials Used in DNA-Biosensors

Nanomaterial	Type of Biosensor	Target Sequence	Detection Limit	Ref.
Nanogold	Electrochemical	Breakpoint cluster region gene and a cellular abl (BCR/ABL) fusion gene in chronic myelogenous leukemia	1.0×10^{-13} M (oligonucleotides); detection of PCR samples	Lin et al. 2010
Cupric hexacyanoferrate (CuHCF) nanoparticles	Electrochemical	Influenza A virus	1.0×10^{-15} M (oligonucleotides)	Chen et al. 2010
Zinc oxide nanowires, multiwalled carbon nanotubes, and gold nanoparticles	Electrochemical	/	3.5×10^{-14} M (oligonucleotides)	Wang et al. 2010a
Multiwalled carbon nanotubes and gold nanoparticles	Electrochemical	/	3.5×10^{-13} M (oligonucleotides)	Zhang, Wang et al. 2010
Gold nanoparticles (GN) embedded silicon nanowires	Electrochemical	Breast cancer DNA	1.0×10^{-12} M (oligonucleotides)	Ryu et al. 2010
Gold nanoparticles	Electrochemical	Hepatitis B virus	9.1×10^{-11} M (oligonucleotides)	Li et al. 2010a

Material	Detection method	Target	Sensitivity	Reference
Gold nanoparticles	Electrochemical	/	1.0×10^{-13} M (oligonucleotides)	Li et al. 2010b
Gold nanoparticles	Electrochemical	/	1.0×10^{-11} M (oligonucleotides)	Liu et al. 2010
Gold nanoparticles	Electrochemical	/	3.5×10^{-14} M (oligonucleotides)	Wang et al. 2010b
Gold-capped nanoparticle array chip	Optical (localized surface plasmon resonance)	BIGH3 gene associated with corneal dystrophies	1.0×10^{-12} M (oligonucleotides)	Yoo et al. 2010
Gold nanorods	Optical (surface plasmon resonance)	*Chlamidya Trachomatis*	2.5×10^{-10} M (oligonucleotides); detection of PCR samples	Parab et al. 2010
Hollow gold nanospheres	Electrochemical	/	1.0×10^{-12} M (oligonucleotides)	Liu et al. 2010
Multiwalled carbon nanotubes	Electrochemical	Genetically modified organisms	4.0×10^{-12} M (oligonucleotides)	Lien et al. 2010
Multiwalled carbon nanotubes	Electrochemical	Phosphinothricin acetyltransferase gene (genetically modified organisms)	3.1×10^{-12} M (oligonucleotides)	Zhou et al. 2010
Multiwalled carbon nanotubes	Electrochemical	/	0.1×10^{-12} M (oligonucleotides)	Zhu et al. 2010
Silicon nanowires	Electrical	/	1.0×10^{-13} M (oligonucleotides)	Zhang, Luo et al. 2010

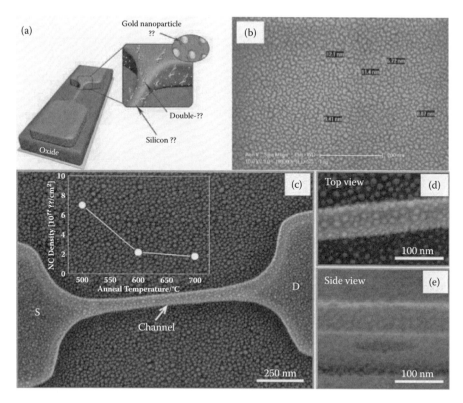

FIGURE 10.1 Scheme (a) and SEM images of the GN-embedded SiNW device. (b) The cracked Au film due to incomplete agglomeration at 400°C, and (c) Au nanoparticles after complete agglomeration at 500°C. The SEM close-up view at the top (d) and the side (e) of the SiNW. (From Ryu, S. et al. 2010. *Biosens Bioelectron* 25:2182–2185. With permission.)

In clinical diagnostics, several DNA-based biosensors have been recently developed for the detection of virus-related sequences, such as hepatitis virus B (Ding et al. 2008; Yao et al. 2008) or papilloma virus (HPV) (Xu et al. 2008; Dell'Atti et al. 2007), or for the recognition of disease-related sequences, such as leukaemia (Wang et al. 2009b), meningitis (Patel et al. 2010), or breast cancer (Castaneda et al. 2007).

In a recent work (Dequaire et al. 2009), a bienzymatic electrochemical biosensor was developed for the detection of human cytomegalovirus (HCMV). 406-Base-pair-amplified HCMV DNA sequence targets were obtained by PCR amplification with biotinylated primers in order to produce biotinylated targets, which were then immobilized onto neutravidin-coated electrodes. The immobilization of the DNA targets and their hybridization with digoxigenin-labeled detection probes were performed in a single step and followed by the use of an antidigoxigenin antibody conjugated with alkaline phosphatase (anti-Dig-AP) and p-aminophenylphosphate (PAPP)/p-aminophenol (PAP) as a substrate/product couple. The sensitivity of the assay was further enhanced in the presence of a second enzyme, Diaphorase (DI), which was used to reduce back the quinonimines produced by the first enzymatic reaction and to regenerate the oxidized form of DI to its reduced native state by its natural substrate, NADH. The obtained detection limit (6×10^7 copies of HCMV-amplified DNA fragments per electrochemical cell) is competitive with other monoenzymatic electrochemical DNA sensors, and the bienzymatic detection enabled a 100-fold lower HCMV DNA detection limit to be estimated.

An interesting paper (Liepold et al. 2008) introduced an electrical detection technology that is easy to handle, easy to integrate into automated diagnosis systems, but can also be applied in a conventional manner, for example, as a sensor to detect polymerase chain reaction (PCR) amplicons directly in the PCR reaction vessel. The work deals with an electrical biosensor detection principle for applications in medical and clinical diagnosis based on the electrically detected displacement assay (EDDA). The electrodes used in the work are shaped as a "dipstick" that can be directly placed into PCR tubes containing the DNA amplicons. The microelectrode array consists of 32 gold electrodes onto which different DNA probes can be immobilized. The detection principle is based on a first step of hybridization between the immobilized capture probe and a signaling probe labeled with an electroactive moiety (ferrocene). The signaling probe is hybridized to the capture probe at the end carrying the label, whereas at the other end it presents an extra sequence that is free for the hybridization with the target PCR amplicon. The amplicon can hybridize to this remote end of the signaling probe that is released in a solution, leading to the decrease in an electrochemical signal. The biosensor array was applied to the detection of specific sequences of HPV (HPV-6) both with standard solutions and PCR amplicons.

Another multielectrode array-based DNA biosensor was recently reported for the detection of human immunodeficiency virus 1 and 2 (HIV1 and HIV2; Zhang, Peng et al. 2010). The array comprised six gold working electrodes and a gold auxiliary electrode, which were fabricated by gold sputtering technology, and a printed Ag/AgCl reference electrode fabricated by screen-printing technology. The DNA biosensor array for simultaneous detection of HIV1 and HIV2 was fabricated in a sequence by self-assembling each of the two thiolated hairpin-DNA probes onto the surfaces of the corresponding three working electrodes (Figure 10.2).

The hybridization events were monitored by square wave voltammetry using methylene blue (MB) as a hybridization redox indicator. The oxidation currents of MB decreased with increasing concentration of HIVs due to the higher affinity of MB for a single strand rather than double strands of DNA. The result showed that the oxidation peak currents were linear over ranges from 20 to 100 nM for HIV1 and HIV2, with the same detection limits of 0.1 nM (S/N = 3). The obtained sensitivity was higher with respect to previously reported works based on label-free single biosensors (Jin et al. 2007) and label-free biosensor arrays (Acero Sanchez et al. 2009).

10.3 DNA-BASED BIOSENSOR FOR ENVIRONMENTAL APPLICATION

Environmental applications of genosensors are in the field of species identification (Palchetti and Mascini 2008). For instance, genosensors have been extensively exploited in the detection of pathogenic microorganisms relevant to food, biodefense, and environmental contamination applications. Genosensors have mainly been

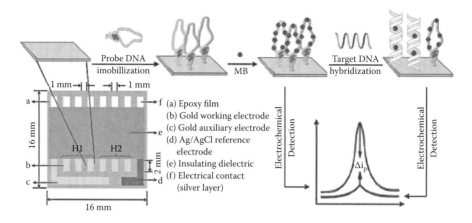

FIGURE 10.2 Schematic diagrams of the multielectrode array (six gold electrodes) and of the biosensor array. (From Zhang, D., Peng, Y., Qi, H. et al. 2010. *Biosens Bioelectron* 25:1088–1094. With permission.)

coupled to nucleic acid amplification techniques such as PCR or nucleic acid sequence based amplification (NASBA), as the specific detection method of the amplified base sequence. Mo et al. (2002) reported the detection of a few *E. coli* cells per 100 mL of water combining the detection of the *lac* gene with PCR amplification and quartz crystal microbalance (QCM). QCM detection of PCR samples of *Aeromonas* strains, isolated from water, vegetables, or human specimens, were reported by Tombelli et al. (2000). The simultaneous electrochemical detection of *Salmonella enterica*, *Lysteria monocytogenes*, *Staphylococcus aureus*, and *E. coli 0157:H7* amplicons in less than 1 h was demonstrated by our group using a disposable electrode array (Farabullini et al. 2007). Wang (2002) described electrochemical genosensors for *Cryptosporidium* as well as *E. coli*, *Giardia*, and *Microbacterium tubercolosis*. A genomagnetic assay for the electrochemical detection of food pathogens based on in situ DNA amplification with magnetic primers was described by Lermo et al. (2007). Electrochemical detection of harmful algae and other microbial contaminants in coastal waters using hand-held biosensors was also demonstrated by LaGier et al. (2007). Baeumner et al. (2003) detected as few as 40 *E. coli* cells per 1 mL sample using a simple optical dipstick-type biosensor coupled to NASBA, emphasizing the fact that only viable cells are detected and no false-positive signals are obtained from dead cells present in a sample, which is important in respect to safety and also food and environmental sample sterilization assessments; similarly, a biosensor for the protozoan parasite *Cryptosporidium parvum* was developed (Esch et al 2001). Baeumner's group also developed a genosensor with an integrated microfluidic system and a minipotentiostat for the quantification of *Dengue virus* RNA (Kwakye et al. 2006). By combining microelectronics and microfluidics with the simple and effective liposome signal enhancement technology and an amperometric transducer, the authors designed a miniaturized electrochemical detection system (miniEC) that was easy to assemble and use. DNA and RNA molecules were quantified using a sandwich hybridization assay similar to the one previously described (Zaytseva et al.

2005), with the transduction mechanism being based on electrochemical rather than fluorescence detection.

A fast procedure that did not require any amplification of the targeted nucleic acids by PCR or NASBA was reported by Elsholz et al. (2006), where a low-density electrical 16S rRNA specific oligonucleotide microarrays and an automated analysis system for the identification and quantization of pathogens, such as *Escherichia coli*, *Pseudomonas aeruginosa*, *Enterococcus faecalis*, *Staphylococcus aureus*, and *Staphylococcus epidermidis* were described.

Genosensors were also applied to the detection of genetically modified plants or foods, such as the surface plasmon resonance (SPR) genosensors described by Feriotto et al. (2002) for the detection of Roundup Ready soybeans in which PCR-amplified sequences from transgenic and wild-type soybeans were investigated by the SPR sensor principle. Examples of QCM or electrochemical biosensors are reported by Mannelli et al. (2003) and Meric et al. (2004).

However, in the literature, many other examples of genosensor technology are reported. Although not specifically developed for environmental applications, the described technologies are prerequisites for the development of DNA-based biosensors for environmental applications.

10.4 NEW FRONTIERS IN NUCLEIC BIOSENSORS: APTAMER-BASED BIOSENSORS

In biosensors, where the detection system requires a biomolecular recognition event, antibody-based detection methodologies are still considered the standard assays, especially in clinical analysis. These assays are well established, and they have been demonstrated to reach the desired sensitivity and selectivity. However, the use of antibodies in multianalyte detection methods and in the analysis of very complex samples could encounter some limitations, mainly deriving from the nature and synthesis of these protein receptors. In order to circumvent some of these drawbacks, other recognition molecules are being explored as alternatives.

The awareness that nucleic acids can assume stable secondary structures and that they can be easily synthesized and functionalized has led to the idea of selecting new nucleic acids ligands called aptamers. Aptamers are artificial single-stranded DNA or RNA ligands that can be generated against aminoacids, drugs, proteins, and other molecules (Tombelli et al. 2005). Their name derives from the Latin word *aptus*, which means "to fit." They are generated by exploiting combinatorial chemistry technology, from very large random sequence oligonucleotide libraries, through an iterative process of absorption, recovery, and reamplification, called SELEX (Systematic Evolution of Ligands by Exponential enrichment) (Tuerk et al. 1990; Ellington et al. 1990).

With respect to their application, aptamers were selected in the past mainly for their use as therapeutic agents; now, for the first time, an aptamer has been recently approved by the US Food and Drug Administration for the clinical treatment of age-related ocular vascular disease (Ng et al. 2006). In addition to the therapeutic field, aptamers have been then used in several analytical methodologies, such as mass spectrometry (Cole et al. 2007) or biosensors (Tombelli et al. 2005). These aptamer-based methods have been mainly employed in the clinical area for the development of diagnostic assays; despite the analytical application of aptamers in this field, they are still under investigation, and the scientific community still needs further research to demonstrate the advances brought about by this new kind of ligand.

The important factors in the success of analytical and diagnostic assays based on aptamers are the affinity and the specificity of the aptamer that provides molecular recognition. The selected aptamers can bind to their targets with affinity ranging from the micro- to the nanomolar level, and they can discriminate between closely related targets.

The fact that some aptamers fold or make a conformational change upon associating with their molecular targets represents an interesting mechanism that can be exploited in the design of new aptamer-based biosensors. Various assays, especially electrochemical sensors, based on this approach have been used for the detection of different targets such as theophylline (Ferapontova et al. 2009a; Ferapontova et al. 2009b), lysozyme (Cheng et al. 2007), Botulinum neurotoxin (Wei et al. 2009), adenosine (Wang et al. 2009a; Wu et al. 2007), cocaine (Swensen et al. 2009, 2010), or thrombin (Tan et al. 2009; Xiao, Lubin et al. 2005a, Xiao, Piorek et al. 2005b). In the electrochemical approach, the interaction of a labeled aptamer with its target can modulate the distance of the electroactive labels from the sensor electrode, thereby altering the redox current.

Despite this high number of published assays based on this approach, however, most of the selected aptamers are well folded and fail to undergo any significant conformational change upon target binding, as recently discussed by Plaxco and co-workers (Xiao et al. 2009; White et al. 2010). Generally, when the conformational change is absent or partial and does not generate any signaling event, a change in the aptamer geometry is necessary, which is achieved by the introduction of an antisense oligonucleotide that hybridizes with the aptamer, keeping it in the unfolded form in the absence of the target (Zuo et al. 2007) or by destabilization of the native aptamer fold by truncation or the introduction of point mutations (Lai et al. 2007). These aptamer engineering approaches (Figure 10.3) have been systematically explored and compared by using two representative aptamers (ATP and IgE aptamer), and it was observed that the relative change in signal upon target binding varies by more than two orders of magnitude across the various investigated constructs and that the optimal geometry is specific to the aptamer sequence upon which the sensor is built (White 2010).

The majority of the reported aptamer-based biosensors, sensors, or assays make use of the thrombin-binding aptamer. Actually, despite nowadays the great number of selected aptamers for a plethora of different molecules, when having a look at the published works on aptamer-based assays, only some specific aptamers have been used, therefore limiting the application of the assays and demonstrating that the proposed approaches often cannot be generalized to all the available aptamers but are strictly related to the aptamer sequence and structure. Among the hundreds (>900) of publications about aptamer-based assays, sensors, or biosensors in the last ten years, almost 60% are dominated by only eight aptamers (thrombin, ATP, platelet-derived growth factor (PDGF), IgE, cocaine, lysozyme, theophylline, and vascular endothelial growth factor [VEGF]), with the thrombin aptamer representing the great part of this number.

(a) Folded aptamer

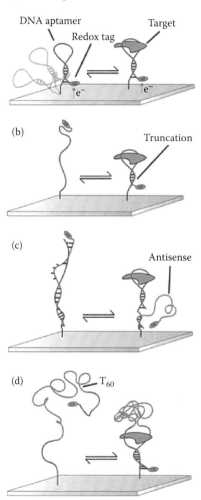

FIGURE 10.3 Approaches to engineer aptamers (A) to undergo large-scale, binding-induced conformational changes: (B) the destabilization of the wild-type aptamer via introduction of sequence truncations or point mutations, (C) the introduction of antisense sequences, or (D) the introduction of long unstructured sequences internal to the aptamer. (From White, R.J. et al. 2010. *Analyst* 135:589–594. With permission.)

The thrombin DNA aptamer (15-mer, 5′-GGTTGGTGTGGTTGG-3') was the first one selected in vitro, specific for a protein without nucleic-acid-binding properties (Bock et al. 1992). Many assays, mainly biosensors, based on the thrombin-binding aptamer for the detection of thrombin have been developed in the last few years (Radi et al. 2006; Zhang et al. 2006) but only few of them have been really used as an analytical method for the detection of thrombin in real samples.

A sandwich assay was developed by using two selected aptamers binding thrombin in two different, not overlapping, sites (Centi et al. 2007). The protein captured by the first aptamer was detected after the addition of the second biotinylated aptamer and of streptavidin labeled with alkaline-phosphatase and the detection of the product generated by the enzymatic reaction was achieved by differential pulse voltammetry (DPV). Good sensitivity and selectivity were demonstrated by the sensor with a detection limit of 0.45 nM in the buffer and negligible signal generated by negative control proteins. The system was also demonstrated to recognize the target analyte in protein-rich media such as thrombin-spiked serum and plasma samples. Moreover, mimicking the physiological clogging event, thrombin was generated in situ by the conversion of its precursor prothrombin present in plasma, and its concentration was measured at different incubation times. The results were well correlated with a thrombogram-mimicking software.

Other interesting aptamer-based biosensors have been developed for the detection of small molecules: small-molecule sensors are more difficult to develop than protein sensors since there are far fewer moieties for aptamer binding. Several examples will be reported here.

A DNA aptamer for oxytetracycline has been selected (Niazi et al. 2008) with a K_d of about 10 nM and good specificity toward oxytetracycline with respect to doxycycline and tetracycline. This aptamer has been recently used for the detection of oxytetracycline coupled to an interdigitated array electrode chip, composed of 65 pairs of generator/collector electrodes (Kim et al. 2009). The aptamer, modified at the 3′ end by a thiol group, was immobilized onto the gold electrode chip. The detection was based on the reduction in current after the binding of the immobilized aptamer to different concentrations of oxytetracycline. Cyclic voltammetry and square wave voltammetry were used to record the current after the binding of the aptamer to the antibiotic, washing, and addition of $K_3Fe(CN)_6$, which was used as a mediator to generate the electron flow between the bulk solution and working electrode. A dynamic range for the detection of oxytetracycline between 1 and 100 nM was observed, and the biosensor could be reused after regeneration with NaCl 2 M about 30 times without losing sensitivity. Current changes for the interaction of the immobilized aptamer with doxycycline and tetracycline were similar to the background signal when testing the complete assay without any chemical target, and they were clearly distinguishable from the current changes obtained when testing oxytetracycline.

In another work, faradic impedance spectroscopy (FIS) was used as a transduction technique for a competitive aptamer-based assay for the detection of neomycin B (De-los-Santos-Alvarez et al. 2007). The interesting feature of this work is the possibility of easily detecting a small molecule such as neomycin B with an electrochemical aptamer-based assay. In the presented work, an aptamer specific for neomycin B (Cowan et al. 2000) was used in a competitive/displacement assay format. In particular, neomycin B was immobilized onto gold electrodes, and this modified surface was saturated with the specific aptamer by affinity binding. By exposing the modified system to different concentrations of neomycin B, a displacement of the bound aptamer was observed, resulting in a drop of the electron-transfer resistance consistent with the reduction of the negative charge of the electrode surface. The competitive assay was very fast (equilibrium in 5 min), with a linear range covering 2 orders of magnitude (0.75–500 μM) and a submicromolar limit of detection.

Very high specificity toward neomycin B was also observed with respect to other very similar antibiotics. The application of the method to the analysis of real samples was also demonstrated by testing neomycin-spiked whole milk with a recovery of 102% and 109% for two different neomycin concentrations.

The same aptamer has been characterized and used for neomycin B detection by using the surface plasmon resonance technique (De-los-Santos-Alvarez et al. 2009).

Cocaine-binding aptamer (Stojanovic et al. 2000) has been widely used for the development of fluorescence, colorimetric, molecular beacon, electrochemical, and chemiluminescence assays.

The detection of cocaine in adulterated samples and biological fluids was performed by the development of an MB-aptamer-based electronic biosensor (Baker et al. 2006). The scheme and operation mode of the sensor is similar to the one developed by the same group for PDGF (Lai et al. 2007). The cocaine aptamer-based sensor detected 500 μM of cocaine in calf serum and human saliva, which represents approximately a 200-fold lower LOD than the Scott Test for cocaine currently used in law enforcement. Furthermore, the sensitivity and selectivity of the sensor were not affected by the presence of cocaine cutting and masking adulterants such as coffee, flour, baking soda, or mustard powder.

The principle of the formation of supramolecular aptamer complexes has been recently used for the detection of cocaine by different research groups (Golub et al. 2009; Sharon et al. 2009; Zuo et al. 2009). The aptamer specific for cocaine has been divided into two fragments which, if incubated together in the presence of cocaine, assemble into a supramolecular aptamer–cocaine complex stabilized both by the aptamer-target interaction and by the base-pairing interaction. This kind of detection principle is very useful since it overcomes some of the drawbacks of the much used classical "sandwich assays": With this format, it is possible to perform the assay for small molecules that could not easily bind simultaneously to two antibodies. Moreover, the selection of two different aptamers for the same molecule is still very rare and so also the development of sandwich assays based on aptamers is very difficult, especially for small molecules.

This principle has been used by the Willner group for the detection of cocaine by using electrochemical (voltammetry, impedance, and ion-sensitive field-effect transistor [ISFET]), photoelectrochemical, and SPR techniques (Golub et al. 2009; Sharon et al. 2009). In a first work (Golub et al. 2009), electrochemical detection was achieved by anchoring on the two fragments of the cocaine aptamer to a gold electrode; the other fragment was functionalized with platinum nanoparticles. In the presence of cocaine, the two fragments self-assembled into the complex, which is fixed onto the gold electrode and labeled with the platinum nanoparticles. The formation of the aptamer–cocaine complex is then detected and amplified by the electrocatalytic reduction of H_2O_2. A detection limit of 1×10^{-5} M for cocaine was reached with an analysis time of approximately 30 min. A similar protocol was followed for photo-electrochemical detection. In this case, the second fragment was labeled with CdS nanoparticles, which enabled the detection of cocaine once the complex was formed, by their photoexcitation. A photocurrent was generated by the concomitant interaction of the CdS particle with the electrode and the interaction of the CdS particle with the electron donor triethanolamine in a solution. A detection limit of 1×10^{-6} M for cocaine was reached with the same analysis time as for electrochemical detection (30 min). For the SPR technique, the second aptamer fragment was labeled with gold nanoparticles. The readout signal (SPR shift) was generated by the coupling between the localized plasmons on the organized gold nanoparticles on the gold chip surface and the surface plasmons. The same detection limit as for photoelectrochemical detection (1×10^{-6} M) was achieved with the same analysis time (30 min). A very interesting feature of this principle is the absence of a background or nonspecific signal when all three techniques were tested in the absence of the target molecule cocaine.

The same principle of self-assembling of the supramolecular aptamer–target complex was used by the same group in a different work (Sharon et al. 2009) for the detection of cocaine by impedimetric or ISFET analysis. For the impedimetric analysis, one of the two fragments of the cocaine aptamer was immobilized via thiol modification onto a gold electrode. The other fragment was labeled with gold nanoparticles. Upon the addition of cocaine, the formation of the supramolecular complex and the presence of the gold nanoparticles increase the negative charge on the electrode. The signal readout was represented by the increase of interfacial electron transfer resistance in the presence of a negatively charged redox label ($Fe(CN)_6^{3-/4-}$). The detection limit for the analysis of cocaine was 1×10^{-5} M. For the ISFET detection, the first fragment of the cocaine aptamer, labeled with amine groups, was fixed onto the Al_2O_3 gate of the FET functionalized with 3′-aminopropyl triethoxysilane. This modified FET was used to detect the formation of the complex between this fragment and the second fragment labeled with gold nanoparticles and cocaine. A detection limit of 1×10^{-6} M was achieved.

It should be noted that all of these proposed techniques could be used to detect cocaine at patophysiological levels (10^{-5}–10^{-7} M); they have very low background signals in the absence of cocaine, and they are easily and fully regenerated by dissociation of the complex.

The self-assembling of the two cocaine aptamer fragments was also used by a different group (Zuo et al. 2009) to develop an electrochemical sensor. One of the fragments was fixed onto a gold electrode via thiol coupling. The other fragment was labeled with an electroactive molecule (methylene blue). The concentration of methylene blue at the electrode surface increases after the addition of cocaine, which contributes to the formation of the fragments–target complex, and this increase can be monitored by voltammetry. A diction limit of 1×10^{-6} M for cocaine was achieved, and the analysis was conducted also on real matrices such as whole blood without any nonspecific matrix effects. A very fast analysis time was presented (1 min after the addition of the target), and the sensor could be regenerated by a simple treatment with distilled water.

Li et al. (2007) reported the development of an electro-generated chemiluminescence (ECL) aptamer-based biosensor to detect cocaine. The ECL-AB biosensor is fabricated by self-assembly of ruthenium complex/NHS-labeled aptamer, which is immobilized onto a gold electrode surface via thiol-Au interactions. In the absence of a target, the aptamer is thought to remain partially unfolded, with only one of its three double-stranded stems intact, resulting in a small ECL signal. In the presence of a target, the aptamer presumably folds into the cocaine-binding three-way junction, and thus a strong ECL signal is generated due to the tag close to the electrode surface. The experimental detection limit was 5.0 nM, and the assay specificity was tested using heroin and caffeine (50 nM) as negative controls.

REFERENCES

Baker, B.R., Lai, R.Y., Wood, M.S. et al. 2006. An electronic aptamer-based small molecule sensor for the rapid, label-free detection of cocaine in adulterated samples and biological fluids. *J Am Chem Soc* 128:3138–3139.

Baeumner, A.J., Cohen, R.N., Miksic, V., and Min, J. 2003. RNA biosensor for the rapid detection of viable *Escherichia coli* in drinking water *Biosens Bioelectron* 18:405–413.

Bock, L.C., Griffin, L.C., Latham, J.A. et al. 1992. Selection of single-stranded DNA molecules that bind and inhibit human thrombin. *Nature* 355:564–566.

Castaneda, M.T., Merkoci, A., Numera, M. et al. 2007. Electrochemical genosensors for biomedical applications based on gold nanoparticles. *Biosens Bioelectron* 22:1961–1967.

Centi, S., Tombelli, S., Minunni, M. et al. 2007. Aptamer-based detection of plasma proteins by an electrochemical assay coupled to magnetic beads. *Anal Chem* 79:1466–1473.

Ceretti, H., Ponce, B., Ramirez, S.A. et al. 2010. Adenosine reagentless electrochemical aptasensor using a phosphorothioate immobilization strategy. *Electroanalysis* 22:147–150.

Chen, H., Zhang, J., Wang, K. et al. 2008. Electrochemical biosensor for detection of BCR/ABL fusion gene using locked nucleic acids on 4-aminobenzenesulfonic acid-modified glassy carbon electrode. *Anal Chem* 80:8028–8034.

Chen, X., Xie, H., Seow, Z. et al. 2010. An ultrasensitive DNA biosensor based on enzyme-catalyzed deposition of cupric hexacyanoferrate nanoparticles. *Biosens Bioelectron* 25:1420–1426.

Cheng, A.K., Ge, B., Yu, H.Z. 2007. Aptamer-based biosensors for label-free voltammetric detection of lysozyme. *Anal Chem* 79:5158–5164.

Clark, L.C. 1956. Monitor and control of blood and tissue oxygen tensions. *Trans Am Soc Artif Intern Organs* 2:41–48.

Cole, J.R., Dick, L.W., Morgan, E.J. et al. 2007. Affinity capture and detection of immunoglobulin E in human serum using an aptamer-modified surface in matrix-assisted laser desorption/ionization mass spectrometry. *Anal Chem* 79:273–279.

Cowan, J.A., Ohyama, T., Wang, D.Q. et al. 2000. Recognition of a cognate RNA aptamer by neomycin B: quantitative evaluation of hydrogen bonding and electrostatic interactions. *Nucleic Acids Res* 28:2935–2942.

Dell'Atti, D., Tombelli, S., Minunni, M. et al. 2006. Detection of clinically relevant point mutations by a novel piezoelectric biosensor. *Biosens Bioelectron* 21:1876–1879.

Dell'Atti, D., Zavaglia, M., Tombelli, S. et al. 2007. Development of combined DNA-based piezoelectric biosensors for the simultaneous detection and genotyping of high risk human papilloma virus strains. *Clin Chim Acta* 383:140–146.

De-los-Santos-Alvarez, N., Lobo-Castañon, M.J., Miranda-Ordieres, A.J. et al. 2007. Modified-RNA aptamer-based sensor for competitive impedimetric assay of neomycin B. *J Am Chem Soc* 129:3808–3809.

De-los-Santos-Alvarez, N., Lobo-Castañon, M.J., Miranda-Ordieres, A.J. et al. 2009. SPR sensing of small molecules with modified RNA aptamers: detection of neomycin B. *Biosens Bioelectron* 24:2547–2553.

Ding, C., Zhao, C., Zhang, M. et al. 2008. Hybridization biosensor using 2,9-dimethyl-1,10-phenantroline cobalt as electrochemical indicator for detection of hepatitis B virus DNA. *Bioelectrochemistry* 72:28–33.

Dzyadevych, S.V., Arkhypova, V.N., Soldatkin, A. et al. 2008. Amperometric enzyme biosensors: past, present and future. *ITBM-RBM* 29:171–180.

Ellington, A.D. and Szostak, J.W. 1990. In vitro selection of RNA molecules that bind specific ligands. *Nature* 346:818–822.

Elsholz, B., Wo, R., Blohm, L., Albers, J., Feucht, H., Grunwald, T., Jurgen, B., Schweder, T., and Hintsche, R. 2006. Automated detection and quantitation of bacterial RNA by using electrical microarrays. *Anal Chem* 78:4794–4802.

Esch, M.B., Baeumner, A.J., and Durst, R.A. 2001. Detection of cryptosporidium parvum using oligonucleotide-tagged liposomes in a competitive assay format. *Anal Chem* 73:3162–3167.

Farabullini, F., Lucarelli, F., Palchetti, I., Marrazza, G., and Mascini, M. 2007. Disposable electrochemical genosensor for the simultaneous analysis of different bacterial food contaminants. *Biosens Bioelectron* 22:1544–1549.

Ferapontova, E.E. and Gothelf, K.V. 2009a. Optimization of the electrochemical RNA-aptamer based biosensor for theophylline by using a methylene blue redox label. *Electroanalysis* 21:261–1266.

Ferapontova, E.E. and Gothelf, K.V. 2009b. Effect of serum on an RNA aptamer-based electrochemical sensor for theophylline. *Langmuir* 25:4279–4283.

Feriotto, G., Borgatti, M., Mischiati, C., Bianchi, M., and Gambari, R. 2002. Biosensor technology and surface plasmon resonance for real-time detection of genetically modified Roundup Ready soybean gene sequences *J Agric Food Chem* 50, 955–962.

Frank, R. and Hargreaves, R. 2003. Clinical biomarkers in drug discovery and development. *Nature Rev* 2:566–580.

Golub, E., Pelossof, G., Freeman, R. et al. 2009. Electrochemical, photoelectrochemical, and surface plasmon resonance detection of cocaine using supramolecular aptamer complexes and metallic or semiconductor nanoparticles. *Anal Chem* 81:9291–9298.

Henry, O.Y.F., Acero Sanchez, J.L., Latta, D. et al. 2009. Electrochemical quantification of DNA amplicons via the detection of non-hybridized guanine bases on low-density electrode arrays. *Biosens Bioelectron* 24:2064–2070.

Jin, Y., Yao, X., Liu, Q. et al. 2007. Hairpin DNA probe based electrochemical biosensor using methylene blue as hybridization indicator. *Biosens Bioelectron* 22:1126–1130.

Kim, Y.S., Niazi, J.H., and Gu, M.B. 2009. Specific detection of oxytetracycline using DNA aptamer-immobilized interdigitated array electrode chip. *Anal Chim Acta* 634:250–254.

Kwakye, S., Goral, V.N., and Baeumner, A.J. 2006. Electrochemical microfluidic biosensor for nucleic acid detection with integrated minipotentiostat. *Biosen Bioelectron* 2:2217–2223.

Labuda, J., Oliveira Brett, A.M., Evtugyn, G., Fojta, M., Mascini, M., Ozsoz, M., Palchetti, I., Paleček, E., and Wang J. 2010. Electrochemical nucleic acid-based biosensors: concepts, terms, and methodology (IUPAC Technical Report). *Pure Appl Chem* 82:1161–1187.

LaGier, M.J., Fell, J.W., and Goodwin, K.D. 2007. Electrochemical detection of harmful algae and other microbial contaminant in coastal waters using hand-held biosensors. *Marine Pollution Bull* 54:757–770.

Lai, R.Y., Plaxco, K.W., and Heeger, A.J. 2007. Aptamer-based electrochemical detection of picomolar platelet-derived growth factor directly in blood serum. *Anal Chem* 79:229–233.

Lermo, A., Campoy, S., Barbe, J., Hernandez, S., Alegret, S., and Pividori, M.I. 2007. In situ DNA amplification with magnetic primers for the electrochemical detection of food pathogens. *Biosens Bioelectron* 22:2010–2017.

Li, F., Feng, Y., Dong, P. et al. 2010a. Gold nanoparticles modified electrode via a mercaptodiazoaminobenzene monolayer and its development in DNA electrochemical biosensor. *Biosens Bioelectron* 25:2084–2088.

Li, X., Fu, P., Liu, J. et al. 2010b. Biosensor for multiplex detection of two DNA target sequences using enzyme-functionalized Au nanoparticles as signal amplification. *Anal Chim Acta* 673:133–138.

Li, Y., Qi, H., Peng, Y. et al. 2007. Electrogenerated chemiluminescence aptamer-based biosensor for the determination of cocaine. *Electrochem Commun* 9:2571–2575.

Lien, T.T.N., Lamb, T.D., An, V.T.H. et al. 2010. Multi-wall carbon nanotubes (MWCNTs)-doped polypyrrole DNA biosensor for label-free detection of genetically modified organisms by QCM and EIS. *Talanta* 80:1164–1169.

Liepold, P., Kratzmüller, T., Persike, N. et al. 2008. Electrically detected displacement assay (EDDA): a practical approach to nucleic acid testing in clinical or medical diagnosis. *Anal Bioanal Chem* 391:1759–1772.

Lin, J. and Ju, H. 2005. Electrochemical and chemiluminescent immunosensors for tumor markers. *Biosens Bioelectron* 20:1461–1470.

Lin, L. Lin, X., Chen, J. et al. 2009. Electrochemical biosensor for detection of BCR/ABL fusion gene based on hairpin locked nucleic acids probe. *Electrochem Commun* 11:1650–1653.

Lin, L., Chen, J., Lin, Q. et al. 2010. Electrochemical biosensor based on nanogold-modified poly-eriochrome black T film for BCR/ABL fusion gene assay by using hairpin LNA. *Probe Talanta* 80:2113–2119.

Liu, S., Liu, J., Han, X. et al. 2010. Electrochemical DNA biosensor fabrication with hollow gold nanospheres modified electrode and its enhancement in DNA immobilization and hybridization. *Biosens Bioelectron* 25:1640–1645.

Liu, S., Liu, J., Wang, L. et al. 2010. Development of electrochemical DNA biosensor based on gold nanoparticle modified electrode by electroless deposition. *Bioelectrochemistry* 79:37–42.

Lucarelli, F., Tombelli, S., Minunni, M. et al. 2008. Electrochemical and piezoelectric DNA biosensors for hybridisation detection. *Anal Chim Acta* 609:139–159.

Mannelli, I., Minunni, M., Tombelli, S., and Mascini, M. 2003. Quartz crystal microbalance (QCM) affinity biosensor for genetically modified organisms (GMOs) detection. *Biosens Bioelectron* 18:129–140.

Meric, B., Kerman, K., Marrazza, G., Palchetti, I., Mascini, M., and Ozsoz, M. 2004. Disposable genosensor, a new tool for the detection of NOS-terminator, a genetic element present in GMOs. *Food Control* 15:621–626.

Mo, X.T, Zhou, Y.P., Lei, H., and Deng, L. 2002. *Enzyme Microbial Technol*, 30:583–589.

Newman, J.D., Tigwell, L.J., Turner, A.P.F. et al. 2004. Biosensors—A Clearer View. Cranfield University Publication, Cranfield, Beds., UK.

Ng, E.W.M., Shima, D.T., Calias, P. et al. 2006. Pegaptanib, a targeted anti-VEGF aptamer for ocular vascular disease. *Nat Rev Drug Discov* 5:123–132.

Niazi, J.H., Lee, S.J., Kim, Y.S. et al. 2008. ssDNA aptamers that selectively bind oxytetracycline. *Bioorg Med Chem* 16:1254–1261.

Obika, S., Nanbu, D., Hari, Y. et al. 1998. Stability and structural features of the duplexes containing nucleoside analogues with a fixed N-type conformation, 2′-O,4′-C-methyleneribonucleosides. *Tetrahedron Lett* 39:5401–5404.

Palchetti, I. and Mascini, M. 2008. Nucleic acid biosensors for environmental pollution monitoring. *Analyst* 133:846–854.

Parab, H.J., Jung, C., Lee, J. et al. 2010. A gold nanorod-based optical DNA biosensor for the diagnosis of pathogens. *Biosens Bioelectron* 26:667–673.

Patel, M.K., Solanki, P.R., Kumara, A. et al. 2010. Electrochemical DNA sensor for *Neisseria meningitidis* detection. *Biosens Bioelectron* 25:2586–2591.

Radi, A.E., Acero Sanchez, J.L., Baldrich, E. et al. 2006. Reagentless, reusable, ultrasensitive electrochemical molecular beacon aptasensor. *J Am Chem Soc* 128:117–124.

Rochelet-Dequaire, M., Djellouli, N., Limoges, B. et al. 2009. Bienzymatic-based electrochemical DNA biosensors: a way to lower the detection limit of hybridization assays. *Analyst* 134:349–353.

Ryu, S., Kim, C., Han, J. et al. 2010. Gold nanoparticle embedded silicon nanowire biosensor for applications of label-free DNA detection. *Biosens Bioelectron* 25:2182–2185.

Sharon, E., Freeman, R., Tel-Vered, R. et al. 2009. Impedimetric or ion-sensitive field-effect transistor (ISFET)-aptasensors based on the self-assembly of Au nanoparticle-functionalized supramolecular aptamer nanostructures. *Electroanalysis* 21:1291–1296.

Stojanovic, M.N., de Prada, P., and Landry, D.W. 2000. Fluorescent sensors based on aptamer self-assembly. *J Am Chem Soc* 122:11547–11548.

Swensen, J.S., Xiao, Y., Ferguson, B.S. et al. 2009. Continuous, real-time monitoring of cocaine in undiluted blood serum via a mocrifluidic, electrochemical aptamer-based sensor. *J Am Chem Soc* 131:4262–4266.

Tan, E.S.Q., Wivanius, R., Toh, C.S. et al. 2009. Heterogeneous and homogeneous aptamer-based electrochemical sensors for thrombin. *Electroanalysis* 21:749–754.

Thevenot, D.R., Toth, K., Durst, R.A. et al. 2001. Electrochemical biosensors: Recommended definitions and classification. *Biosens Bioelectron* 16:121–131.

Tombelli, S., Minunni, M., and Mascini, M. 2005. Analytical application of aptamers. *Biosens Bioelectron* 20:2424–2434.

Tombelli, S., Mascini, M., Sacco, C., and Turner, A.P.F. 2000. A DNA piezoelectric biosensor assay coupled with a polymerase chain reaction for bacterial toxicity determination in environmental samples. *Anal Chim Acta* 418:1–9.

Tuerk, C. and Gold, L. 1990. Systematic evolution of ligands by exponential enrichment: RNA ligands to bacteriophage T4 DNA polymerase. *Science* 249:505–510.

Wang, J. 2006. Electrochemical biosensors: towards point-of-care cancer diagnostics. *Biosens Bioelectron* 21:1887–1892.

Wang, J., Li, S., and Zhang, Y. 2010a. A sensitive DNA biosensor fabricated from gold nanoparticles, carbon nanotubes, and zinc oxide nanowires on a glassy carbon electrode. *Electrochim Acta* 55:4436–4440.

Wang, J., Wang, F., and Dong, S. 2009a. Methylene blue as an indicator for sensitive electrochemical detection of adenosine based on aptamer switch. *J Electroanal Chem* 626:1–5.

Wang, J., Zhang, S., and Zhang, Y. 2010b. Fabrication of chronocoulometric DNA sensor based on gold nanoparticles/poly(L-lysine) modified glassy carbon electrode. *Anal Biochem* 396:304–309.

Wang, K., Chen, J., Chen, J. et al. 2009b. A sandwich-type electrochemical biosensor for detection of BCR/ABL fusion gene using locked nucleic acids on gold electrode. *Electroanalysis* 21:1159–1166.

Wang, J. 2002. Electrochemical nucleic acid biosensors. *Anal Chim Acta* 469, 63–71.

Wei, F. and Ho, C.M. 2009. Aptamer-based electrochemical biosensor for *Botulinum* neurotoxin. *Anal Bioanal Chem* 393:1943–1948.

Wengel, J. 1999. Synthesis of 3′-C- and 4′-C- Branched Oligodeoxynucleotides and the Development of Locked. Nucleic Acid (LNA). *Acc Chem Res* 32:301–310.

White, R.J., Rowe, A.A., and Plaxco. K.W. 2010. Re-engineering aptamers to support reagentless, self-reporting electrochemical sensors. *Analyst* 135:589–594.

Wu, Z.S., Guo, M.M., Zhang, S.B. et al. 2007. Reusable electrochemical sensing platform for highly sensitive detection of small molecules based on structure-switching signalling aptamers. *Anal Chem* 79:2933–2939.

Xiao, Y., Lubin, A.A., Heeger, A.J. et al. 2005a. Label-free electronic detection of thrombin in blood serum by using an aptamer-based sensor. *Angew Chem Int Ed* 44:5456–5459.

Xiao, Y., Piorek, D., Plaxco, K.W. et al. 2005b. A reagentless signal-on architecture for electronic, aptamer-based sensors via target-induced strand displacement. *J Am Chem Soc* 127:17990–17991.

Xiao, Y., Uzawa, T., White, R.J. et al. 2009. On the signaling of electrochemical aptamer-based sensors: collision- and folding-based mechanisms. *Electroanalysis* 21:1267–1271.

Xu, L., Yu, H., Akhras, M.S. et al. 2008. Giant magnetoresistive biochip for DNA detection and HPV genotyping. *Biosens Bioelectron* 24:99–103.

Yang, C.Y.J., Wang, L., Wu, Y.R. et al. 2007. Synthesis and investigation of deoxyribonucleic acid/locked nucleic acid chimeric molecular beacons. *Nucl Acids Res* 35:4030–4041.

Yao, C., Zhu, T., Tang, J. et al. 2008. Hybridization assay of hepatitis B virus by QCM peptide nucleic acid biosensor. *Biosens Bioelectron* 23:879–885.

Yoo, S.Y., Kim, D., Park, T.J. et al. 2010. Detection of the most common corneal dystrophies caused by $BIGH_3$ gene point mutations using a multispot gold-capped nanoparticle array chip. *Anal Chem* 82:1349–1357.

Zhang, D., Peng, Y., Qi, H. et al. 2010. Label-free electrochemical DNA biosensor array for simultaneous detection of the HIV-1 and HIV-2 oligonucleotides incorporating different hairpin-DNA probes and redox indicator. *Biosens Bioelectron* 25:1088–1094.

Zhang, Z., Zhou, J., Tang, A. et al. 2010. Scanning electrochemical microscopy assay of DNA based on hairpin probe and enzymatic amplification biosensor. *Biosens Bioelectron* 25:1953–1957.

Zhang, Y., Wang, J., and Xu, M. 2010. A sensitive DNA biosensor fabricated with gold nanoparticles/ploy (p-aminobenzoic acid)/carbon nanotubes modified electrode. *Colloids Surf B: Biointerfaces* 75:179–185.

Zhang, G., Luo, Z.H.H., Huang, M.J. et al. 2010. Morpholino-functionalized silicon nanowire biosensor for sequence-specific label-free detection of DNA. *Biosens Bioelectron* 25:2447–2453.

Zhang, H., Wang, Z., Li, X.F. et al. 2006. Ultrasensitive detection of proteins by amplification of affinity aptamers. *Angew Chem Int Ed* 45:1576–1580.

Zhou, N., Yang, T., Jiao, K. et al. 2010. Electrochemical deoxyribonucleic acid biosensor based on multiwalled carbon nanotubes/Ag-TiO_2 composite film for label-free phosphinothricin acetyltransferase gene detection by electrochemical impedance spectroscopy. *Chin J Anal Chem* 38:301–306.

Zhu, N., Gao, H., Xu, Q. et al. 2010. Sensitive impedimetric DNA biosensor with poly(amidoamine) dendrimer covalently attached onto carbon nanotube electronic transducers as the tether for surface confinement of probe DNA. *Biosens Bioelectron* 25:1498–1503.

Zuo, X., Song, S., Zhang, J. et al. 2007. A target-responsive electrochemical aptamer switch (TREAS) for reagentless detection of nanomolar ATP. *J Am Chem Soc* 129:1042–1043.

Zuo, X., Xiao, Y., and Plaxco, K.W. 2009. High specificity, electrochemical sandwich assays based on single aptamer sequences and suitable for the direct detection of small-molecule targets in blood and other complex matrices. *J Am Chem Soc* 131:6944–6945.

Zaytseva, N.V., Montagna, R.A., and Baeumner, A.J. 2005. Microfluidic biosensor for the serotype-specific detection of Dengue virus RNA. *Anal Chem* 77:7520–7527.

11 Materials Science of DNA—Conclusions and Perspectives

James G. Grote

CONTENTS

References.. 316

It has been 11 years since I attended that first seminar on purified deoxyribonucleic acid (DNA) given by Professor Naoya Ogata from the Chitose Institute of Science and Technology. At that seminar, Ogata discussed a new process that he developed for deriving DNA from fish milt and roe sacs, waste products of the salmon industry in Hokkaido, Japan. He purified it by removing 98% of the proteins and stabilizing it with a cationic surfactant to render it water insoluble, but dissolvable in alcohols. You have read about the process in depth in Chapter 7. He displayed a sample of a purified DNA-surfactant free-standing film that he had cast to illustrate the optical transparency of DNA. Working with nonlinear optical (NLO) polymer electro-optic (EO) modulators at the time, I was looking for optical cladding materials with an electrical resistivity one to three orders of magnitude lower than that of the NLO core material, an optical propagation loss no more than two times that of the core material, and resistance to the organic solvents used to dissolve the core material. In order to optimize the performance of NLO polymer EO modulators, top and bottom cladding materials with all of the foregoing desired properties listed were needed. However, all the polymers that were identified with the desired electrical resistivity had more than 10 times the desired optical loss, and polymers that had the desired optical loss had 10–1000 times the desired electrical resistivity. In addition, the solvents used to dissolve the NLO core material would also dissolve the cladding materials, rendering stacking these materials to fabricate waveguide structures problematic to say the least.

In fact, during the course of that research, it became apparent that the solvent issue was the most important issue to solve. Even though cladding materials with the desired optical and electrical properties were never realized, trade-offs could be made in order to use materials lacking the desired properties. But the problem still remained: these materials all dissolved with the same organic solvents, such as cylopentanone, dichloroethane, etc., which resulted in either completely dissolving the layer underneath or creating a rough interface between the two layers, significantly increasing the optical insertion loss. So, when Ogata conveyed to the audience that his DNA-surfactant complex, or biopolymer, was only dissolvable in alcohols and not dissolvable in the typical organic solvents used to dissolve other polymers, my interest was sparked. This suggested that there was potential for this new DNA-based biopolymer to work as both the top and bottom cladding layer. Was this biopolymer material optical waveguide quality? Did the material have low enough electrical resistivity? These, I believed, could prove easier to achieve than the solvent issue.

This was my entry point to using DNA. It was discovered, through further processing, that the DNA-based biopolymer developed by Ogata not only demonstrated low optical absorption, low optical propagation, and low optical scattering loss, but it demonstrated low electrical resistivity as well, which is one of the unique properties of DNA. In addition, we discovered that the electrical resistivity could be reduced by as much as two orders of magnitude by simply reducing the molecular weight, without increasing the optical losses. Using a lower-molecular-weight DNA-based biopolymer as the top and bottom cladding layers in an NLO polymer EO modulator, we were able to demonstrate a 90% poling efficiency and a 3X decrease in the optical insertion loss of the device, from 15 to 10 dB (Grote et al. 2006).

Since that first application, this new DNA-based biopolymer has been used for many types of electronic and photonic applications. New DNA-based complexes have been developed with new material properties. This has led to an international research effort in what we call biotronics, using biotechnology and biopolymers for electronic and photonic applications.

Even with growing research in DNA-based biopolymers and the achievements presented in this book and elsewhere, there is still much about DNA that we do not know. Use of DNA-based biopolymers in such devices as organic light-emitting diodes and organic field-effect transistors has resulted in significant improvements in device performance, all with an unoptimized material. In order to improve and optimize the performance of these types of devices, we must first optimize the properties of the material.

We know through the literature, and by experimentation, that reducing the molecular weight of the DNA reduces the electrical resistivity; however, we are not certain why that is the case. As the DNA is sonicated, the overall molecular weight is reduced. Since the adenine-thymine (A-T) bond is weaker than the guanine-cytosine (G-C) bond, the DNA will typically break at the A-T bond during sonication, thus increasing the G-C to A-T content. Increased G-C to A-T content in dry double-stranded DNA of the same length has been reported to have lower electrical resistivity (Iqbal et al. 2005). In addition, the lengths of the DNA base pairs are also reduced during sonication, which has been also suggested to reduce the electrical resistivity. In addition, with shorter DNA lengths, we have the potential for more DNA strands to be in parallel with the applied electric field, which could also be a reason for the reduced electrical resistivity. So, is the decrease in electrical resistivity due to the shorter length, increased G-C to A-T content, more DNA stands in parallel with the electric field, or a combination of all three? When the DNA is sonicated, we end up with many different base pair lengths, or molecular weights, but with the majority being near a certain molecular weight. There are still some low- and high-molecular-weight DNAs. A way to better control the base pair lengths would be to use enzymes;

however, the yields are low and the costs are high. We have been somewhat successful using polyethylene glycol (PEG) to reduce the distribution of the base pairs by reducing the high- and low-molecular-weight DNA; however, the yield is low for this method as well. So, we have not yet been able to determine the effect molecular weight has on device performance, other than reduced electrical resistivity.

It has been shown that adding the hexadecyltrimethyl ammonium chloride (CTMA) surfactant to the DNA, to render it water insoluble, increases the electrical resistivity by several orders of magnitude. Since CTMA is a long molecule, the DNA strands would be further apart. Since charge flow between DNA molecules, as well as through them, has been reported (Schuster and Angelov 2004), could the electrical resistivity be reduced by decreasing the distance between DNA base pairs?

Ogata's DNA has a reduced protein content totaling around 2%. Is that low enough? What effect do proteins have on device performance? Sonicating the DNA produces numerous sequences of DNA all blended together. What role does sequence play and could particular sequences be identified that would be best for particular applications?

We have begun investigation of plasmid-DNA. Since it can be produced in large quantities having one molecular weight, no proteins attached and specific genomic sequences, it may hold the answers we are looking for, as well as help solve forthcoming production, yield, and cost issues.

Some other properties of DNA and DNA-CTMA we have measured include high thermal conductivity, high optical damage threshold, and high photochemical stability.

We recently measured the thermal conductivity of both DNA and DNA-CTMA and compared it with the thermal conductivity of polymethylmethacrylate (PMMA). The thermal conductivity of DNA and DNA-CTMA measured 0.82 and 0.6 W/mK, respectively. Compared with the thermal conductivity of PMMA, 0.12 W/mK, reported in the literature (Kodama et al. 2009), we find that the thermal conductivity of DNA is 7 times higher than the thermal conductivity of PMMA and the thermal conductivity of DNA-CTMA is 5 times higher. This suggests that if DNA is introduced into the device, the potential exists for removing heat from the device more efficiently than using other polymer materials.

We measured the optical damage threshold of DNA using ultrafast Z-Scan at λ = 750 nm. We found the optical damage threshold for DNA-CTMA to be 2.3–2.6 J/cm^2. This is comparable to the 2.7–4.0 J/cm^2 that we measured for fused silica. We found silk to have an optical damage threshold 2× higher than that of fused silicon.

The photochemical stability was measured by Rau's group at the Universitatea Politehnica din Bucuresti. They found that the kinetic degradation constant for both DNA and DNA-CTMA measured 2.78×10^{-6}/m and 2.57×10^{-6}/m, respectively, at λ = 312 nm (UV), at room temperature. The kinetic degradation constant for polycarbonate measured 3.13×10^{-6}/m under the same conditions, which is about 12% higher than for the DNA and DNA-CTMA samples. At 85°C, the kinetic degradation constant for DNA and DNA-CTMA measured 6.68×10^{-6}/m and 36.6×10^{-6}/m, respectively, where the kinetic degradation constant for polycarbonate measured 11000×10^{-6}/m, or ~2000 times higher than DNA and ~200 times higher than DNA-CTMA. This indicates that DNA-based materials are more photochemically stable than other polymers, especially at elevated temperatures.

An interesting property demonstrated by Subramanyam's group at the University of Dayton was a nearly 50% tunability in which the dielectric constant of DNA-CTMA could be induced by an applied electric field in the 10–20 GHz range (Bartsch et al. 2007). This is typically found in inorganic ferroelectric materials, such as barium strontium titanate (BST), but not in polymers. Electric-field-induced refractive index tuning is present in poled polymer complexes; however, at 10–20 GHz, the change in the dielectric constant is several orders of magnitude lower than that observed in BST and what we have seen in DNA-CTMA.

In addition to its unique electromagnetic and optical properties, DNA possesses a self-assembly property and an "ordered" type nature that have been used to enhance the properties of several material complexes and electronic and photonic devices.

Prasad's group at the University at Buffalo demonstrated an enhancement in two-photon lasing using DNA as a host, compared to other polymer hosts; they found DNA could be doped much more heavily without aggregation than the other polymer hosts (He et al. 2006).

Jin and Choi's group at Korea University used DNA as a host material to demonstrate a significant enhancement in the photoluminescence of both fluorescent- and phosphorescent-type guest materials for sensing and organic light-emitting diode applications (Lee et al. 2008; Cho et al. 2010).

Sotzing's group at the University of Connecticut recently demonstrated nearly exact white color, light-emitting diode by blending donor and acceptor fluorescent dyes with DNA-CTMA at a donor:acceptor ratio of 1:20 (Ner et al. 2009). Illuminating the DNA-CTMA-donor:acceptor complex at $\lambda = 365$ nm, they achieved a chromaticity of (0.35, 0.34). Exact white has a chromaticity of (0.33, 0.33). The donor dye was 4-[4-dimethylaminostyryl]-1-docosylpyridinium bromide, and the acceptor dye was Coumarin 102. Comparing the DNA host with other polymer hosts, they found that more energy is transferred from the donor to the acceptor when DNA is used and that the observed enhancement in the energy transfer efficiency between the donor and acceptor can be attributed to self-assembly, the well-defined organization of the dyes by association within the DNA. An actual device was also fabricated by electrospinning the DNA-CTMA-donor–acceptor complex inside the lens of a 400 nm light-emitting diode.

Samoc's group at the Australian National University has performed many refractive index studies of DNA and DNA-CTMA complexes. They have measured a large second-order nonlinearity associated with DNA and have observed a liquid crystalline-like behavior (Samoc et. al. 2006).

Kajzar's group at the Université d'Angers demonstrated a fourfold increase in third-order nonlinearity using a disperse red one-DNA complex when compared with a polymethylmethacrylate polymer host (Krupka et al. 2008). This increase is thought to be due, in part, to the highly polarizable conjugated π electrons in DNA and that, again, there appears to be a self-assembly or alignment taking place in the DNA that is not present in amorphous polymers.

Naik's group at the Air Force Research Laboratory discovered yet another interesting property of DNA. They were able to achieve an 86% selectivity of (6,5) single-walled carbon nanotubes (SWNTs) from a metallic and semiconducting mixture by sonicating a blend of the SWNT mixture with DNA. This compares to 50%

selectivity using d(GT)$_{20}$, a synthetic oligo-DNA. High selectivity is important since it has been shown that one needs at least 86% selectivity of semiconducting SWNTs for high-mobility transistors. Another advantage of using DNA is that the selectivity of (6,5) SWNTs is $1000 less per gram to process than using d(GT)$_{20}$.

In addition to the 7x decrease in gate voltage achieved by Sariciftci's group at the Johannes Kepler University of Linz, using a DNA-based gate dielectric in a pentacene-based organic field effect transistor (Singh et al. 2006), Jin and Choi's group recently increased transistor mobility by using a photoreactive DNA (Kim et al. 2010). Peyghambarian's group at the University of Arizona has developed a DNA-sol-gel complex with enhanced dielectric properties potentially suitable for both capacitors and gate dielectrics (Norwood et. al. 2008). They have achieved a nearly 900 V/µm dielectric breakdown voltage with a dielectric constant of 7–8.

In order to use any new material, device fabrication processes also need to be taken into account. We have demonstrated that DNA films can be applied via spin deposition, casting, flow coating, ink jet printing, vapor deposition, spray deposition, and pulsed laser deposition. Photoresist can be deposited without dissolving the DNA-based materials, and these materials can be exposed to UV and etched. We have also demonstrated that they can be ablated using an excimer laser and can be nano-imprinted. So, are we close to having DNA-based commercial products in the near future?

From the first point contact transistor, demonstrated by Brattain and Bardeen in 1947, to the p-n junction (Shockley 1949), to the announcement of the first commercially feasible silicon transistors in 1954 (Warner and Grung 1983), it took until the mid-1970s to actually realize commercial silicon devices (Teal 1976). Silicon had to be a nearly defect-free crystalline material and ultra-purified to 99.999999999% to become electronic grade (Hashim et al. 2007). Diffusion and implantation techniques had to be developed. Optimum diffusion and implantation materials had to be identified. Photoresist and lithography technologies also had to be developed.

The silicon microelectronics we know today took the combined efforts of physics, engineering, and metallurgy pioneers such as Bardeen, Shockley, Teal, Ebers, Early, Ohl, Scaff, Theurer, Schumacher, Buehler, Sparks, Hall, Taylor, Noyce, Moore, Galvin, and Kilby. It also took the commitments of companies such as Bell, Texas Instruments, Motorola, Fairchild, RCA, General Electric, Sylvania, Raytheon, Westinghouse and, later on, Mostek, AMI, and Intel. With decades of research and engineering, it has been an internationally driven endeavor.

Based on this past history for silicon, it would seem that we are probably at silicon's mid-1950s point of development with DNA. We may be able to shorten time a little by taking advantage of past work with silicon, organics, and DNA, but there is still much more to be done and much more to learn before we can realize commercial devices. We need to maximize device performance through better understanding of material properties and the effects of protein content, molecular weight, and genomic sequence. We need to investigate new sources of DNA such as plasmid-DNA and new guest materials or dopants. We need to better understand and optimize the binding processes for these guest materials. We need to better understand and optimize the alignment of DNA utilizing electrical, magnetic, mechanical, and self-alignment. We need to better understand and optimize the deposition and fabrication processes.

All of the research and development that was needed to realize silicon-based micro-electronics also applies to DNA.

In conclusion, as the authors of this book, and other researchers from around the world, have demonstrated, biotechnology and DNA-based biopolymers are not only applicable for genomic sequencing and clinical diagnosis and treatment, but can also have a major impact on nonbiotech applications as well, such as electronics and photonics, opening up a whole new field for bioengineering. Where silicon has been the building block of inorganic electronics and photonics, DNA holds promise to become the building block for organic electronics and photonics. We hope that many more of you will join us in this exciting new field of biotronics.

As editors of this book on Materials Science of DNA, both Jung-Il Jin and I wish to acknowledge the pioneering research on DNA materials performed by Professor Naoya Ogata. His breakthrough inspired all of the authors of this book, and many other researchers from around the world, to perform their own research using Ogata's DNA, which has, in turn, led to their own breakthroughs in DNA. Finally, we want to thank the Air Force Research Laboratory, the Korea Science and Engineering Foundation, the Air Force Office of Scientific Research, the Asian Office of Aerospace Research and Development, the European Office of Aerospace Research and Development and the South American Office of Aerospace Research and Development, the Defense Advanced Research Projects Agency and the Chitose Institute of Science and Technology. This work would not have been possible without their support.

REFERENCES

Bartsch, C., Subramanyam, G., Axtell, H., Grote, J., Hopkins, F., Brott, L. and Naik, R. 2007. A new capacitive test structure for microwave characterization of biopolymers "temperature and bias dependant microwave dielectric properties of new biopolymers." *Microwave Opt. Technol. Lett.*, 49(6), 1261–1265.

Cho, M., Lee, U., Kim, Y., Shin, J., Kim, Y., Park, Y., Ju, B., Jin, J.-I. and Choi, D. 2010. Organic soluble deoxyribonucleic acid (DNA) bearing carbazole moieties and its blend with phosphorescent Ir(III) complexes. *J. Polym. Sci. Part A-Polym. Chem.,* 48(9), 1913–1918.

Grote, J., Heckman, E., Hagen, J., Yaney, P., Diggs, D., Subramanyam, G., Nelson, R., Zetts, J., Zang, D., Singh, B., Sariciftci, N., and Hopkins, F. 2006. DNA—new class of polymer. *Proc. SPIE,* 6117, 0J1–0J6.

Hashim, U., Ehsan, A., and Ahmad, I. 2007. High purity polycrystalline silicon growth and characterization, *Chiang Mai J. Sci.*, 34(1), 47–53.

He, G., Zheng, Q., Prasad, P., Grote, J., and Hopkins, F. 2006. Toward biological laser: IR two-photon excited visible lasing from a DNA-chromophore-surfactant complex. *Opt. Lett.*, 31(3), 359–361.

Iqbal, S., Balasundaram, G., Ghosh, S., Bergstrom, D. and Bashir, R. 2005. Direct current electrical characterization of ds-DNA in nanogap junctions. *Appl. Phys. Lett.*, 86, 153901.

Kim, S., Kuang, Z., Grote, J., Farmer, B., and Naik, R. 2008. Enrichment of (6,5) single wall carbon nanotubes using genomic DNA. *Nano Lett.*, 8(12), 4415–4420.

Kim, Y., Jung, K., Lee, U., Kim, K., Hoang, M., Jin, J. and Choi, D. 2010. High-mobility bio-organic field effect transistors with photoreactive DNAs as gate insulators. *Appl. Phys. Lett.,* 96(10), 103307.

Kodama, T. et al. 2009. Heat conduction through a DNA-gold composite. *Nano Lett.*, 9, 2005.

Krupka, O., El-Ghayoury, A., Rau, I., Sahraoui, B., Grote, J. and Kajzar, F. 2008. NLO properties of functionalized DNA thin films. *Thin Solid Films*, 516, 8932–8936.

Lee, J., Do, E., Lee, U., Cho, M., Kim, K., Jin, J.I., Shin, D., Choi, S. and Choi, D. 2008. Effect of binding mode on the photoluminescence of CTMA-DNA doped with (E)-2-(2-(4-(diethylamino)styryl)-4H-pyran-4-ylidene)malononitrile. *Polymer,* 49(25), 5417–5423.

Ner, Y., Grote, J., Stuart, J. and Sotzing, G. 2009. White luminance from multi-dye doped electrospun DNA nanofibers via fluorescence resonance energy transfer. *Angewandte Chem.*, 48, 1–6.

Norwood, R., DeRose, C., Himmelhuber, R., Peyghambarian, N., Wang, J., Li, L., Ouchen, F. and Grote, J. 2008. Dielectric and electrical properties of Sol-Gel/DNA blends. *Proc. SPIE*, 7403, 74030A.

Samoc, M., Samoc, A., and Grote, J. 2006. Complex nonlinear refractive index of DNA. *Chem. Phys. Lett.*, 431, 132–134.

Schuster, G., ed. 2004. DNA electron transfer processes: Some theoretical notions. In *Long Range Charge Transfer in DNA II—Topics in Current Chemistry,* 237, p. 18, Berlin, Springer-Verlag.

Shockley, W. et al. 1976. The path to the conception of the junction transistor. *IEEE Trans. Electron Devices*, 23, 597.

Shockley, W. 1949. The theory of p-n junctions in semiconductors and p-n junction transistors. *Bell Syst. Tech. J.*, 28, 435.

Singh, B., Sariciftci, S., Grote, J., and Hopkins, F. 2006. Bio organic-semiconductor field-effect transistor (BiOFET) based on deoxyribonucleic acid (DNA) gate dielectric. *J. Appl. Phys.*, 100, 024514.

Teal, G. 1976. Single crystals of germanium and silicon-basic to the transistor and integrated circuit. *IEEE Trans. Electron Devices*, 23, 621.

Warner, R. and Grung, B. 1983. *Transistors Fundamentals for the Integrated-Circuit Engineer*, Chapter 1. New York, John Wiley & Sons.

Index

A

Acridine orange, 54
    structure, 51
Adenine
    ligands, 147
    thymine bond, 312
All-DNA-based electro-optic waveguide
    modulator, 256–263
    device testing and performance, 259–261
    electro-optic coefficient, 257–258
    fabrication, 256
Alternating current (AC) modulation, 262
Amine-modified DNA, 125
Ammonium tetrachloroplatinate dopant behavior,
    220
Amorphous polycarbonate (APC), 196, 211,
    214–215
    capacitive test structure, 197
Anthraquinone, 54
Applied Wave Research's (AWR) Microwave
    Office, 194, 196
Aptamer
    approaches, 301
    biosensors, 300
    cocaine-binding, 303
As-received DNA
    dopant Baytron P iron behavior, 221
    sodium counter ion, 216
Atomic force microscope (AFM), 201
    images, 16

B

Barium strontium titanate (BST), 314
    frequency response, 200
Baseline green device
    luminance and current density, 284
Baytron P (BP)
    behavior, 221
    doped film, 219
B-DNA, 67
Biocompatible field-effect transistors (BioFETs),
    267–272
Biocompatible light emitting diodes (BioLED)
    EBL, 274–275
    emitting molecules, 273
    energy level diagram, 276
    ETL, 274
    hole blocking layers, 274

    hole transport layers, 273
    lifetime efficiency curves, 285
    luminance and current density, 285
    luminance *vs.* current density, 280
    materials used for fabrication, 273–275
Biocompatible light emitting diodes fabrication,
    276–277
    anode patterning and deposition, 276
    molecular beam deposition, 277
    solvent-based deposition of HTL and EBL,
    277
Biomedical application of DNA films, 245–252
    cell culture on DNA films, 246–249
    UV cross-linking of DNA films, 245
    wound-healing effect, 247–249
Biopolymers, 2
    initial bottom-gate BioFET, 267–270
    measurement setup, 266
    semiconducting layer, 266–270
Biosensors
    array schematic diagrams, 298
    nanomaterials, 294–295
Bisintercalation, 53
    structure, 69
Bottom-gate BioFET
    drain current, 269
    FETs, 267–270
    saturation mobility, 270
    structure, 268

C

Capacitive test structure, 193
Carbon-based nanomaterials and DNA, 77–112,
    79
    advantages and disadvantages of synthetic
    approaches, 109
    CNT, 80–81
    DNA-CNT hybrids, 82–101
    DNA-graphene hybrids, 104–108
    GRP, 102–103
    preparation and properties of composites,
    77–112
Carbon nanotubes (CNT), 78
    articles published, 80
    attachment to DNA, 87
    covalent linkage, 83–90
    dispersion, 97
    DNA and carbon-based nanomaterials, 80–81
    DNA hybrid, 82–101

DNA hybrid materials, 93, 99, 101
DNA-programmed assembly, 35
DNA wrapped, 39, 91
   encapsulation, 91–92
   formation of supramolecular conjugates, 95
   gold-thiol bonding, 39
   interacting with DNA oligonucleotide, 92
   nanomaterials, 82–101
   noncovalent interactions, 93–101
   polymers bundles, 82
   preparing, 95
   properties, 99
   strategy to prepare, 93
   synthesize, 81, 82
   system, 98
   wrapped by DNA, 94
Cationic surfactant complex. *see* cetyltrimethyl
     ammonium chloride (CTMA) DNA
Cetyltrimethyl ammonium chloride (CTMA)
     DNA, 313
   blend film, 238
   blending of dye-intercalated with synthetic
     polymers, 236
   capacitance, 201
   capacitive test structure, 197
   chromophore film, 184
   chromophore films, 184
   core and cladding materials, 257
   dc conductivities, 223
   dielectric constant, 198
   dissolved in ethanol, 234
   DNA-surfactant, 185
   dye complexes, 235
   dye-PMBA films, 237
   fabrication, 184
   film doped with Eu-FOD, 240
   film effect of long-term exposure to humidity,
     217
   fluorescence intensities, 237, 238
   fluorescence quantum yields, 239
   frequency response, 199
   index of refraction with Cauchy fit, 188
   PBMA blend film, 236
   PEDOT based blends output characteristics,
     271
   PPIF cross-linker, 189
   preparation, 182–184
   refractive indices, 257
   resistivity, 214
   structure, 185–186
   UV-Vis spectra, 235
   water absorption, 238
Chelation metal compounds, 233, 237–241
Chiral DNA nanotubes, 17
Chiral pyramids, 34
Circular dichroism (CD) spectroscopy, 55–56
$C_n$mim-DNA, 170. *see also* Ionic liquidized DNA

CD spectra, 173
   effect of added ionic liquid, 175
   FT-IR spectra, 172
   Raman spectra, 174
   temperature dependence on ionic
     conductivity, 175
Cocaine binding aptamer and detection, 303
Computed density of states, 130
   simulated structures, 132
Conjugated carbon nanotubes tail strategy, 36
Coplanar waveguide (CPW) probes, 190, 192
Coralyne, 54
Corner temperature, 211
Crosslinkable organo polysilicates preparation,
   233
Cross-linked DNA-CTMA film
   DNA-surfactant, 183
   preparation, 184
CTMA. *see* Cetyltrimethyl ammonium chloride
     (CTMA) DNA
Current-voltage (I-V)
   characteristics of DNA, 275
   conductance curves from STS measurement,
     131
Cyclic voltammetry (CV), 90
Cytosine neutralization, 166

D

Daunomycin, 59
DC. *see* Cetyltrimethyl ammonium chloride
     (CTMA) DNA; Direct current (DC)
Density of states, 130
   simulated structures, 132
Deoxyribonucleic acid (DNA). *see also* specific
     type, combination or property
   absorption, 57
   applications, 231–253
   array seven unique tiles, 23
   articles published, 80
   attachment to carbon nanotubes, 87
   based biopolymer, 312
   base pairing, 6
   base pairs arrangement, 62
   bases, 57, 240
   biocompatible and renewable material, 79
   CD spectra, 186
   chelating cites, 240
   chelation with novel metals or rare earth
     metal compounds, 233
   chemical modification, 9
   components, 2, 3
   conductivity measurements, 133
   conductivity values, 123
   confirmation analysis using SQUID, 172
   cube schematic synthetic process, 24
   DNA and carbon-based nanomaterials, 79

double helix structure, 5
DSC curves, 191
electrical and magnetic properties, 121–154
EMR signals, 138
fibril birefringence, 141
immobile four-way branched junction, 15
intercalation of ligand, 50
lattices construction, 20
link with SWNT, 84
melting curves, 187
molecular structure, 4
molecular weight, 181
octahedron, 25
oligonucleotide interacting with carbon
        nanotube, 92
rule of sequence symmetry, 21
salmon sperm, 8
STM image, 129
stretching curves, 66
TGA curves, 190
threefold symmetric three-point-star motifs,
        26
three-point-star and four-point-star motifs,
        22
Deoxyribonucleic acid molecules
    charge transfer, 209
    crystalline inorganic inclusions, 144
    electric resistance, 122
    force spectroscopy, 66
    ligand redistribution, 61
    obtaining DOS, 130
    ratios, 245
    transformer, 143
Deoxyribonucleic acid origami
    DNA nanotube, 29
    gate oxide, 37
    nanostructures and nanomaterials via self-
        assembly, 27–29
    power demonstration, 28
    scaffolded, 27
    tensegrity objects, 29
    two single-walled carbon nanotubes, 37
Deoxyribonucleic acid surfactant, 180–226
    absorption loss, 187–188
    capacitance measurements, 200
    capacitive test structure, 190–200
    compared to nonbiopolymers, 212–213
    conductive dopants, 218–219
    cross-linked DNA-CTMA films, 183
    CTMA, 185
    CTMA-chromophore films, 184
    CTMA structure, 185–186
    data, 212–223
    data analysis, 208–211
    DC resistivity studies, 203–224
    effect of humidity and measurement
        accuracy, 216

experimental procedure, 194–195
index of refraction, 187
material characterization, 185–189
measurement technique, 205–224
molecular weight, 180–182
non-cross-linked DNA-CTMA films, 182
optical loss, 187–188
PEDOT, 185
precipitation with CTMA surfactant, 181
preparation of DNA-CTMA films, 182–184
processing, 180–185
propagation loss, 188
results and analysis, 196–199
RF electrical characterization, 190–202
silk, 215
thin-film processing, 180–226
Diamidino phenylindole (DAPI) structure, 51
Dielectrophoresis (DEP) field, 88
Differential scanning calorimetry (DSC), 189
Direct current (DC), 262
Discotic liquid crystal (DLC)
    iron phthalocyanine, 152
DNA. *see* Deoxyribonucleic acid (DNA)
Dopant Baytron P
    behavior, 219
    iron behavior as-received DNA, 221
Double-crossover molecules (DX), 14
Double helix
    stereochemical structure, 7
    structure, 5
Double-stranded DNA (dsDNA)
    cyclotron motion of spin carriers, 141
    EMR signals, 139
    EMR spectra, 146
    force extension curve, 67
    helix, 50, 67
    magnetization-magnetic field strength (M-H)
        curves, 140
    motional dimension analyses of spin, 140
Double-walled carbon nanotubes (DWNT), 101
Dual-polarization interferometry, 68
Dye complexes, 235
Dysonian lineshape, 148

E

Electrical conductivity of DNA, 130–133
Electrically detected cyclotron resonance
        (EDCR), 142
Electrical properties of DNA, 121–133
    charge transport in dry DNA, 121–129
    electrical conductivity of DNA, 130–133
Electric force microscopy, 203, 204
Electric resistance DNA molecules, 122
Electromagnetic radiation (EMR)
    absorption derivatives, 134
    A-DNA, 138

dsDNA, 139
   signals, 138, 139
   spectra dsDNA, 146
Electron blocking layer (EBL) DNA, 273
   current density *vs.* voltage, 282
   luminous efficiency *vs.* luminance, 283
Electro-optic waveguide modulator, 256–263
   coefficient, 257–258
   device testing and performance, 259–261
   fabrication, 256
Engineer aptamers
   approaches, 301
Environmental application of DNA-based
       biosensors
   nucleic acids-based biosensors, 297–298
Ethidium bromide, 54, 59
   intercalation processes, 61
   structural changes of B-DNA upon
       intercalation, 65
   structure, 51
Ethyl carbodiimide hydrochloride (EDC), 85
Ethylimidazolium cation, 174
Europium DNA
   CD spectra, 241
   CTMA and PMMA fluorescence quantum
       yields, 239
   fluorescence intensity, 242, 243

F

Faradic impedance spectroscopy (FIS), 302
Field-effect transistors (FETs), 263–271
   all-polymer FETs, 265
   DNA-based thin-film devices, 266–270
   DNA biopolymer as semiconducting layer,
       266–270
   improvements to BioFET, 271–272
   initial bottom-gate BioFET, 267–270
   measurement setup, 266
   output characteristics, 272
   polymer FETs, 264–265
   principles of operation, 263
   sensor applications, 265
Films DNA
   gas permeation, 252
   infrared spectrum between nonirradiated and
       irradiated, 247
   resistivities, 212
   thickness spin-coated, 183
   UV-curing conditions, 246
Fluorescence intensity
   DNA-CTMA blend film, 238
   DNA-CTMA film, 240
   DNA molar ratios, 245
Fluorescence quantum yields, 239
Fluorinated PMMA, 237
Fluorination of polymers, 163

Fourier transform infrared spectroscopy (FTIR),
       89
Freeze-dried zinc-DNA, 145
Fully dissociated DNA, 67

G

Gamma-DNA
   M-H curves, 145
   toroids cryoelectron micrographs, 142
Gate
   oxide DNA origami, 37
   voltage, 127
Genome DNA, 14
Gold complexes
   CD spectra, 241
   mono-functionalized DNA, 31
   preparation, 240
Gold nanoparticles
   chiral pyramids, 34
   embedded SiNW device, 296
   graphene oxide and reduced graphene oxide,
       40
   sequence-specific one-dimensional assembly,
       38
Gold substrate DNA nanostructure, 41
Gold-thiol bonding
   DNA-wrapped carbon nanotubes, 39
   interaction, 38
Graphene, 38
   2D building material, 103
   DNA and carbon-based nanomaterials,
       102–103
   DNA hybrids, 104–108
Graphene oxide (GO)
   DNA coating and aqueous dispersion, 107
   DNA-mediated assembly of gold
       nanoparticles, 40
Green-emitting BioLED device
   current density *vs.* voltage, 282
   DNA-based thin-film devices, 278–279
   efficiency *vs.* voltage, 282
   luminance current-voltage characteristics,
       279
   luminous efficiency *vs.* luminance, 283
   operation, 281
Groove-binding DNA, 53
Guanine
   cytosine bond, 312
   ligands, 147

H

Half-intercalation model, 51
Helical DNA wrapping of nanotubes
   HR-TEM images, 100
Helical twist, 63

Helix, 50
  force extension curve, 67
  molecular packing structure, 139
  structure, 5
Hewlett-Packard 8720B Microwave Network
      Analyzer, 194
Hexadecyltrimethyl ammonium chloride. *see*
      cetyltrimethyl ammonium chloride
      (CTMA) DNA
High-resolution transmission electron
      microscopy (HR-TEM), 81
Holliday, Robin, 14
Holliday junction, 14, 19
Hydrophobic forces, 93

I

Imidazolium cation, 171
Induced CD (ICD) signal, 55–56
Intercalation
  DNA structure, 52, 54, 64
  ligand DNA, 53
  processes, 61
  properties DNA photoinduced switching, 52
Intercalation of organic ligands to modify
      properties of DNA, 49–69
  determination of intercalator-DNA
        association, 55–56
  dynamic aspects of intercalation, 60–61
  intercalation general principles, 54–61
  structural changes of DNA upon
        intercalation, 62–68
  thermodynamics, 57–59
Ionic conductivity of DNA films, 165
Ionic liquidized DNA, 163–176
  inner column, 164–169
  ionic liquidized bases in DNA, 167–169
  ionic liquidized DNA-inner column, 164–169,
        167–169
  ionic liquidized DNA-outer column, 170–175
  low-molecular-weight model compounds,
        164–166
  mixture, 164–169
  preparing, 165
  structure, 164
Iron policium M-H curves, 152
Irradiated DNA, 248
Isothermal titration calorimetry (ITC), 106

L

Ligands
  adenine, 147
  redistribution DNA molecules, 61
Linear dichroism (LD) spectroscopy, 55–56
Line-Reflect-Relfect-Match (LRRM) calibration,
      195

Lipid DNA complex film
  preparation, 233, 234
  synthetic polymers blending, 235
Luminance current-voltage characteristics, 279

M

Magnetic properties of DNA, 134–151
  discotic liquid crystals as DNA-mimicking
        compounds, 152
  DNA magnetism, 135–151
  historical recount, 134
  mesoscopic ring flux, 137
  M-H curves, 137
Magnetization-magnetic field strength (M-H)
      curves, 136
  dsDNA, 140
  iron, 152
Manganese DNA
  EMR spectra, 150, 151
  magnetic susceptibility, 151
  single-line shape, 150
Materials science of DNA, 1–9
  conclusions and perspectives, 311–316
  naturally occurring organic polymers, 1
  nucleic acids structures, 2–6
Metal compounds
  DNA, 233
  photonic applications of DNA, 237–241
  UV-VIS spectra, 234
Metal-insulator-semiconductor field-effect
      transistor (MISFET), 263
Methylene blue, 54
Microwave conductivity, 149
Microwave probe station, 195
Minimized sequence symmetry (MSS) criteria,
      14
Minor groove binders, 50
Molecular lithography, 41
Mono-functionalized DNA, 31
Multiwalled carbon nanotubes (MWNT), 81
  SEM images and length distribution, 96

N

Nanogrids, 21
Nanomaterials, 294–295
Nano (biomacromolecular) or micro (cellular)
      levels, 13
Nanostructure gold substrate, 41
Nanostructures and nanomaterials via DNA-
      based self-assembly, 13–43
  carbon hybrid nanostructures, 37–49
  carbon nanotubes, 34–36
  DNA origami, 27–29
  molecular lithography, 41
  nanomaterials, 30–43

nanoparticles in one and two dimensions, 30–31
one-dimensional DNA nanostructures, 16–17
three dimensional ordering of gold nanoparticles with DNA, 32–33
three dimensional self-assembly of DNA polyhedra, 22–24
three dimensional self-assembly of periodical DNA crystals, 25–26
two-dimensional DNA self-assembly, 18–21
Nanotubes
  DNA origami, 29
  from DX molecules, 17
  guided assembly of gold nanoparticles, 35
Near-infrared Raman spectra, 68
Neighbor exclusion principle, 65, 68
Neodymium DNA solution fluorescence intensity, 244
Neutralized DNA
  ionic conductivity, 168
  temperature dependence of ionic conductivity, 167
Nitrogen ligands, 147
Nogalamycin structure, 59
Non-cross-linked DNA-CTMA films
  absorption loss spectrum, 188
  DNA-surfactant, 182
Non-DNA polymer films, 213
Nonlinear optical (NLO) polymer electro-optic (EO) modulators, 311
Novantrone structure, 59
Nucleic acid (NA), 2, 49
  structures, 2–6
Nucleic acid (NA) biosensor, 291–305
  aptamer-based, 299–304
  defined, 292
  diagnostics, 293–296
  environmental application, 297–298
  neutralized, 166

O

Ogata, Naoya, 311
One-dimensional DNA nanostructures, 16–17
Organic field-effect transistor (OFET), 264
Organic light emitting diodes (OLED)
  with HTLs, 274
  with HTLs and HBL, 275
  with HTLs and HBL and ETL, 275
Organic polymers naturally occurring, 1
Oxytetracycline, 302

P

Palladium DNA
  CD spectra, 241
  fluorescence intensity, 243

Peptide nucleic acid (PNA), 86
Phase images, 203, 204
Phosphoric acid di-n-butyl ester (PDE), 170
Photochemical stability, 313
Photonic applications of DNA, 232–244
  chelation of DNA, 237–241
  experimentals, 233–234
  results and discussion, 235
  stability improvements of DNA photonic devices, 233–235
Plasmid-DNA
  investigation, 313
Poly ethylenedioxythiophene (PEDOT)
  CTMA DNA based blends output characteristics, 271
  DNA-surfactant, 185
Polyethylene glycol (PEG), 313
Polymer field-effect transistor, 264–265
  all-polymer FETs, 265
  sensor applications, 265
Polymers
  CNT bundles, 82
  typical alternating polarity, 207
Polymethylmethacrylate (PMMA), 313
  current density *vs.* voltage, 282
  EBL, 280, 282, 283
  fluorescence quantum yields, 239
  luminous efficiency *vs.* luminance, 283
Polynucleotides. *see* Nucleic acid (NA)
Poly phenyl isocyanate co formaldehyde (PPIF)
  cross-linker, 183–184
  DNA-CTMA films, 189
Poly(G)-poly(C) DNA molecule current-voltage and conductance curves, 131
Poly(A)-poly(T) DNA molecule values of gate voltage, 124–125
Polysaccharides, 2
Polyvinyl carbazole, 280
  current density *vs.* voltage, 282
  luminous efficiency *vs.* luminance, 283
Proflavine, 54, 59
  structure, 51
Programmed assembly
  CNT building blocks, 35
  two-dimensional assembly of gold nanoparticles, 32
Protein, 2
Pseudotetragonal symmetry, 20
Pulse voltammetry (PV), 90
Pyrene-tagged single strand DNA (Py-ssDNA), 108, 109
  GRP hybrid, 108

Q

Quartz crystal microbalance (QCM), 298

R

Rare earth metal compounds
DNA, 233
photonic applications of DNA, 237–241
Reduced graphene oxide (RGO)
DNA coating and aqueous dispersion, 107
DNA-mediated assembly of gold
nanoparticles, 40
nanosheets, 106
Resistivity measurement, 207
Rhombohedral DNA crystals, 26
Rhombus-like DNA motif, 19
Ribonucleic acid (RNA), 78
Rolling circle amplification (RCA) technique, 30
Rothemund, Paul, 15

S

Scanning tunneling spectroscopy (STS), 127
statistical analysis, 132
Self-assembly
carbon nanotubes, 34–36
DNA, 13–43
gold-carbon hybrid nanostructures, 37–49
gold nanoparticles in one and two
dimensions, 30–31
nanomaterials, 30–43
Silicon device
GN-embedded, 296
Silicon microelectronics, 315
Silk films resistivities, 215
Single-layer DNA-based films
poling results, 259
Single-stranded DNA (ssDNA) oligonucleotides,
87, 90
attached to functionalized SWNT electrodes,
127
GRP stable dispersion, 105
noncovalent dispersant, 99
wrapped SWNT, 97
Single-walled carbon nanotubes (SWNT) DNA,
35, 81, 314
composite formation, 86
connected with DNA, 126
device characteristic, 126
device morphology, 128
with DNA strands, 91, 126
electrical characterization, 128
functionalized point contacts, 91
hybrids covalently linked, 85
method to cut and functionalize, 126
multilayer film fabrication, 88
multilayer films, 89
wrapped, 99
Sodium counter ion
as-received DNA, 216

Spin-coated DNA-CTMA film thickness, 183
Stability improvements of DNA photonic
devices, 233–235
Stretched DNA, 67
Superconducting quantum interference device
(SQUID) measurement, 139
Supramolecular aptamer-target complex, 304
Supramolecular functionalization of CNT, 93
Surface plasmon resonance (SPR), 60
genosensors, 299

T

TappingMode topography, 203, 204
Temperature dependence on ionic conductivity,
175
Tensegrity objects
DNA origami, 29
DNA triangle, 20, 26
Terbium DNA solution fluorescence intensity,
244
Tetrafluoroborate, 164, 170
effect of added ionic liquid, 169
Thermogravimetric analysis (TGA), 189
Thin-film devices DNA, 256–286
all-DNA-based electro-optic waveguide
modulator, 256–263
comparison of DNA-CTMA, 280–282
development of BioLED DNA, 273–286
fabrication of BioLEDs, 276–277
field-effect transistors, 266–270
green (Alq3)-emitting BioLED results,
278–279
lifetime of BioLED and baseline devices,
283–286
materials used for fabrication of BioLEDs,
273–275
Thin-film transistor (TFT), 263
Three-dimensional gold nanoparticle supra-
crystals, 33
Three dimensional ordering of gold nanoparticles
with DNA, 32–33
Three dimensional self-assembly of DNA
polyhedra, 22–24
Three dimensional self-assembly of periodical
DNA crystals, 25–26
Threefold symmetric three-point-star, 26
Three-layer poled DNA-based waveguide, 261
Thrombin-binding aptamer, 300
Thrombin DNA aptamer, 302
Thymine ligands, 147
Toroids cryoelectron micrographs gamma-DNA,
142
Transistors probe station setup, 267
Transmission electron microscope (TEM), 17
Transverse electro-optic polymer-phase
modulator, 260

Tricationic ligand, 51
Trifluoromethane sulfonylimide (TFSI), 165,
          166, 168
    effect of added ionic liquid, 169
Trioxatriangulenium cation, 63
Trisintercalation, 53
Two-dimensional DNA lattices, 18
Two-dimensional DNA self-assembly, 18–21
Two single-walled carbon nanotubes, 37

U

Ultraviolet curved DNA film
    cell culture, 249
    growth number as functions of incubation
          days, 249
Unwinding, 66

V

Vacuum-dried lithium DNA activation energy,
          222
Volume resistivity measurements, 205

W

Waveguide modulator, 258
Wounded skin recovery, 250
Wound of rat skin, 251

X

X-ray photoelectron spectroscopy (XPS), 86

Z

Zinc-DNA
    with Curie law, 148
    EMR spectrum, 147, 150
    freeze-dried, 145
    microwave conductivity, 149
    temperature dependence, 148